CHEMISTRY RESEARCH AND APPLICATIONS

BENZOFURAN

PRODUCTION AND APPLICATIONS

Chemistry Research and Applications

Additional books and e-books in this series can be found on Nova's website under the Series tab.

CHEMISTRY RESEARCH AND APPLICATIONS

BENZOFURAN

PRODUCTION AND APPLICATIONS

MURILO SOUZA BARROS
EDITOR

nova
science publishers
New York

Library of Congress Cataloging-in-Publication Data

ISBN: 978-1-53617-873-9
Library of Congress Control Number: 2020936718

Published by Nova Science Publishers, Inc. † New York

CONTENTS

Preface		**vii**
Chapter 1	Origins and Biological Applications of Benzofurans *Jorddy Neves Cruz, Sebastião Gomes Silva, Jose de Arimateia Rodrigues do Rego, Joyce Karen Lima Vale, Fernanda Wariss Figueiredo Bezerra, Wanessa Almeida da Costa, Davi do Socorro Barros Brasil, Cleydson Breno Rodrigues dos Santos, Mozaniel Santana de Oliveira and Eloisa Helena de Aguiar Andrade*	**1**
Chapter	Dihydrobenzofuran Neolignans: An Overview of Their Chemistry and Biological Activities *Herbert J. Dias, Eliane O. Silva, Tatiana M. Vieira and Antônio E. M. Crotti*	**31**
Chapter 3	Palladium or Rhodium Catalyzed Benzofuran and Derivatives Synthesis via C-H Functionalizations *Bo Sun, Tian-Yu Yao, Lu-Lu Zhang, Ming-Li and Lin-Bao Zhang*	**117**

Chapter 4 Benzofurans Scaffolds: Applications in Anticancer Drugs Discovery and Development **199**
Jorddy Neves Cruz, Kauê Santana da Costa, Marcia Moraes Cascaes, Fernanda Wariss Figueiredo Bezerra, Ingryd Nayara de Farias Ramos, Mozaniel Santana de Oliveira, Eloisa Helena de Aguiar Andrade and Cleydson Breno Rodrigues dos Santos

Index **219**

PREFACE

Benzofurans have garnered the attention of medicinal chemists due to their broad pharmacological activity. *Benzofuran: Production and Applications* provides insight into the natural origin of these compounds, reported biological activities for various analogs, synthetic routes to obtain benzofurans substituted at different heterocycle sites, and controlled release encapsulation methods.

Dihydrobenzofuran neolignans are secondary metabolites arising from the coupling of two phenylpropanoid units during the shikimate biossynthetic pathway. In this compilation, the natural sources and synthetic routes from which dihydrobenzofuran neolignans can be obtained are reviewed, and their biological potential is highlighted.

Additionally, the authors focus on the palladium and rhodium catalyzed multiple C-H functionalizations in the synthesis of benzofurans and their derivatives. The use of Ir and Ru salts is also explored.

A review of the aspects related to cancer initiation and progress, as well as the applications of benzofurans as structural frameworks for the development of new drugs against cancer, is provided.

Chapter 1 - Benzofurans have gained the attention of medicinal chemists due to their broad pharmacological activity. Such compounds may be synthesized or obtained from natural sources. Currently, about 30 drugs used in pharmacotherapy have the benzofuran nucleus, and there are also several drugs being developed with this nucleus as a privileged structure. Heterocyclic-ring systems occupy a relevant position in organic chemistry as they are an integral part of chemistry and life sciences. These systems have been sought in modern research worldwide because they are central structures in molecules that have relevant biological activities. Several drugs containing the benzofuran ring have been developed from compounds of natural and synthetic origin, due to their recognized biological and pharmacological interests. The diversity of biological effects attributed to the benzofuran nucleus is associated with the radical change that can occur in the heterocyclic structure. The present review will provide insight into the natural origin of these compounds, reported biological activities for various analogs, synthetic routes to obtain benzofurans substituted at different heterocycle sites, and controlled release encapsulation methods.

Chapter 2 - Dihydrobenzofuran neolignans (DBNs) are secondary metabolites arising from the coupling of two phenylpropanoid (C_6C_3) units during the shikimate biossynthetic pathway. These compounds are ubiquotous in the plant kingdom, especially in higher plants, and display several biological activities that are traditionally exploited by the pharmaceutical and food industries. Because of the interest in the DBN structure, chemistry, and bioactivities, the current literature contains numerous studies on their preparation by both the phytochemical and synthetic approaches. Here, the authors present the natural sources and synthetic routes from which DBNs can be obtained and highlight their biological potential.

Chapter 3 - Benzofurans and their derivatives are a class of important structural motifs owing to their wide existence in natural products, pharmaceuticals and biologically active compounds, thus inspiring extensive research for their synthesis under mild and simple reaction conditions. During the past decades, various strategies have been developed to synthesize such useful heterocyles. Among these methods, the annulation reaction represents one of the most straightforward ways to produce the target molecules. In this context, great attention has been paid to the protocols that possess the atom and step efficiency, with which multiple C-H functionalizations could be accomplish the expectations. This review is mainly focused on the palladium and rhodium catalyzed multiple C-H functionalizations of the development in the field of benzofurans and their derivatives' synthesis. The use of Ir or Ru salts will be slightly referred to at the end of the context.

Chapter 4 - In this review, the authors exhibit a general view of aspects related to cancer initiation and progress, as well as, the applications of benzofurans as structural frameworks for the development of new drugs against cancer. Using recent studies in the scientific literature, the authors discuss the cytotoxic activities of compounds containing benzofurans as structural core against different cancer cell lines and their applications to the design and discovery of new benzofuran derivatives with biological activities against interesting molecular targets for cancer therapy.

In: Benzofuran: Production and Applications ISBN: 978-1-53617-873-9
Editor: Murilo Souza Barros

Chapter 1

ORIGINS AND BIOLOGICAL APPLICATIONS OF BENZOFURANS

Jorddy Neves Cruz[1], Sebastião Gomes Silva[1], Jose de Arimateia Rodrigues do Rego[2], Joyce Karen Lima Vale[3], Fernanda Wariss Figueiredo Bezerra[4], Wanessa Almeida da Costa[3], Davi do Socorro Barros Brasil[3], Cleydson Breno Rodrigues dos Santos[5], Mozaniel Santana de Oliveira[1] and Eloisa Helena de Aguiar Andrade[1]

[1]Botany Coordinating, Paraense Emilio Goeldi Museum, Belém, PA, Brazil

[2]Department of Natural Science, Pará State University, Belém, PA, Brazil

[3]Department of Chemical Engineering, University of Pará, Belém, PA, Brazil

[4]Faculty of Food Engineering, University of Pará, Belém, PA, Brazil

[5]Department of Biological and Health Sciences, Federal University of Amapá, Macapá, AP, Brazil

ABSTRACT

Benzofurans have gained the attention of medicinal chemists due to their broad pharmacological activity. Such compounds may be synthesized or obtained from natural sources. Currently, about 30 drugs used in pharmacotherapy have the benzofuran nucleus, and there are also several drugs being developed with this nucleus as a privileged structure. Heterocyclic-ring systems occupy a relevant position in organic chemistry as they are an integral part of chemistry and life sciences. These systems have been sought in modern research worldwide because they are central structures in molecules that have relevant biological activities. Several drugs containing the benzofuran ring have been developed from compounds of natural and synthetic origin, due to their recognized biological and pharmacological interests. The diversity of biological effects attributed to the benzofuran nucleus is associated with the radical change that can occur in the heterocyclic structure. The present review will provide insight into the natural origin of these compounds, reported biological activities for various analogs, synthetic routes to obtain benzofurans substituted at different heterocycle sites, and controlled release encapsulation methods.

Keywords: benzofurans, natural product, synthesis, biological activity

1. INTRODUCTION

Compounds that have the benzofuran nucleus as a privileged structure have emerged as promising pharmacologically active molecules.

Benzofurans are heterocyclic nuclei that act as scaffolds for the binding of radicals, which provide selectivity and potency for the molecule (Scheme 1). Thus, this pharmacophoric group has been used in new drug development projects around the world (Masubuchi et al. 2003; Hayakawa et al. 2004).

Compounds that have the benzofuran nucleus can be obtained from natural or synthetic sources. These compounds are found in different plant species, being obtained from extracts or through the extraction of essential oils (EOs). It is also possible to obtain them as metabolites of bacteria and fungi (Chand et al. 2017).

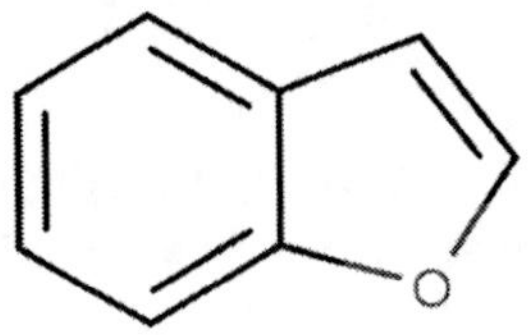

Scheme 1. Benzofuran nucleus.

Isolated compounds of natural and synthetic origin, EOs and extracts containing aromatic compounds have shown larvicidal, antimicrobial, anti-inflammatory, anti-hyperlipidemic and anti-depressant activities, and potential against cardiovascular and skin diseases (Chand et al. 2017; Oramas-Royo et al. 2017).

In addition to the broad biological activity presented by different analogs, these scaffolds draw the attention of medicinal chemists, also due to their physicochemical properties. The ability of heterocyclic nuclei to have good solubility and salt formation is recognized in the literature, characteristics that are appreciated by pharmacologists because they help in optimizing properties, such as bioavailability, and drug excretion. Isolated molecules and EOs that have molecules of this class can be encapsulated for controlled release (Munin and Edwards-Lévy 2011; Miao et al. 2019).

Currently, 30 drugs that have the benzofuran nucleus as a preferred structure have been released for pharmacotherapeutic use in the US, Europe and Canada (Radadiya and Shah 2015). Many other molecules are protected by patents because they are promising chemicals for pharmacological use but have not yet been approved for therapeutic use. So far, if a search for the term "Benzofuran" is performed on the platforms World Intellectual Property Organization (WIPO) ("World Intellectual Property Organization" n.d.) and European Patent Office (EPO) ("European Patent Office" n.d.), 4,552 and 4,515 results of patent documents will be found, respectively.

In this work, the obtaining of benzofuran compounds from natural products, and from different synthetic routes; the main advances in the use of these compounds in research against various diseases; and different encapsulating methods of bioactive compounds were highlighted.

2. Natural Origin of Benzofuran Compounds

Knowledge involving the chemistry of natural products, especially the chemistry of secondary metabolism, is a multidisciplinary science that involves several areas, because the chemical composition of these products is directly influenced by biotic factors (herbivorism, anthropic action, allelopathy, and others), and abiotic factors (edaphoclimatic conditions, light radiation, and others) (Verma and Shukla 2015).

The secondary metabolites are the chemical compounds that suffer most changes in plant metabolism, either qualitatively or quantitatively, in order to provide a better adaptation of the plant to the exposed conditions. Thus, the researcher who ventures into these studies should understand the need to seek in other sciences the explanations for their questions. It is known that since before the Christian era, the collection and use of plants for medicinal purposes had already met certain criteria, such as the time of collection. However, plants do not always follow the same pattern of behavior, which means that those cultivated for commercial or subsistence purposes may have a different metabolism than those found in their natural habitat. It justifies the application of knowledge involving the agrarian sciences, such as soil chemical composition, weather and climate factors, and anthropic action (Jäger 2005; Soejarto et al. 2005).

Benzofurans, heterocyclic compounds with benzene condensed to the furan ring, are precursors of many compounds of simple and complex molecular structure.

These compounds have various pharmacological activities that can be explored for the production of analog series, which lead to a high degree of structural diversity (Khanam and Shamsuzzaman 2015b).

An alternative in the search for benzofurans and their derivatives are the natural products, where these compounds are found in large numbers, in various organs of many species scattered in different parts of the world, as shown in Table 1.

Table 1. Compounds that contain benzofuran nucleus derived from natural plant products

Species	Collection Site	Vegetable part	Constituents	Chemical structure[a]	Reference
Coriandrum sativum	India	Flowers/ methanol: water (80:20) extract	2,3-dihydro Benzofuran (15.4%)	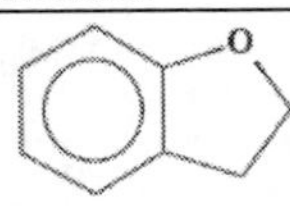	(Dharmalingam and Nazni 2013)
Mariscus pedunculatus	Brazil	Rhizomes and roots/ hydrodistillation	1-(4-hydroxy-2-isoprenyl-6-methoxy-2,3-dihydro-benzofuran-5-yl)-ethanone (Remirol – 56%) and 1-(5-hydroxy-7-methoxy-2,2-dimethyl2H-chromen-6-yl)-ethanone (Evodionol – 7%)	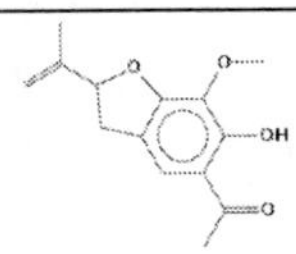 Remirol 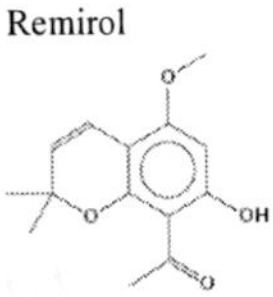 Evodionol	(Siani et al. 2001)
Cinnamomum sp	Seoul, Korea	Leaf/essential oil	4,7-dimethyl-benzofuran (10.6%)	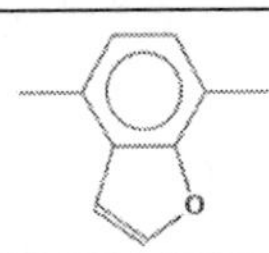	(Kim et al. 2015)
Commiphora myrrha	Sudan	Resins/essential oil	6-ethenyl-4,5,6,7-tetrahydro-3,6-dimethyl-5-isopropenyl-, trans- benzofuran (26.63%)	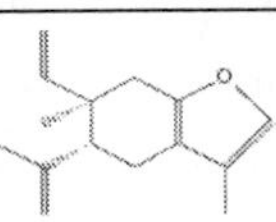	(Gadir and Ahmed 2014)
Curcuma zedoaria	Taichung, Taiwan	Rhizomes/ essential oil	6-ethenyl-4,5,6,7-tetrahydro-3,6-dimethyl-5-isopropenyl-trans-benzofuran (12%)	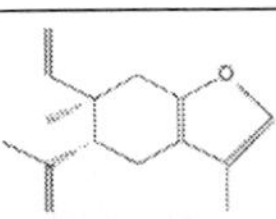	(Chen et al. 2013)
Aristolochia longa	Tunisia	Roots/essential oil	2,3-dihydro-benzofuran (6.0 and 7.81%)	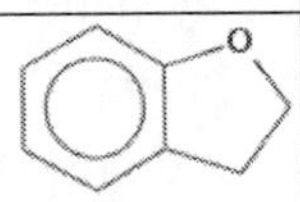	(Dhouioui et al. 2016)
Otostegia michauxii	Fars, Iran	Aerial parts/ essential oil	Benzofuran, 2-methyl (12.9%)	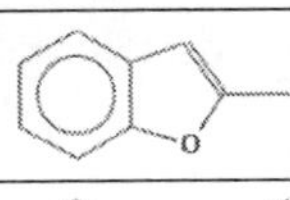	(Javidnia et al. 2010)
Cinnamomum zeylanicum	-	-	benzofuran, 3-methyl (17.97%)	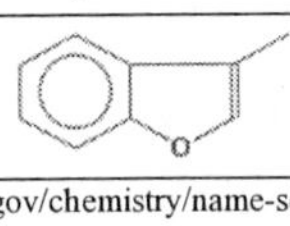	(Perumal et al. 2016)

a = The chemical structures were obtained at the Nist platform (https:// webbook.nist.gov/chemistry/name-ser/).

3. Synthesis and Semi-Synthesis of Benzofurans

Heterocyclic-ring systems occupy a relevant position in organic chemistry as they are an integral part of chemistry and life sciences. These systems have been sought in modern research worldwide because they are central structures in molecules that have relevant biological activities (Saini et al. 2013; Martins et al. 2015)

In the pharmaceutical industry, heterocyclic systems are especially prevalent, being present in over 60% of the drugs used in pharmacotherapy. These compounds may be present in the biomolecules as radicals or as components of the pharmacophoric group (Khanam and Shamsuzzaman 2015a).

Benzofurans are an important class of heterocycles present in various natural products and bioactive molecules, as well as pharmaceuticals and polymers. Several benzofuran derivatives have been recognized as biologically and pharmacologically relevant molecules (Nevagi, Dighe, and Dighe 2015).

Benzofuran derivatives are versatile agents that can be used to design and develop new therapies because they exhibit potent biological properties, such as antihyperglycemic, analgesic, antiparasitic, antimicrobial, and antitumor activities (Dawood 2013). In addition, substituted benzofurans may exhibit fluorescent sensor, oxidizer, and antioxidant activities, and be present in a variety of drugs, and other chemical and agricultural domains (Khanam and Shamsuzzaman 2015b; Structure et al. 1995).

To obtain compounds with benzofuran nucleus, many synthesis methods were developed. Some of these techniques are introduced below.

3.1. Ether Synthesis

Benzofurans were synthesized via Friedel-Crafts intramolecular reaction. Aryloxyacetaldehyde diethyl acetals were obtained from arylphenol according to the methodology described in the literature by Sun et al. (2015) (Sun et al. 2015), in the presence of phosphoric acid and chlorobenzene (PhCl), at 130°C, under constant stirring, for 12 h. Wang et al. (2019) (Ma et al. 2019), after screening the catalysts and solvents, verified the effect of the catalyst concentration on the reaction yield, in which 300 µL (4.8 equiv.) of phosphoric acid is the most suitable amount for 1 mol of the initial reagent, with 96% yield. The reaction was robust and versatile, and allowed the synthesis of unsubstituted 2,3 benzofuran derivatives, with satisfactory yields. This reaction proves to be effective for aromatic ring-substituted electron donating or withdrawing groups, such as Me, tBu, OMe, Ph, F, Cl, Br, I, and CO_2iPr. This mechanism can be viewed in Scheme 2.

3.2. Synthesis of Benzofurans in Aqueous Medium

Products and chemical reactions that do not produce environmentally harmful compounds, or that harm the environment less than traditional processes are the focus of green chemistry. Due to this, the development of ecological synthetic routes has gained more prominence and importance in the last few years.

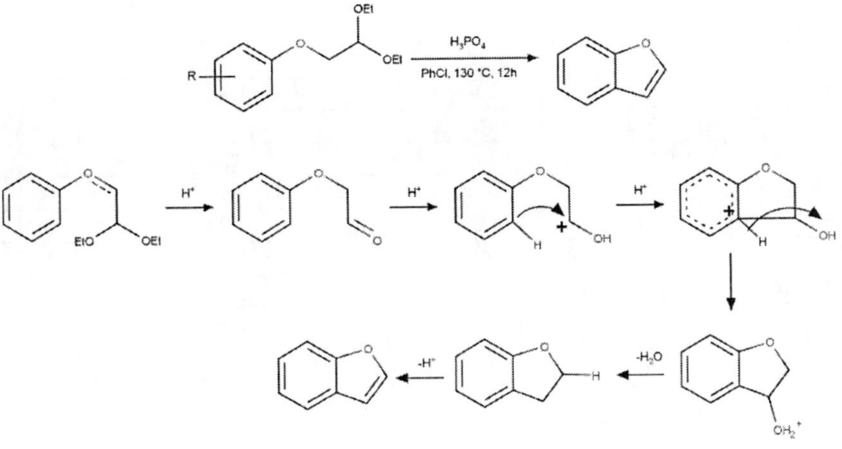

Scheme 2. Mechanism proposal.

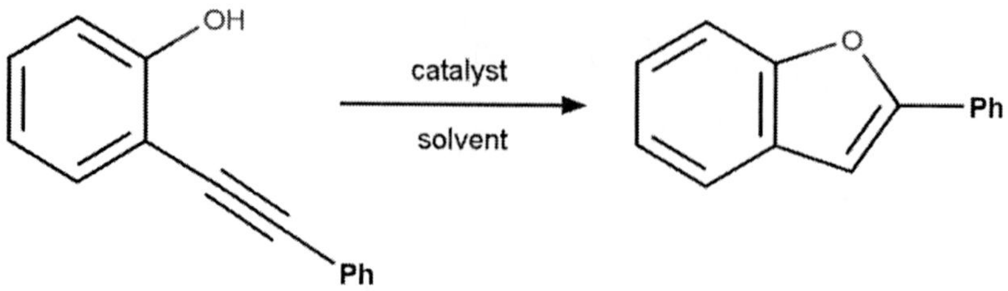

Scheme 3. Cyclization of 2-(phenylethynyl)phenol (1a) to 2-phenylbenzofuran.

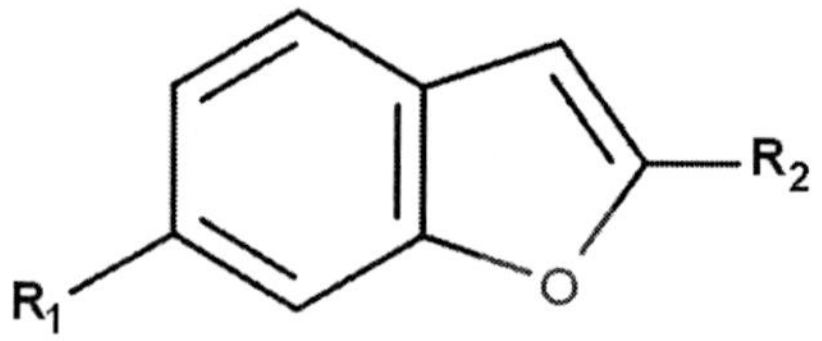

Entry	**R1**	**R2**	**Yield (%)**
1	H	C_6H_5	98
2	H	2-Thienyl	87
3	H	4-MeC_6H_5	90
4	H	4-FC_6H_5	92
5	H	4-ClC_6H_5	90
6	H	4-BrC_6H_5	84
7	H	3-MeC_6H_5	81
8	H	3-FC_6H_5	98
9	H	3-ClC_6H_5	98
10	H	3-BrC_6H_5	97
11	H	*n*-Bu	29
12	Me	C_6H_5	94
13	Me	4-MeC_6H_5	43
14	*t*-Bu	C_6H_5	34
15	*t*-Bu	4-MeC_6H_5	20
16	*t*-Bu	4-ClC_6H_5	32
17	Cl	C6H_5	65
18	Cl	4-MeC_6H_5	70
19	Cl	4-ClC_6H_5	44

Scheme 4. Benzofuran nucleus and radicals substituted in R1 and R2, with their respective reaction yields.

Thus, Sun et al. (2014) developed an intermediate strategy for the synthesis of benzofurans in water. The heterogeneous synthesis of these compounds occurs with the cyclization reaction of 2-(phenylethynyl) phenol to 2-phenylbenzofuran, without using transition metal as catalyst (Scheme 3).

This efficient and promising methodology in the field of green chemistry has produced compounds that can be easily isolated and purified, and through this technique it has been possible to produce molecules with a wide variety of functional groups substituted on R1 and R2 (Scheme 4).

3.3. Cyclization

Andiappan et al. (2017) synthesized a series of new derivatives of 3-(glycinamide)-benzofuran-2-carboxamide, and 3-(b-alanamide) – benzoderivatives of furan-2-carboxamide (5a-o), with antimicrobial, anti-inflammatory, and DPPH free radical scavenging activities.

Compounds 5k and 5o (Scheme 5) had the best anti-inflammatory activity, however they showed low antimicrobial capacity, and low potential for free radical sequestration.

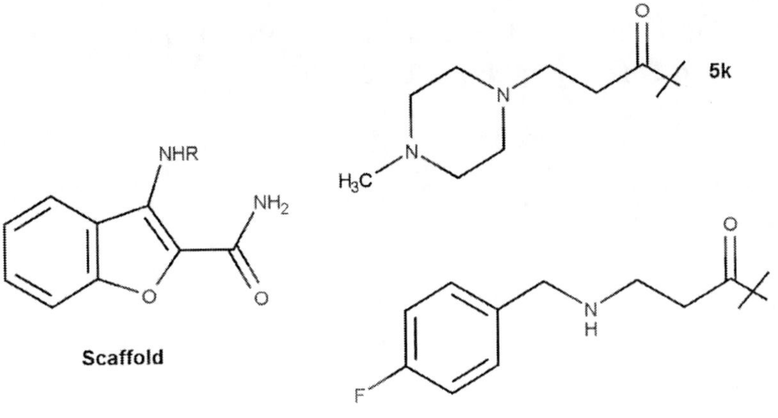

5k

5o

Scheme 5. Synthesized compounds.

The general procedure for the synthesis of these compounds occurred as follows: a solution of 3-[(chloroacetyl)amino]-1-benzofuran-2-carboxamide 4 (0.0012 mol) in dichloromethane (10 mL), triethylamine (0.0024 mol), and corresponding amines (0.0014 mol) were added at ambient temperature. After 6 h, the reaction mixture was diluted with dichloromethane (50 mL), washed with water (2 × 25 mL), and brine (25 mL). The organic layer was dried over anhydrous sodium sulphate, filtered and evaporated to get the expected compound (Lavanya, Sribalan, and Padmini 2017).

3.4. Sonogashira Reaction

The Sonogashira reaction usually occurs in the presence of Pd (II) catalyst together with copper or under copper-free conditions. Several heterogeneously supported palladium catalysts have been developed to produce benzofuran derivatives under different conditions, such as zeolites $[Pd(NH_3)_4]/(NH_4)Y$, and mixed oxide of $[Pd(NH_3)_4]/NaY$, Pd/MgLa, PS-PdO NPs, Pd(OAc)2-nanometric carbon balls (NCB), Pd(II,0)/Cu-Fe_3O_4, CuO-Fe_3O_4, and Pd-Cu/C (Chinchilla and Nájera 2011).

Djakovitch and Rollet (2004) developed a benign heterogeneous zeolite catalyst of $[Pd(NH_3)4]^{2+}/(NH_4)Y$, used for benzofuran synthesis by reaction between o-iodophenol and terminal alkyne, under copper-free condition in the presence of triethylamine in DMF: H2O (4: 1), at 80°C, for 6 h (Scheme 6) (Djakovitch and Rollet 2004).

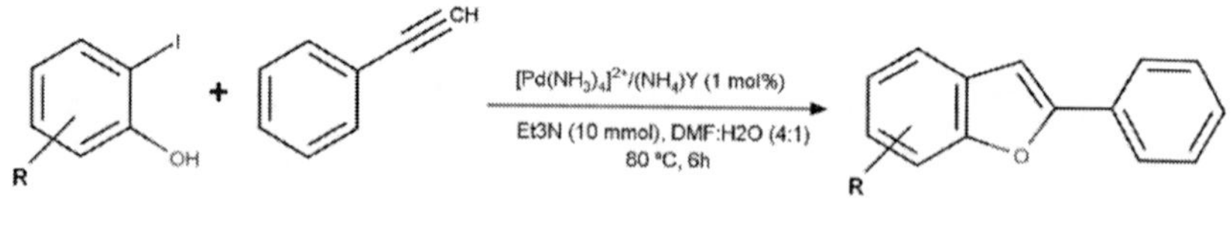

Scheme 6. Synthesis of 2-substituted benzofurans catalyzed by copper-free palladium zeolites.

The Pd catalyst supported by MgLa mixed oxide was used, which is air and moisture stable, and highly efficient for the synthesis of benzofuran derivatives. The reaction was performed in distilled DMF and Et3N, at 80°C for 6h to form product with more than 60% yield (Scheme 7). Although reactions with o-iodophenol provided moderate to high yields, in the case of o-iodine, and o-chlorine, very low phenol yields (less than 10%) were obtained (Yum et al. 2013).

The reaction between o-iodophenol and terminal alkynes produced the expected product in the presence of $Pd(OAc)_2$-NCBs, with additive LiCl and Cs_2CO_3 base in DMF, at 110°C, for 15 h (Scheme 8).

3.5. Ruthenium-Catalyzed Synthesis

The highly selective C-H alkylation reaction occurs between phenols and 1,2 diols, in the presence of cyclopentene and toluene using a complex cationic ruthenium hydride catalyst [(C6H6)(PCy3)(CO)RuH] + BF4. This C-H coupling method shows a wide scope of the high chemoselective substrate, providing an efficient protocol for the preparation of various substituted benzofuran derivatives, with yields ranging from 60 to 95% (Scheme 9) (Lee, Kwon, and Yi 2012; Grimshaw and Thompson 1987).

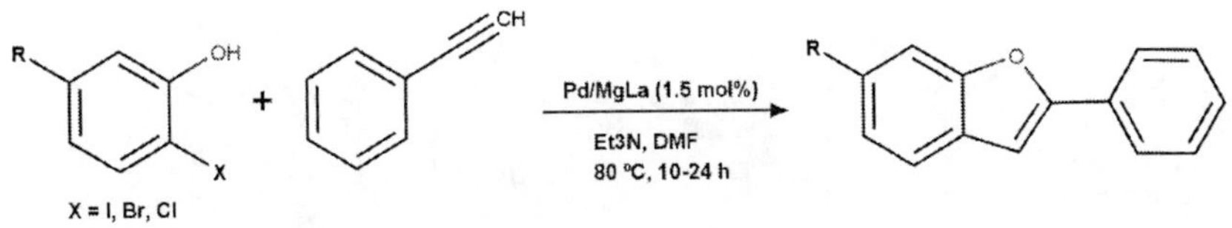

Scheme 7. Synthesis of 2-substituted benzofurans.

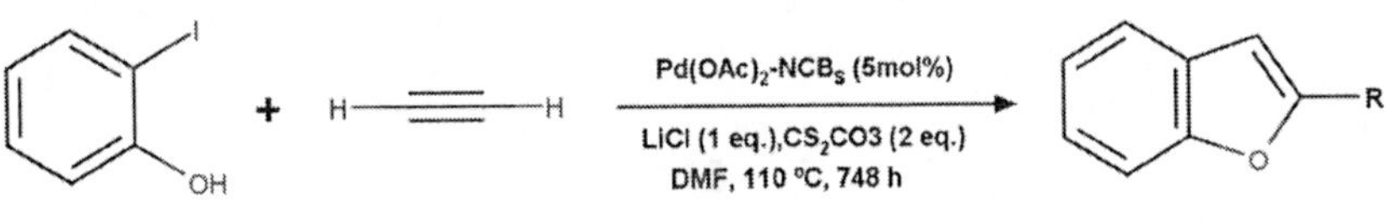

Scheme 8. $Pd(OAc)_2$-NCB_S catalyzed synthesis of 2-substituted benzofurans.

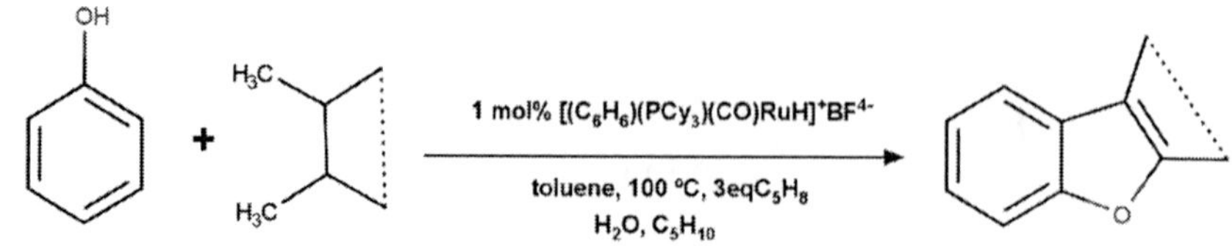

Scheme 9. Production reaction of benzofurans.

Formed Products (Scheme 10):

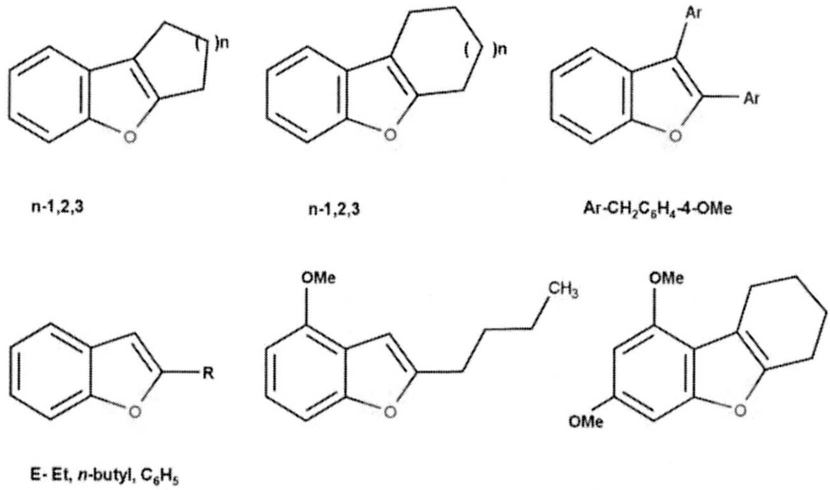

Scheme 10. Benzofurans formed from the alkylation reaction.

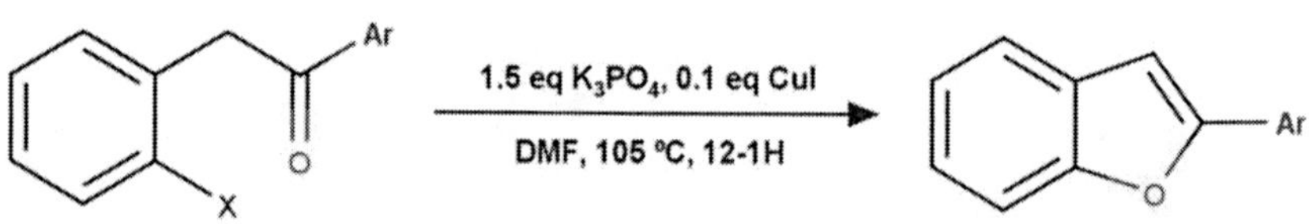

X	Ar
Br	p-SMeC6H4
Br	p-MeC6H4
I	OMe
I	Mesitylene

Scheme 11. Synthesis of 2-aryl substituted benzofurans via ring closure.

3.6. Synthesis Catalyzed by Palladium and Copper

α-substituted 2-halodesoxybenzoines and ketones were heated with a mixture of DMF bases with K_3PO_4, in the presence of CuI catalyst to obtain substituted benzofurans. The products obtained were isolated by crystallization by addition of water, with yields of 70 to 90%, and excellent purity (Scheme 11) (Lu et al. 2007).

Ketones may also be used as reagents for benzofuran production. The reaction occurs between 1-bromo-2-iodobenzene and α-keto ester in the presence of THF, at 100°C (Scheme 12). The intermediate is formed by C-C intermolecular bonding followed by an intramolecular C-O bond, originating the 2,3-substituted benzofurans (Torborg, Zapf, and Beller 2008).

2,5-Substituted benzofurans are produced by reacting phenol-type 1-bromo-benzene with α-keto esters, using 0.5 mol% Na_2PdCl_4 in TMEDA, in the presence of 2- (di-tert-butylphosphine)-N-phenylindole, at 80°C. Phenols in the presence of AcBr in TMEDA first form an intermediate, which will later be coupled with alkynes, forming 2,5-substituted benzofurans (Scheme 13). Products are obtained with yields of up to 99% (Hirano, Satoh, and Miura 2011).

3.7. Photochemical Reaction

Ghosh et al. (2012) performed the synthesis of 2-arylbenzofurans using the Sonogashira reaction induced by visible light, and cyclization of 5-endo-dig in orthohalophenols water, and palladium-catalyzed terminal alkynes. The synthesis presented by these authors has the advantage of obtaining products with satisfactory yield and, besides, there is no formation of complexes with Ru and Ir, or with any other additives (Ghosh, Das, and Saikh 2012).

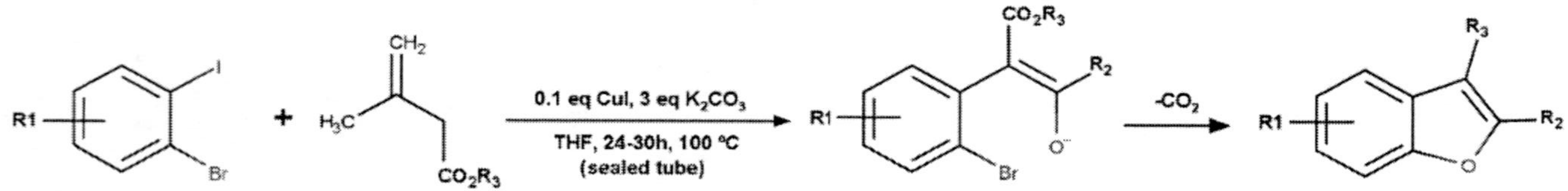

R1 – Me, Cl, CF_2, $COCH_3$, CO_2CH_3, CH_2OH, CH_2OTBS, OCH_2, NO_2
R2 – Et, Bn, ME
R3 – Me, C_4H_{17}, C_2H_4, C_6H_5, *i*Pr, C_6H_5

Scheme 13. Reaction between 1-bromo-2-iodobenzene and α-keto ester.

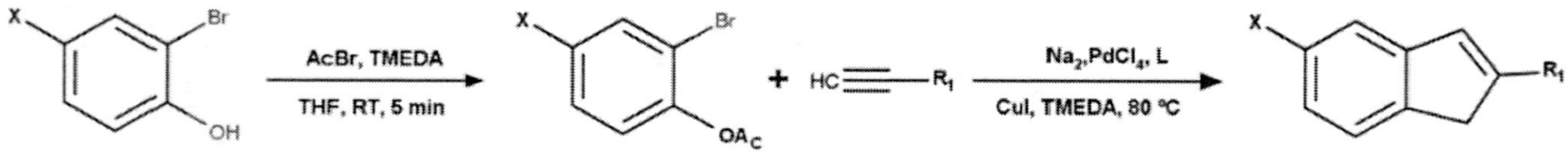

X	NO_2	NO_2	NO_2	NO_2	CN	CN	CN	CHO	H	OH
R1	C_6H_5	3,5-dimethoxyphenyl	$C(CH_3)_2OH$	C_6H_{12}	C_6H_5	$C(CH_3)_2OH$	C_6H_{12}	C_6H_5	C_6H_5	C_6H_5

Scheme 13. Synthesis of 2,5-substituted benzofurans from 2-bromo-4-substituted phenols.

This class of benzofurans can be obtained from natural products, and are scaffolds that have great biological activity (Hou et al. 2008). Thus, as proposed by the authors, obtaining it from a rapid synthesis process, which occurs with two reactions, besides using water as a solvent, makes this synthesis route very attractive for researchers who want efficiency, and speed in obtaining these compounds.

The synthesis of 2-arylbenzofurans is accomplished by the palladium-free copper-catalyzed Sonogashira reaction in water, induced by visible light. The reaction occurs with Sonogashira coupling reaction mediated by $PdCl_2$ and PPh_3, followed by 5-endo-dig cyclization to originate 2-aryl/alkylbenzofurans. The procedure occurs with photo-irradiation of substituted ortho-iodine bromophenols, and terminal alkynes in the presence of palladium chloride, triphenylphosphine, and triethylamine in aqueous medium. The products are then obtained in 49-77% yields by highly pure silica gel chromatography (Scheme 14).

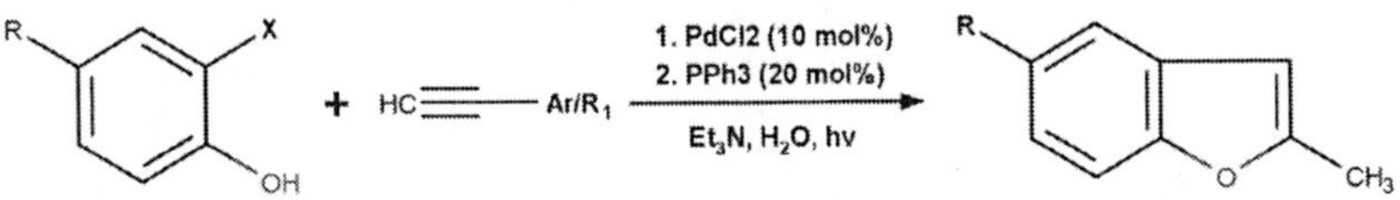

a : R = H, X = I/Br, Ar = C_6H_5
b : R = CH3, X = Br, Ar = C_6H_5
c : R = Cl, X = Br, Ar = C_6H_5
d : R = H, X = I, Ar = (p-OCH_3) C_6H_5
e : R = CH_3, X = Br, Ar = (p-OCH_3) C_6H_5
f : R = Cl, X = Br, Ar = (p-OCH_3) C_6H_5
g : R = H, X = I, Ar = (p-Cl) C_6H_5
h : R= Cl, X = Br, R1 = C_6H_{13}

Scheme 14. Visible light Sonogashira coupling, and cyclization reaction of ortho-halophenols, and terminal alkynes in water.

Table 2. Encapsulated EO and their characteristics

Encapsulated EO	Mold	Method	Polymers	Size	Application/ Activity	Reference
Atractylodes macrocephala Koidz	Liposomes	RESS	-	173 nm	-	(Wen et al. 2010)
Citronella	Microcapsules	Complex coacervation	Gelatin CTLW and Sodium Alginate	434 µm	-	(de Matos, Scopel, and Dettmer 2018)
Curcumin	Nanoparticles	Nanoprecipitation	PEG-PLA and PVP	70-85 nm	-	(Chow et al. 2015)
Lippia sidoides	Nanoparticles	Spray Drying	Cashew Gum and Alginate	223-399 nm	Larvicidal	(de Oliveira, Paula, and Paula 2014)
Mustard	Microcapsules	Complex coacervation	Genipine	5-10 µm	Antimicrobial	(Peng et al. 2014)
Nin	Microcapsules	Spray Drying	Polyvinyl Alcohol, Arabic Gum, and Whey Protein Isolate/ Maltodrextine	28,84 µm, 32,43 µm e 52,88 µm	-	(Sittipummongkol and Pechyen 2018)
Thyme	Nanoparticles and Nanocapsules	Nanoprecipitation and Nanoencapsulation	Chitosan	6,4 nm and 9,1 nm	Antimicrobial	(Sotelo-Boyás et al. 2017)

4. Biological Activity of Benzofurans

Several drugs containing the benzofuran ring have been developed from compounds of natural and synthetic origin, due to their recognized biological and pharmacological interests.

The diversity of biological effects attributed to the benzofuran nucleus is associated with the radical change that can occur in the heterocyclic structure. Thus, several synthetic routes can be applied to its synthesis. The most well-known benzofurans compounds are ailanthoidol, amiodarone, and bufuralol (Khanam and Shamsuzzaman 2015b; Naik et al. 2015).

Benzofuran compounds have been studied for the treatment of cancer, viruses, microbes, fungus, and diseases such as depression, asthma, hypertension, heart failure, skin diseases (psoriasis, eczema, vitiligo, cutaneous lymphomas), gout, tuberculosis, etc. (Nevagi, Dighe, and Dighe 2015; Dawood 2013).

Studies report the wide variety of biological activities that benzofuran compounds, isolated from plant species, have (Hiremathad et al. 2015). Cao et al. (2015) (Cao et al. 2015) isolated 14 compounds from stem bark of *Styrax obassia*. The authors evidenced that the new isolated benzofuran (2-(3,4-dimethoxyphenyl)-5-(1,3-dihydroxypropyl)-7-methoxybenzofuran) presented anti-inflammatory activity, evaluated by the ability of reduction of lipopolysaccharide (LPS)-induced nitric oxide in RAW264.7 macrophages. In addition, pretreating cells along with the compound significantly eliminated their LPS-induced expression of COX-2 protein.

In the study of Liu et al. (2011) (Liu et al. 2011), extracts obtained with ethyl acetate (EtOAc) from ripe fruits of *Styrax agrestis* presented a new series of acetylcholinesterase (AChE)-inhibiting benzofurans, which can also act in prevention of peptide amyloid beta (Aβ) aggregation. The study suggested the influence that the structure that contains the furan ring, the ester group, and the length of hydrocarbon side chain, have on the inhibitory potential of benzofuran derivatives against AChE, the enzyme responsible for the metabolic and catalytic hydrolysis of acetylcholine, and related to Alzheimer's disease.

Timmers et al. (2015) (Timmers et al. 2015) attributed to benzofuran egonol, isolated from the fruits of *Styrax ramirezii*, inhibition of lipopolysaccharide (LPS)-induced nitric oxide (NO) production, release of reactive oxygen species (ROS), and inflammation biomarkers. In addition, egonol could reduce the messenger RNA (mRNA), levels of inducible nitric oxide synthase (iNOS), cyclooxygenase-2 (COX-2), interleukin-1β (IL-1β), and interleukin-6 (IL-6); evidencing the anti-inflammatory activity of the fruit.

Benzofurans from euparin extracted of *Petasites hybridus* showed moderate antioxidant activity by 2,2-diphenyl-1-picrylhydrazyl (DPPH) radical scavenging, and ferric (Fe^{3+}) reducing anti-oxidant power (FRAP); and high reduction activity when compared to standards butyl-hydroxytoluene (BHT) and t-butyl-hydroxyquinone (TBHQ) (Ezzatzadeh and Hossaini, 2018). Mohammed et al., (2016) (Mohammed, Omran, and Hussein 2016) associated to compound 2H-benzo[f]oxireno[2,3-E]benzofurano-8,(9H)-one,9-[[(1,3-benzodio, found in the methanolic extract of the fruits of *Piper nigrum*, its antimicrobial and antifungal activities.

Luo et al., (2012) (Luo et al. 2012) isolated two new 2-arylbenzofuran neolignans from methanolic extracts of *Zanthoxylum capense* roots; these compounds were named zanthocapensol (compound 1) and zanthocapensate (compound 2). The *in vitro* antibacterial activity of the compounds was tested against three gram-positive bacterial strains (*Staphylococcus aureus* ATCC 6538 and 25923, and *Enterococcus faecalis* ATCC 51299), and two gram-negative (*Pseudomonas aeruginosa* ATCC 27853 and *Escherichia coli* ATCC 25922). Compound 2 showed inhibitory activity against *Staphylococcus aureus* strains. Although the compounds are analogs, the antibacterial activity was reduced in presence of compound 1 due to a hydroxyl group terminal, which in compound 2 contained a methyl carboxylate group.

5. Controlled Release Techniques

5.1. Encapsulation

The development of formulations of EO microspheres, microcapsules, nanoparticles, and liposomes aims at the controlled release of bioactive compounds. In food products, they preserve substances from moisture, light, heat, oxygen, besides allowing practicality during handling. And in drugs, they act as substitutes for capsules, solutions, tablets, and conventional medicines, which have their activities limited to their physicochemical properties (Eratte et al. 2018; Rivas et al. 2017). Different encapsulation techniques were investigated in order to contribute to the conservation of bioactive compounds present in EOs. Nanoprecipitation, coacervation, spray drying, and rapid expansion of supercritical solutions (RESS) are examples of techniques applied in the synthesis of EO particles. Advantages of these methods include increased biological activity of EOs (Asbahani et al. 2015). Table 2 shows the characteristics of encapsulated EO, with their respective applications.

5.1.1. Nanoprecipitation

Developed by Fessi et al. (Fessi et al. 1989), nanoprecipitation or solvent displacement method is based on the interfacial deposition of a polymer, after the displacement of a water-miscible semi-polar solvent from a lipophilic solution, which allows the synthesis of narrowly distributed, low-energy monodisperse nanoparticles, applicable to lipidic substances, such as EOs. This technique is simple, quick, and easy to experiment. The mechanism requires the presence of at least two miscible phases (organic and aqueous). The organic phase is composed of EO and solvated polymers, whereas the aqueous phase is formed by a non-solvent or mixture of non-solvents for a given polymer, which may be supplemented with surfactants (Asbahani et al. 2015; Rivas et al. 2017).

5.1.2. Coacervation

Coacervation consists of a phase separation process induced by medium modification (temperature, pH, solubility, ionic strength), under controlled conditions. Two phases make up this method: coacervate (rich in colloids), and equilibrium fluid (low number of colloids). Coacervation can be classified as simple or complex. In simple coacervation method, a single polymer is involved, and the coacervates are formed by dehydration caused by the addition of a desolvating salt or liquid to the reaction medium, known as coacervation agent or inducing agent. In complex coacervation mechanisms, the electrostatic interactions between two or more polymers with opposite charges (usually polysaccharides and proteins), lead to coacervate grouping, and phase separation (Timilsena et al. 2019; Eratte et al. 2018).

5.1.3. Spray Drying

Spray drying is a simple and fast technique that allows the transformation of liquids (solutions, emulsions, dispersions, and pastes) into particles with adjustable size, distribution, shape, porosity, density, and chemical composition. Spray drying steps consist of: (1) heating of the drying gas; (2) droplet generation; (3) droplet drying, and (4) particle collection. In practice, the liquid is atomized in a certain nozzle, where the solvent is quickly removed from the droplets by the action of hot air, occurring then the spraying. The particles formed are then separated and collected (Arpagaus et al. 2018).

5.1.4. Rapid Expansion of Supercritical Solutions (RESS)

The process of rapid expansion of supercritical solutions (RESS) is divided into one extraction step, and one of rapid expansion. In extraction step, the solute saturates the supercritical fluid, whereas in rapid expansion, the saturated supercritical solution is depressurized through a heated nozzle to prevent clogging by solute precipitation into a low-pressure vessel. Due to the rapid depressurization, the density of the supercritical fluid decreases and, consequently, its solubility power also decreases, which leads to solvent supersaturation, and formation of small particles (Bagheri, Hashemipour, and Mirzaie 2019).

Conclusion

Compounds that have the benzofuran nucleus as preferred structure exhibit potent pharmacological activity against various targets that are involved in diseases caused by bacteria, fungi, viruses, and other microorganisms. Many articles have reported lead molecules that have this heterocycle as scaffold. These compounds can be obtained from natural marine and plant sources, and are found in essential oils and extracts so that they can be isolated from them. They can also be obtained by green synthetic route with high yields and purity. For the wide pharmacological activity of benzofurans, the ability of this heterocycle to support different radicals that give the molecule high selectivity and potency can be highlighted, according to the pharmacological mechanism investigated.

References

Arpagaus, Cordin., Andreas, Collenberg., David, Rütti., Elham, Assadpour. & Seid, Mahdi Jafari. (2018). "Nano Spray Drying for Encapsulation of Pharmaceuticals." *International Journal of Pharmaceutics*, *546* (1–2), 194–214. https://doi.org/10.1016/j.ijpharm.2018.05.037.

Asbahani, A. El., Miladi, K., Badri, W., Sala, M., Aït Addi, E. H., Casabianca, H., El Mousadik, A., et al. (2015). "Essential Oils: From Extraction to Encapsulation." *International Journal of Pharmaceutics*, *483* (1–2), 220–43. https://doi.org/10.1016/j.ijpharm.2014.12.069.

Bagheri, Hamidreza., Hassan, Hashemipour. & Maryam, Mirzaie. (2019). "Investigation on Hydrodynamic and Formation of Nano Particle by RESS Process: The Numerical Study." *Journal of Molecular Liquids*, *281*, (May), 490–505. https://doi.org/10.1016/j.molliq.2019.02.108.

Cao, Thao Quyen., Manh, Hung Tran., Jeong, Ah Kim., Phuong, Thao Tran., Jeong, Hyung Lee., Mi, Hee Woo., Hyeong, Kyu Lee. & Byung, Sun Min. (2015). "Inhibitory Effects of Compounds from Styrax Obassia on NO Production." *Bioorganic and Medicinal Chemistry Letters*, *25* (22), 5087–91. https://doi.org/10.1016/j.bmcl.2015.10.020.

Chand, Karam., Rajeshwari, Asha Hiremathad., Mahak, Singh., Amelia Santos, M. & Rangappa, S. Keri. (2017). "A Review on Antioxidant Potential of Bioactive Heterocycle Benzofuran: Natural and Synthetic Derivatives." *Pharmacological Reports*. Elsevier B.V. https://doi.org/10.1016/j.pharep.2016.11.007.

Chen, Chien-chang., Yuhsin, Chen., Yi-ting, His., Chih-Sheng, Chang., Li-Fen, Huang., Chi-tang, Ho., Tzong-der, Way. & Jungyie, Kao. (2013). "Chemical Constituents and Anticancer Activity of Curcuma Zedoaria Roscoe Essential Oil against Non-Small Cell Lung Carcinoma Cells *in Vitro* and *in Vivo*." *Journal of Agricultural and Food Chemistry*, *61* (47), 11418–27. https://doi.org/10.1021/jf4026184.

Chinchilla, Rafael. & Carmen, Nájera. (2011). "Recent Advances in Sonogashira Reactions." *Chemical Society Reviews*. https://doi.org/10.1039/c1cs15071e.

Chow, Shing Fung., Ka, Yee Wan., Kwok, Kin Cheng., Ka, Wai Wong., Changquan, Calvin Sun., Larry, Baum. & Albert, Hee Lum Chow. (2015). "Development of Highly Stabilized Curcumin Nanoparticles by Flash Nanoprecipitation and Lyophilization." *European Journal of Pharmaceutics and Biopharmaceutics*, *94*, (August), 436–49. https://doi.org/10.1016/j.ejpb.2015.06.022.

Dawood, Kamal M. (2013). "Benzofuran Derivatives: A Patent Review." *Expert Opinion on Therapeutic Patents*, *23* (9), 1133–56. https://doi.org/10.1517/13543776.2013.801455.

Dharmalingam, R. & Nazni, P. (2013). "Phytochemical Evaluation of Coriandrum L Flowers." *International Journal of Food and Nutritional Sciences*, *2*, 34–39.

Dhouioui, Mouna., Abdennacer, Boulila., Hedia, Chaabane., Mongia, Saïd Zina. & Hervé, Casabianca. (2016). "Seasonal Changes in Essential Oil Composition of *Aristolochia Longa* L. Ssp. Paucinervis Batt.

(Aristolochiaceae) Roots and Its Antimicrobial Activity." *Industrial Crops and Products*, *83*, (2016), 301–6. https://doi.org/10.1016/j.indcrop.2016.01.025.

Djakovitch, Laurent. & Patrick, Rollet. (2004). "Sonogashira Cross-Coupling Reactions Catalysed by Copper-Free Palladium Zeolites." *Advanced Synthesis and Catalysis*, *346* (13–15), 1782–92. https://doi.org/10.1002/adsc.200404141.

Eratte, Divya., Kim, Dowling., Colin, J. Barrow. & Benu, Adhikari. (2018). "Recent Advances in the Microencapsulation of Omega-3 Oil and Probiotic Bacteria through Complex Coacervation: A Review." *Trends in Food Science & Technology*, *71*, (January), 121–31. https://doi.org/10.1016/j.tifs.2017.10.014.

"European Patent Office." n.d. Accessed September 3, 2019. https://worldwide.espacenet.com/searchResults?ST=singleline&locale=en_EP&submitted=true&DB=&query=benzofuran.

Fessi, H., Puisieux, F., Devissaguet, J. Ph., Ammoury, N. & Benita, S. (1989). "Nanocapsule Formation by Interfacial Polymer Deposition Following Solvent Displacement." *International Journal of Pharmaceutics*, *55* (1), R1–4. https://doi.org/10.1016/0378-5173(89)90281-0.

Gadir, Suad A. & Ibtisam, M. Ahmed. (2014). "Commiphora Myrrha and Commiphora Africana Essential Oils." *Journal of Chemical and Pharmaceutical Research*, *6* (7), 151–56.

Ghosh, Somnath., Jhantu, Das. & Forid, Saikh. (2012). "A New Synthesis of 2-Aryl/Alkylbenzofurans by Visible Light Stimulated Intermolecular Sonogashira Coupling and Cyclization Reaction in Water." *Tetrahedron Letters*, *53* (44), 5883–86. https://doi.org/10.1016/j.tetlet.2012.08.078.

Grimshaw, James. & Norris, Thompson. (1987). "Cyclization of 2′-Bromodeoxybenzoins: A New Synthesis of 2-Phenylbenzofurans." *Journal of the Chemical Society, Chemical Communications*, no. 4, 240–41. https://doi.org/10.1039/C39870000240.

Hayakawa, Ichiro., Rieko, Shioya., Toshinori, Agatsuma., Hidehiko, Furukawa. & Yuichi, Sugano. (2004). "Thienopyridine and

Benzofuran Derivatives as Potent Anti-Tumor Agents Possessing Different Structure-Activity Relationships." *Bioorganic and Medicinal Chemistry Letters*, *14* (13), 3411–14. https://doi.org/10.1016/j.bmcl.2004.04.079.

Hirano, Koji., Tetsuya, Satoh. & Masahiro, Miura. (2011). "Copper-Catalyzed Annulative Amination of Ortho -Alkynylphenols with Hydroxylamines: Synthesis of 3-Aminobenzofurans by Umpolung Amination Strategy." *Organic Letters*, *13* (9), 2395–97. https://doi.org/10.1021/ol200651r.

Hiremathad, Asha., Mahadeo, R. Patil., Chethana, K. R., Karam, Chand., Amelia Santos, M. & Rangappa, S. Keri. (2015). "Benzofuran: An Emerging Scaffold for Antimicrobial Agents." *RSC Advances*, *5* (117), 96809–28. https://doi.org/10.1039/C5RA20658H.

Hou, Xue-Long., Zhen, Yang., Kap-Sun, Yeung. & Henry, N. C. Wong. (2008). "Chapter 5.3 Five-Membered Ring Systems: Furans and Benzofurans." In, 176–207. https://doi.org/10.1016/S0959-6380(08)80009-8.

Jäger, Anna K. (2005). "Is Traditional Medicine Better off 25 Years Later?" *Journal of Ethnopharmacology*. Elsevier Ireland Ltd. https://doi.org/10.1016/j.jep.2005.05.034.

Javidnia, K., Miri, R., Soltani, M. & Khosravi, A. R. (2010). "Chemical Constituents of the Essential Oil of Otostegia Michauxii Briq. From Iran." *Journal of Essential Oil Research*, *22* (1), 1–2. https://doi.org/10.1080/10412905.2010.9700252.

Khanam, Hena. & Shamsuzzaman. (2015a). "Bioactive Benzofuran Derivatives: A Review." *European Journal of Medicinal Chemistry*, *97* (1), 483–504. https://doi.org/10.1016/j.ejmech.2014.11.039.

———. (2015b). "Bioactive Benzofuran Derivatives: A Review." *European Journal of Medicinal Chemistry*, *97*, (June), 483–504. https://doi.org/10.1016/j.ejmech.2014.11.039.

Kim, Jun-Ran., In-Hong, Jeong., Young, Su Lee. & Sang-Guei, Lee. (2015). "Insecticidal Activity of Cinnamon Essential Oils, Constituents, and (E)- Cinnamaldehyde Analogues against Metcalfa Pruinosa Say (Hemiptera: Flatidae) Nymphs and Adults." *Korean*

Journal of Applied Entomology, December, 375–82. https://doi.org/10.5656/KSAE.2015.10.0.056.

Lavanya, Andiappan., Rajendran, Sribalan. & Vediappen, Padmini. (2017). "Synthesis and Biological Evaluation of New Benzofuran Carboxamide Derivatives." *Journal of Saudi Chemical Society*, *21* (3), 277–85. https://doi.org/10.1016/j.jscs.2015.06.008.

Lee, Dong Hwan., Ki, Hyeok Kwon. & Chae, S. Yi. (2012). "Dehydrative C-H Alkylation and Alkenylation of Phenols with Alcohols: Expedient Synthesis for Substituted Phenols and Benzofurans." *Journal of the American Chemical Society*, *134* (17), 7325–28. https://doi.org/10.1021/ja302710v.

Liu, Jiawei., Vincent, Dumontet., Anne, Laure Simonin., Bogdan, I. Iorga., Vincent, Guerineau., Marc, Litaudon., Van, Hung Nguyen. & Françoise, Gueritte. (2011). "Benzofurans from Styrax Agrestis as Acetylcholinesterase Inhibitors: Structure-Activity Relationships and Molecular Modeling Studies." *Journal of Natural Products*, *74* (10), 2081–88. https://doi.org/10.1021/np200308j.

Lu, Biao., Bao, Wang., Yihua, Zhang. & Dawei, Ma. (2007). "CuI-Catalyzed Domino Process to 2,3-Disubstituted Benzofurans from 1-Bromo-2-Iodobenzenes and β-Keto Esters." *Journal of Organic Chemistry*, *72* (14), 5337–41. https://doi.org/10.1021/jo070729r.

Luo, Xuan., Liliana, Pedro., Vladimir, Milic., Silva, Mulhovo., Aida, Duarte., Noélia, Duarte. & Maria, José U. Ferreira. (2012). "Antibacterial Benzofuran Neolignans and Benzophenanthridine Alkaloids from the Roots of Zanthoxylum Capense." *Planta Medica*, *78* (2), 148–53. https://doi.org/10.1055/s-0031-1280289.

Ma, Weimin., Jiawei, Huang., Xianyu, Huang., Shulin, Meng., Zhengwei, Yang., Chao, Li., Yue, Wang., Ting, Qi. & Baolin, Li. (2019). "Direct Construction of 2,3-Unsubstituted Benzofurans and Benzothiophenes: Via a Metal-Free Catalyzed Intramolecular Friedel-Crafts Reaction." *Organic Chemistry Frontiers*, *6* (4), 493–97. https://doi.org/10.1039/c8qo01204k.

Martins, Pedro., João, Jesus., Sofia, Santos., Luis, R. Raposo., Catarina, Roma-Rodrigues., Pedro, Viana Baptista. & Alexandra, R. Fernandes.

(2015). "Heterocyclic Anticancer Compounds: Recent Advances and the Paradigm Shift towards the Use of Nanomedicine's Tool Box." *Molecules, 20* (9), 16852–91. https://doi.org/10.3390/ molecules 200916852.

Masubuchi, Miyako., Hirosato, Ebiike., Ken, Ichi Kawasaki., Satoshi, Sogabe., Kenji, Morikami., Yasuhiko, Shiratori., Shinji, Tsujii., et al. (2003). "Synthesis and Biological Activities of Benzofuran Antifungal Agents Targeting Fungal N-Myristoyltransferase." *Bioorganic and Medicinal Chemistry, 11* (20), 4463–78. https://doi.org/10.1016/ S0968-0896(03)00429-2.

Matos, Eric F. de., Bianca, S. Scopel. & Aline, Dettmer. (2018). "Citronella Essential Oil Microencapsulation by Complex Coacervation with Leather Waste Gelatin and Sodium Alginate." *Journal of Environmental Chemical Engineering, 6* (2), 1989–94. https://doi.org/10.1016/j.jece.2018.03.002.

Miao, Yu-hang., Yu-heng, Hu., Jie, Yang., Teng, Liu., Jie, Sun. & Xiao-jing, Wang. (2019). "Natural Source, Bioactivity and Synthesis of Benzofuran Derivatives." *RSC Advances, 9* (47), 27510–40. https://doi.org/10.1039/c9ra04917g.

Mohammed, Ghaidaa Jihadi., Aseel, Mohammad Omran. & Haider, Mashkoor Hussein. (2016). "Antibacterial and Phytochemical Analysis of Piper Nigrum Using Gas Chromatography – Mass Spectrum and Fourier-Transform Infrared Spectroscopy." *International Journal of Pharmacognosy and Phytochemical Research, 8* (6), 977–96.

Munin, Aude. & Florence, Edwards-Lévy. (2011). "Encapsulation of Natural Polyphenolic Compounds; a Review." *Pharmaceutics*. https://doi.org/10.3390/pharmaceutics3040793.

Naik, Ravi., Dipesh, S. Harmalkar., Xuezhen, Xu., Kyusic, Jang. & Kyeong, Lee. (2015). "Bioactive Benzofuran Derivatives: Moracins A-Z in Medicinal Chemistry." *European Journal of Medicinal Chemistry, 90*, 379–93. https://doi.org/10.1016/j.ejmech.2014.11.047.

Nevagi, Reshma J., Santosh, N. Dighe. & Satish, N. Dighe. (2015). "Biological and Medicinal Significance of Benzofuran." *European*

Journal of Medicinal Chemistry, *97* (1), 561–81. https://doi.org/10.1016/j.ejmech.2014.10.085.

Oliveira, Erick F. de., Haroldo, C. B. Paula. & Regina, C. M. de Paula. (2014). "Alginate/Cashew Gum Nanoparticles for Essential Oil Encapsulation." *Colloids and Surfaces B: Biointerfaces*, *113*, (January), 146–51. https://doi.org/10.1016/j.colsurfb.2013.08.038.

Oramas-Royo, Sandra., Kriss, Dayana Pantoja., Ángel, Amesty., Carmen, Romero., Isabel, Lorenzo-Castrillejo., Félix, Machín. & Ana, Estévez-Braun. (2017). "Synthesis and Antibacterial Activity of New Symmetric Polyoxygenated Dibenzofurans." *European Journal of Medicinal Chemistry*, *141*, (December), 178–87. https://doi.org/10.1016/j.ejmech.2017.09.062.

Peng, Chao., Su-Qing, Zhao., Jun, Zhang., Gui-Ying, Huang., Lan-Ying, Chen. & Feng-Yi, Zhao. (2014). "Chemical Composition, Antimicrobial Property and Microencapsulation of Mustard (Sinapis Alba) Seed Essential Oil by Complex Coacervation." *Food Chemistry*, *165*, (December), 560–68. https://doi.org/10.1016/j.foodchem.2014.05.126.

Perumal, Anand Babu., Periyar, Selvam Sellamuthu., Reshma, B. Nambiar. & Emmanuel, Rotimi Sadiku. (2016). "Antifungal Activity of Five Different Essential Oils in Vapour Phase for the Control of Colletotrichum Gloeosporioides and Lasiodiplodia Theobromae *in Vitro* and on Mango." *International Journal of Food Science & Technology*, *51* (2), 411–18. https://doi.org/10.1111/ijfs.12991.

Radadiya, Ashish. & Anamik, Shah. (2015). "Bioactive Benzofuran Derivatives: An Insight on Lead Developments, Radioligands and Advances of the Last Decade." *European Journal of Medicinal Chemistry*. Elsevier Masson SAS. https://doi.org/10.1016/j.ejmech.2015.01.021.

Rivas, Claudia Janeth Martínez., Mohamad, Tarhini., Waisudin, Badri., Karim, Miladi., Hélène, Greige-Gerges., Qand, Agha Nazari., Sergio, Arturo Galindo Rodríguez., Rocío, Álvarez Román., Hatem, Fessi. & Abdelhamid, Elaissari. (2017). "Nanoprecipitation Process: From Encapsulation to Drug Delivery." *International Journal of*

Pharmaceutics, *532* (1), 66–81. https://doi.org/10.1016/ j.ijpharm. 2017.08.064.

Saini, Mukhtyar. S., Aran, Kumar., Jaya, Dwivedi. & Rakesh, Singh. (2013). "A Review: Biological Significances of Heterocyclic Compounds." *International Journal of Pharma Sciences and Research*, *4* (3), 66–77.

Siani, Antonio C., Ana, M. P. Silva., Marcos, J. Nakamura., Márcia, V. De Carvalho., Maria, G. M. O. Henriques., Ramos, F. S. & Carlos, R. Kaiser. (2001). "Chemical Composition and Anti-Inflammatory Activity of the Hydrodistillate from Mariscus Pedunculatus." *J. Braz. Chem. Soc*., Vol. *12*.

Sittipummongkol, Kanyarat. & Chiravoot, Pechyen. (2018). "Production, Characterization and Controlled Release Studies of Biodegradable Polymer Microcapsules Incorporating Neem Seed Oil by Spray Drying." *Food Packaging and Shelf Life*, *18*, (December), 131–39. https://doi.org/10.1016/j.fpsl.2018.09.001.

Soejarto, D. D., Fong, H. H. S., Tan, G. T., Zhang, H. J., Ma, C. Y., Franzblau, S. G., Gyllenhaal, C. et al. (2005). "Ethnobotany/ Ethnopharmacology and Mass Bioprospecting: Issues on Intellectual Property and Benefit-Sharing." *Journal of Ethnopharmacology*. Elsevier Ireland Ltd. https://doi.org/10.1016/j.jep.2005.05.031.

Sotelo-Boyás, M., Correa-Pacheco, Z., Bautista-Baños, S. & Gómez y Gómez. Y. (2017). "Release Study and Inhibitory Activity of Thyme Essential Oil-Loaded Chitosan Nanoparticles and Nanocapsules against Foodborne Bacteria." *International Journal of Biological Macromolecules*, *103*, (October), 409–14. https://doi.org/10.1016/ j.ijbiomac.2017.05.063.

Structure, The., Alan, R. Katritzky., Charles, W. Rees., Eric, F. V. Scriven. & Clive, W. Bird. (1995). "Comprehensive Heterocyclic Chemistry II: A Review of the Literature 1982-1995." In *Comprehensive Heterocyclic Chemistry II: A Review of the Literature, 1982-1995*, 11628.

Sun, Nan., Peng, Huang., Yifan, Wang., Weimin, Mo., Baoxiang, Hu., Zhenlu, Shen. & Xinquan, Hu. (2015). "Zeolite-Catalyzed Synthesis of 2,3-Unsubstituted Benzo[b]Furans via the Intramolecular Cyclization of 2-Aryloxyacetaldehyde Acetals." *Tetrahedron, 71* (29), 4835–41. https://doi.org/10.1016/j.tet.2015.05.029.

Timilsena, Yakindra Prasad., Taiwo, O. Akanbi., Nauman, Khalid., Benu, Adhikari. & Colin, J. Barrow. (2019). "Complex Coacervation: Principles, Mechanisms and Applications in Microencapsulation." *International Journal of Biological Macromolecules*, 121 (January), 1276–86. https://doi.org/10.1016/j.ijbiomac.2018.10.144.

Timmers, Michael A., Jorge, L. Guerrero-Medina., Debora, Esposito., Mary, H. Grace., Octavio, Paredes-López., Pedro, A. García-Saucedo. & Mary, Ann Lila. (2015). "Characterization of Phenolic Compounds and Antioxidant and Anti-Inflammatory Activities from Mamuyo (Styrax Ramirezii Greenm.) Fruit." *Journal of Agricultural and Food Chemistry, 63* (48), 10459–65. https://doi.org/10.1021/ acs.jafc. 5b04781.

Torborg, Christian., Alexander, Zapf. & Matthias, Beller. (2008). "Palladium Catalysts for Highly Selective Sonogashira Reactions of Aryl and Heteroaryl Bromides." *Chem Sus Chem, 1* (1–2), 91–96. https:// doi.org/10.1002/cssc.200700004.

Verma, Nidhi. & Sudhir, Shukla. (2015). "Impact of Various Factors Responsible for Fluctuation in Plant Secondary Metabolites." *Journal of Applied Research on Medicinal and Aromatic Plants*. Elsevier GmbH. https://doi.org/10.1016/j.jarmap.2015.09.002.

Wen, Zhen., Bo, Liu., Zongkun, Zheng., Xinkui, You., Yitao, Pu. & Qiong, Li. (2010). "Preparation of Liposomes Entrapping Essential Oil from Atractylodes Macrocephala Koidz by Modified RESS Technique." *Chemical Engineering Research and Design, 88* (8), 1102–7. https:// doi.org/10.1016/j.cherd.2010.01.020.

"World Intellectual Property Organization." n.d. Accessed September 3, 2019. https://patentscope.wipo.int/search/en/result.jsf?_vid=P22-JZZP4Q-17262.

Yum, Eul Kgun., Ok Kyung, Yang., Ji Eun, Kim. & Hee, Jung Park. (2013). "Synthesis of 2-Substituted Benzofurans from o-Iodophenols and Terminal Alkynes with a Recyclable Palladium Catalyst Supported on Nano-Sized Carbon Balls under Copper- and Ligand-Free Conditions." *Bulletin of the Korean Chemical Society*, *34* (9), 2645–49. https://doi.org/10.5012/bkcs.2013.34.9.2645.

In: Benzofuran: Production and Applications ISBN: 978-1-53617-873-9
Editor: Murilo Souza Barros

Chapter 2

DIHYDROBENZOFURAN NEOLIGNANS: AN OVERVIEW OF THEIR CHEMISTRY AND BIOLOGICAL ACTIVITIES

Herbert J. Dias[1,*], Eliane O. Silva[2,*], Tatiana M. Vieira[1,*] and Antônio E. M. Crotti[1,†]
[1]Departamento de Química, Faculdade de Filosofia, Ciências e Letras de Ribeirão Preto, Universidade de São Paulo, Ribeirão Preto, SP, Brazil
[2]Instituto de Química, Universidade Federal da Bahia, Salvador, BA, Brazil

ABSTRACT

Dihydrobenzofuran neolignans (DBNs) are secondary metabolites arising from the coupling of two phenylpropanoid (C_6C_3) units during the shikimate biossynthetic pathway. These compounds are ubiquotous in the plant kingdom, especially in higher plants, and display several biological activities that are traditionally exploited by the pharmaceutical and food industries. Because of the interest in the DBN structure, chemistry, and

* Authors contributed equally to this work.
† Corresponding Author's E-mail: millercrotti@ffclrp.usp.br.

bioactivities, the current literature contains numerous studies on their preparation by both the phytochemical and synthetic approaches. Here, we present the natural sources and synthetic routes from which DBNs can be obtained and highlight their biological potential.

Keywords: dihydrobenzofuran neolignans, natural products, chemical synthesis, biological activity

INTRODUCTION

According to the International Union of Pure and Applied Chemistry (IUPAC), lignans and neolignans are compounds that are characterized by the coupling of two phenylpropanoid (C_6C_3) units [1]. For nomenclature purposes, the C_6C_3 phenylpropanoid unit is numbered from 1 to 6 in the aromatic ring, starting from the carbon attached to the propyl group, and the propyl group is numbered from 7 to 9, starting from the carbon attached to the benzene ring. Currently, the term "lignan" refers to structures that contain two C_6C_3 units bearing a β,β' (8-8') linkage (Scheme 1a), whereas the term "neolignan" must be used for compounds that originate from carbon-carbon coupling other than 8-8' (Scheme 1b) [1].

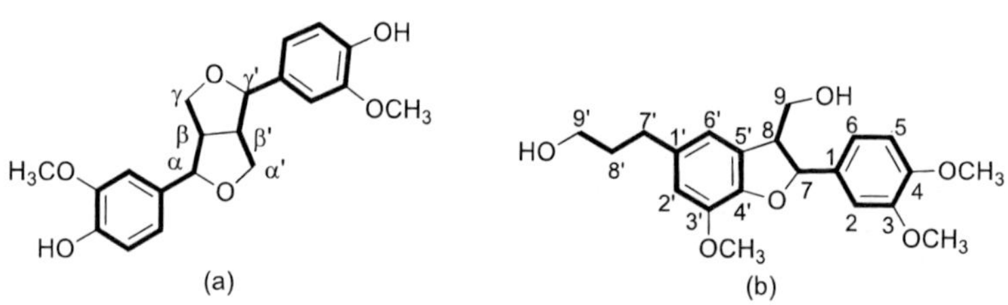

Scheme 1. Differences between the lignan (a) and the neolignan (b) chemical structures.

Dihydrobenzofuran neolignans (DBNs) contain two C_6C_3 units displaying 7.O.4' and 8.5' linkages (Scheme 1b). The term "neolignan" was formally introduced by Gottlieb only in 1978 [2]. Before that year, the general term "lignoids", as described by Haworth, was extensively employed to define all the coupled C_6C_3 units, including lignans and neolignans [3]. Although the distinction between these two independent groups of lignoids has been based on the evolution of biosynthetic knowledge [4], the scientific community has changed the use of these terms periodically over the years. This fact not only makes the search for dihydrobenzofuran neolignans in scientific databases a major challenge, but also induces the student/researcher to underestimate the current literature on this class of compounds. For example, our search for DBN substructures in the SciFinder® database retrieved about 1828 publications (letters, books or papers), 1355 of which were grouped according to the subjects "natural products", "biological activities", and "synthesis" (Figure 1). In contrast, the keywords "dihydrobenzofuran neolignan" matched only 43 publications. This enormous discrepancy comes from the fact that the terms "lignin", "lignan", "phenylbenzofuran", "benzofuran", and others have also been applied to refer to DBN-type structures.

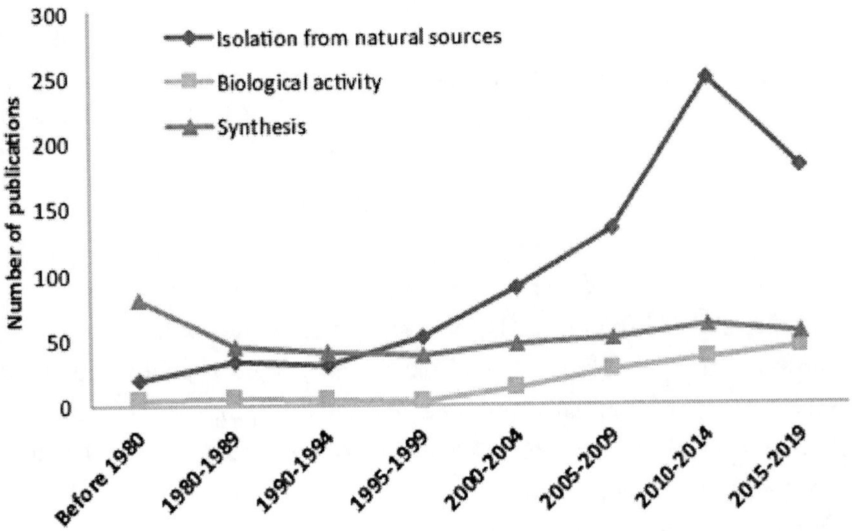

Figure 1. Number of publications on natural and synthetic DBNs and their biological activities between 1990 and 2019.

Most studies on DBNs have been published over the last fifteen years (from 2004 to 2019), as depicted in Figure 1. The increasing number of reports on natural DBNs is directly associated with advances in the isolation and characterization techniques, whilst the search for lead compounds has increased the number of DBNs that have been investigated for their biological actions over this period. Despite the lower number of publications on DBN synthesis and/or structural modification as compared to the number of publications that deal with their isolation from plants and/or evaluate their biological properties, about 30% of the chemical structures reported in this chapter have been obtained by synthesis and/or structural modification.

Even though the DBN chemical structures consist of two phenylpropanoid units only, these compounds present interesting structural diversity, which contributes to a broad spectrum of biological activities. The plant kingdom remains the main source of DBN structural diversity. However, in the last two decades, various synthetic strategies have emerged to build and to modify the structural core of these compounds. Moreover, some DBNs have served as lead compounds for the chemical synthesis of derivatives to improve their biological actions. In this chapter, we provide an overview of the DBN structures, natural sources, synthesis, and biological activities covering the literature since 1970.

Natural Dihydrobenzofuran Neolignan Sources

In terms of biosynthesis, DBNs are secondary metabolites originating from oxidative coupling of hydroxycinnamyl-type alcohol monomers (Scheme 2) [5, 6]. Peroxidase plays a key role in this biosynthetic pathway because it promotes one-electron oxidation of the monomers [7]. The most important monomers are 4-hydroxycinnamyl alcohol (p-coumaryl alcohol), coniferyl alcohol, and sinapyl alcohol, but they can vary depending on the plant taxon. For instance, Gymnosperms dimerize coniferyl alcohol mainly; dicotyledonous plants dimerize coniferyl alcohol and sinapyl alcohol, while monocotyledons use p-coumaryl, coniferyl, and sinapyl alcohols [5]. In Scheme 2, one-electron oxidation involving the coniferyl alcohol phenol hydroxyl results in a resonance-stabilized radical species. Electron pairing through coupling of two resonance structures provides a range of dimeric systems containing reactive quinonemethides, which are susceptible to nucleophilic attack by hydroxyl groups in the same system, or by water molecules from the medium. Coupling of the coniferyl alcohol monomers, as exemplified in Scheme 2, yields dehydrodiconiferyl alcohol and represents the general biosynthetic pathway for DBNs, which are 7.O.4'- and 8.5'- linked dimmers.

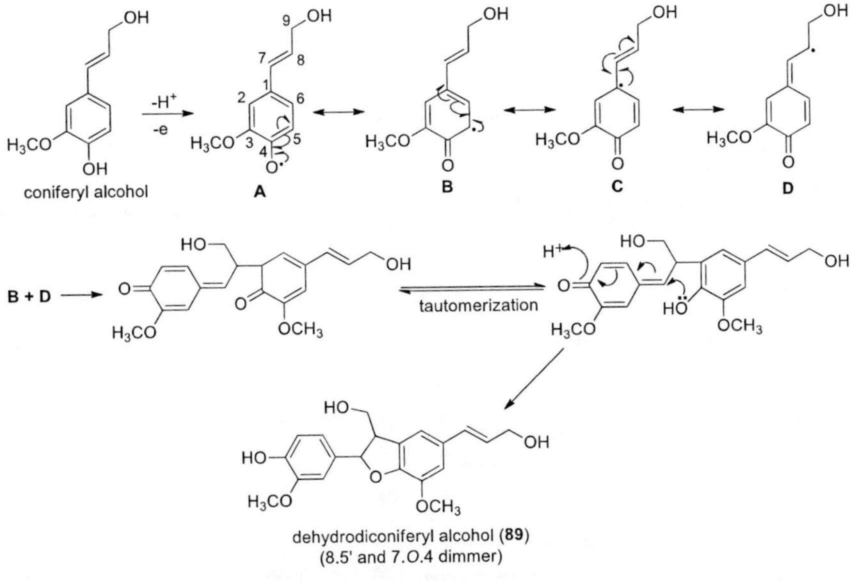

Scheme 2. Biosynthesis of dehydrodiconiferyl alcohol (89) [5, 6].

Coupling of the B- and D-type resonance forms can also produce other DBN classes derived from burchellin by Cope rearrangement and oxy-Cope rearrangements. Nevertheless, studies have reported that DBNs can be converted into other neolignans, as shown in Scheme 3 [8]. Although these neolignans are not the focus of this chapter, they will also be included when the occurrence of these compounds and DBNs are described in the same plant species.

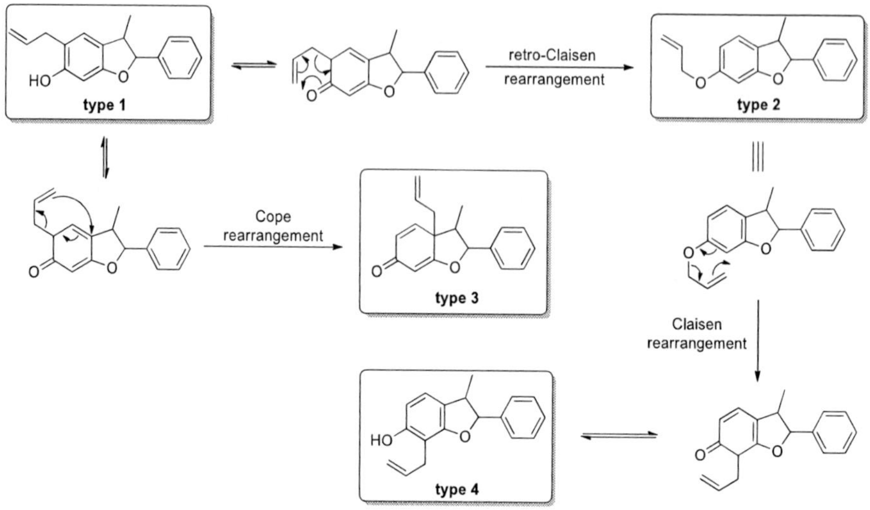

Scheme 3. Dihydrobenzofuran neolignan conversion into other dihydrobenzofuran-like neolignans [8].

With respect to the DBN stereochemistry, the biosynthetic approach based on oxidative coupling of phenylpropanoids implies that four diasteroisomers can emerge due to the stereocenters at C-7 and C-8 (7*R*,8*R*; 7*S*,8*S*; 7*R*,8*S*; and 7*S*,8*R*), as illustrated in Scheme 4. Nonetheless, only in some cases has the absolute stereochemistry of DBNs isolated from plants been exploited. This fact may be associated with difficulties in isolating DBNs as pure enantiomers to determine their absolute stereochemistry by circular dichroism [9] or X-ray crystallography [10]. Additional handicaps can arise from limitations to generating a crystal for X-ray crystallography analysis from small quantities of the isolated compounds, or even from the oily or gummy aspect of some DBNs. As a result of the difficulties in obtaining DBNs as pure enantiomers and thus investigate their absolute stereochemistry, only the C-7 and C-8 relative stereochemistry has been reported, and the compounds are referred to as *cis* (I and II) or *trans* (III and IV) (Scheme 4). The value of the coupling constant between H-7 and H-8 ($J_{7,8}$), which is measured from the ^{1}H Nuclear Magnetic Resonance (NMR) spectrum, usually helps to differentiate between *cis* and *trans* DBNs. According to Li and co-workers, the coupling constant $J_{7,8}$ in the neolignan cores is higher for *cis* isomers (8.2-8.4 Hz) than for *trans* isomers (6.5-7.3 Hz) [11, 12]. In some cases, though, not even the DBN relative stereochemistry has been assigned. On the other hand, mixtures of two enantiomers have been represented by only one of the enantiomeric structures (e.g., III or IV). In this chapter, although we use a standard way to represent all the compounds, we have respected the enantiomer that has been used to represent the DBN in the original papers. Besides that, for compounds that have been isolated as racemic mixtures of *cis* or *trans* enantiomers and which have been represented by two distinct enantiomeric structures in different papers, we have randomly chosen one of them.

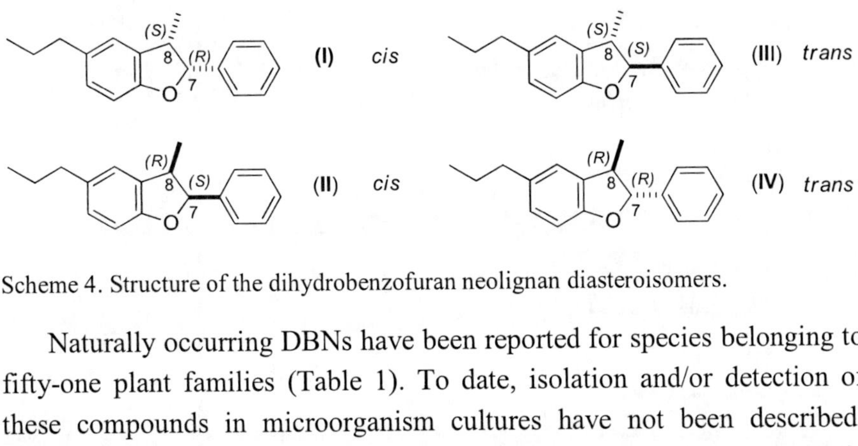

Scheme 4. Structure of the dihydrobenzofuran neolignan diasteroisomers.

Naturally occurring DBNs have been reported for species belonging to fifty-one plant families (Table 1). To date, isolation and/or detection of these compounds in microorganism cultures have not been described. Species belonging to the families Lauraceae (148), Rosaceae (31), Krameriaceae (25), Pinaceae (19), Illiciaceae (17), and Aristolochiaceae (16) contain the highest number of compounds as depicted in Figure 2. Iwashina reported that some chalcones can be specifically found in the families Rosaceae, Lauraceae, and Aristolochiaceae [13]. Therefore, the high occurrence of DBNs and chalcones in the aforementioned plant families could indicate not only a plausible biosynthetic relationship between these compounds as derived from the shikimate pathway, but also some chemotaxonomic relevance. Furthermore, the fact that about 82% of the species are trees or shrubs corroborates the hypothesis that phenylpropanoids in trees or shrubs come from the shikimate pathway, whereas these compounds originate from the acetate pathway in herbaceous Angiosperm species. DBN chemotaxonomic implications have been exploited since the seminal report by Otto Gottileb, published in 1972 [14]. Up to 1965, only one account of the chemistry of the family Lauraceae was available, so Gottlieb's study highlighted various chemical constituents of this family and discussed them in relation to the taxonomy of Lauraceae.

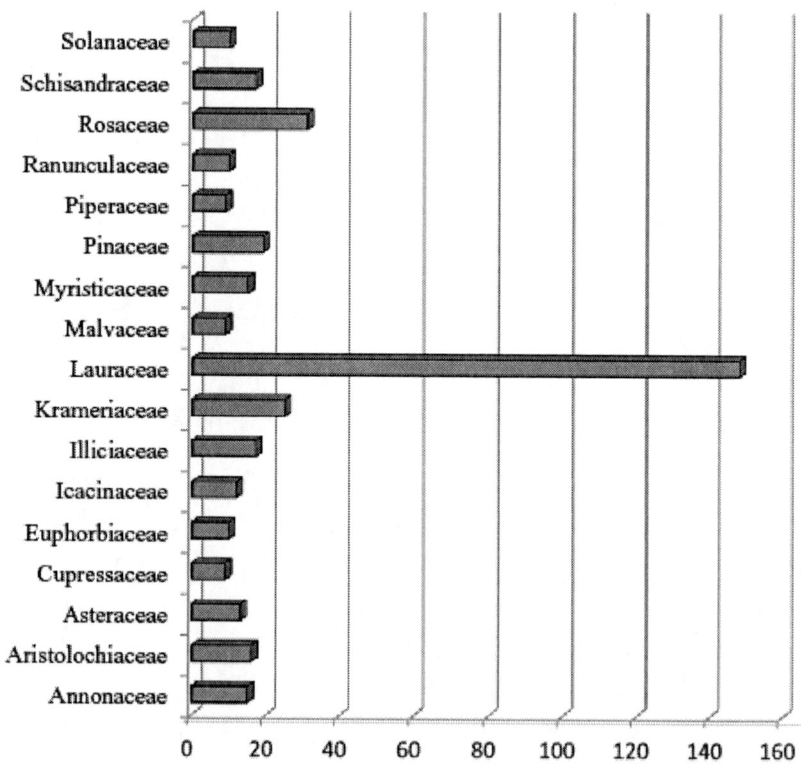

Figure 2. Dihydrobenzofuran neolignan distribution among the different plant families.

Table 1. Natural occurrence of dihydrobenzofuran neolignans

Plant species	Family	Compound	Ref.
Aniba. ferrea Kubitzki	Lauraceae	**366, 368, 374**	[15, 16]
Aniba burchellii	Lauraceae	**(338, 347, 348, 400, 401)**^{PhH}	[17, 18]
Aniba guianensis	Lauraceae	**358**$^{w\text{-}EtOH}$	[19]
Aniba sp	Lauraceae	**9, 8, 77, 292, 296, 294, 297, 295, 228, (233, 234, 346)**^{PhH} **337, 338, 339, 340, 341, 342, 345, 347, 350, 351, 354, (349, 357, 358, 362, 365, 369)**^{tw} **356, 368, (367)**$^{tw\text{-}PhH}$, **(359)**$^{tw\text{-}PhH}$, **(363)**^{PhH}, **294, 295, 297, 298, 299, (390)**^{PhH}, **391, 392, 393, 395, 397, 398, 379, 381, 382 (400)**^{PhH}, **406**	[14, 16, 18, 20-25]
Aniba riparia	Lauraceae	**(342, 345, 346, 356)**^{EtOH}	[16, 26]
Aniba simulans	Lauraceae	**(324, 325, 248, 352, 348, 392, 402, 403, 405)**$^{tw\text{-}PhH}$	[16, 27]
Aniba terminalis Ducke	Lauraceae	**(326, 327, 328, 337, 338, 342, 344, 343, 348, 390)**^{PhH}	[28]
Aquilaria sinensis (Lour.) Spreng	Thymelaeaceae	**(26, 28)**$^{st\text{-}EtOH}$	[29]
Arctium lappa L.[a]	Asteraceae	**(313, 314)**$^{sd\text{-}nd}$	[30]
Arenaria serpyllifolia L.[a]	Caryophyllaceae	**(218, 220)**$^{wp\text{-}EtOH\ 70\%}$	[31]
Aristolochia birostris Duchtre	Aristolochiaceae	**16**$^{ap\text{-}EtOH\ 95\%}$	[32]
Aristolochia fordiana Hemsley	Aristolochiaceae	**(2, 3, 5, 7, 8, 9, 45, 47, 48, 71, 72,)**$^{st\text{-}EtOH\ 95\%}$	[33]
Aristolochia pubescens Willd	Aristolochiaceae	**(74, 77, 138, 139)**$^{st+r\text{-}Hex}$	[34]
Bombax malabaricum DC	Malvaceae	**192, 420, 421**$^{EtOH\ 70\%}$	[35]
Brachanthemum gobicum	Asteraceae	**91, 123, 125, 126, 127**^{AW}	[36]
Breynia fruticosa (L.) Hook	Hippocrateaceae	**(154, 155)**$^{wp\text{-}Me_2CO\ 75\%}$	[37]
Bruguiera gymnorhiza[b,c]	Rhizophoraceae	**120**$^{sb\text{-}EtOH\ 95\%}$	[38]
Buddleja davidii[c]	Loganiaceae	**(230, 235, 236, 315)**$^{st\text{-}MeOH}$	[39]
Callistemon lanceolatus DC.	Myrtaceae	**60**$^{l\text{-}DCM}$	[40]
Canan galatifolia Finet & Gagnep.	Annonaceae	**25**$^{sb\text{-}EtOAc}$	[41]
Caryodaphnopsis baviensis (Lec.) A. Shaw	Lauraceae	**224**$^{fr\text{-}MeOH}$	[42]
Cedrus deodara[c]	Pinaceae	**146, 244**	[43]
Chilean própolis		**(122, 123)**^{MeOH}	[44]
Canarium album L. (Chinese olive)	Burseraceae	**12, 23, 210, 419**$^{EtOH\ 70\%}$	[45]

Table 1. (Continued)

Plant species	Family	Compound	Ref.
Cistus laurifolius L.	Cistaceae	**(165, 204)**$^{l+br\text{-}MeOH}$	[46]
Cleistopholis glauca Pierre ex Engl. & Diels	Annonaceae	**147**$^{sb\text{-}PE}$	[47]
Conocarpus erectus	Combretaceae	**41**^{CHCl_3}	[48]
Coptis japonica[a]	Ranunculaceae	**(93, 94, 95, 102, 103)**^{EtOH}	[49]
Coptis japonica var. *dissecta*[a]	Ranunculaceae	**(85, 87, 93, 94, 95, 102, 103)**$^{rh\text{-}EtOH}$	[50]
Crataegus pinnatifida Bge.[a]	Rosaceae	**(91, 144, 158, 193, 196, 197, 198, 205, 423, 424)**$^{sd\text{-}EtOH\ 70\%}$	[51, 52]
Croton erytrochilus[c]	Euphorblaceae	**(152)**^{lat,Et_2O}	[53]
Croton spp	Euphorblaceae	**(130, 152)**^{Et_2O}	[54, 55]
Dipteronia dyeriana Henry	Aceraceae[d]	**440**$^{br\text{-}nd}$	[56]
Disporum viridescens[c]	Liliaceae[e]	**(129, 160)**$^{l\text{-}MeOH}$	[57]
Durio zibethinus Murr	Bombacaceae	**195**$^{wb\text{-}MeOH}$	[58]
Epimedium diphyllum	Berberidaceae	**192, 193, 204**^{EtOH}	[59]
Eucommia ulmoides Oliv. V.	Eucommiaceae	**(146, 147, 168, 169)**$^{MeOH\ 50\%}$	[60]
Euryale ferox	Nymphaeaceae	**(146, 274, 275)**$^{EtOH\ 85\%}$	[61]
Euonymus alatus (Thunb.) Sieb.	Celastraceae	**(22, 26, 85, 89, 86, 147)**$^{l+t\text{-}MeOH\ 80\%}$	[62]
Euphrasia rostkoviana Hayne[a]	Orobanchaceae	**(30, 29, 317)**$^{wp\text{-}MeOH}$	[63]
Euterpe oleracea Mart.	Arecaceae	**(206, 209)**$^{fp\text{-}MeOH}$	[64]
Helicteres hirsuta Lour	Sterculiaceae	**(193, 194, 195)**$^{st\text{-}MeOH}$	[65]
Heracleum dissectum Ledeb.	Apeace	**(24, 136, 156)**$^{r\text{-}MeOH}$	[66]
Herpetospermum caudigerum[c]	Cucurbitaceae	**(23)**$^{sd\text{-}EtOH\ 95\%}$	[67]
Hibiscus cannabinus var. Salvador[a]	Malvaceae	**(104, 105, 194, 195, 106, 107)**$^{cs\text{-}Me_2CO}$	[68-70]
Hyoscyamus niger L.	Solanaceae	**108**$^{sd\text{-}EtOH\ 95\%}$	[71]
Illicium difengpi K.I.B et K.I.M.	Illiciaceae	**(246, 251, 252, 253, 254,** 257)$^{b\text{-}MeOH}$	[72]
Illicium henryi[c]	Illiciaceae	**183**$^{bk\text{-}EtOH\ 95\%}$	[73]
Illicium oligadrum	Illiciaceae	**(159, 163, 164, 167, 193)**$^{EtOH\ 95\%}$	[74]
Illicium wardii	Illiciaceae	**(266, 267, 268, 269, 309)** $^{EtOH\ 95\%}$	[75]
Juniperus communis L.	Cupressaceae	**278**$^{rbe\text{-}MeOH}$	[76]
Juniperus communis var. *depressa*	Cupressaceae	**(131, 134, 166, 204)** ^{MeOH}	[77]
Krameria cystisoides	Krameriaceae	**(41, 42, 72, 74, 174, 175, 226, 227, 242, 431, 432, 433, 435, 436, 437, 438)**^{MeOH}	[21, 78]
Krameria interior	Krameriaceae	**41, 42, 74, 227, 431, 432, 434, 435, 436, 439**^{MeOH}	[79]

Plant species	Family	Compound	Ref.
Krameria ixina	Krameriaceae	(**41, 43**)^{DCM}	[80]
Krameria lanceolata	Krameriaceae	(**226, 227, 411, 412, 431, 432, 433, 434, 435, 436, 437, 438**)^{MeOH}	[81]
Krameria lappacea	Krameriaceae	(**44, 227, 239, 240, 434, 437, 438, 439**)^{DCM}	[82]
Krameria parvifolia	Krameriaceae	(**41, 176**)^{DCM}	[83]
Larix leptolepis Gord.	Pinaceae	**312**$^{w\text{-}EtOH\ 95\%}$	[84]
Litsea chinpingensis	Lauraceae	**206**$^{EtOH\ 95\%}$	[16, 85]
Leea aequata L.	Vitaceae	(**207, 208**)^{EtOH}	[86]
Leontopodium leontopodioides (Willd.) Beauv	Asteraceae*	**191**$^{ap\text{-}EtOH\ 70\%}$	[87]
Licaria armeniaca	Lauraceae	(**346, 348, 349, 365, 397, 402**)$^{tw\text{-}PhH}$	[88, 89]
Licaria canella	Lauraceae	**369**^{PhH}	[16, 90]
Licaria aritu	Lauraceae	(**8, 9, 77, 76**)^{PhH}	[21]
Licaria chrysophylla	Lauraceae	(**379, 380, 454, 455**)^{EtOH}	[91, 92]
Lindera glauca (Siebold et Zucc.) Blume	Lauraceae	(**209, 221**)$^{twi\text{-}MeOH\ 80\%}$	[93]
Linum usitatissimum L.	Linaceae	(**231, 232**)$^{MeOH\ 80\%}$	[94, 95]
Litsea chinpingensis	Lauraceae	(**9, 177, 206**)$^{t\text{-}EtOH\ 95\%}$	[85]
Lycopodiella cernua (L.) Pic. Serm. (II)	Lycopodiaceae	(**249, 250**)$^{t\text{-}EtOH\ 95\%}$	[96]
Machilus obovatifolia	Lauraceae	(**6, 13, 15, 77, 137, 138, 139, 141, 186, 428, 429**)^{MeOH}	[97]
Machilus odoratissima	Lauraceae	(**9, 13, 15, 83**)^{EtOH}	[98]
Machilus thunbergii	Lauraceae	(**9, 77, 427**)^{DCM}	[99-102]
Magnolia liliflora^{c}	Magnoliaceae	(**232, 300, 301, 355, 407, 408, 444, 445**)$^{l\text{-}MeOH}$	[103]
Mappianthus iodoies^{c}	Icacinaceae	(**41, 49, 50, 75, 73, 82, 80, 89, 90, 91, 199**)$^{s+l\text{-}EtOH}$	[104]
Melia toosendan Sieb. et Zucc.	Meliaceae	(**148, 149, 150, 151**)$^{fr\text{-}EW}$	[105]
Miliusa fragrans Chaowasku and Kessler	Annonaceae	**74**$^{l\text{-}MeOH}$	[106]
Miliusa mollis Pierre	Annonaceae	(**225, 226, 241, 280, 281, 282, 283, 284, 285, 288**)$^{l\text{-}MeOH}$	[107]
Mitrephora wangii HU.	Annonaceae	(**2, 3, 5, 14, 409, 410**)$^{t+l\text{-}EtOAc}$	[108]
Myristica fragrans Houtt.	Myristicaceae	(**8, 9, 10, 13, 14, 46, 71, 74**)$^{st\text{-}MeOH}$	[109, 110]
Nectrandra amazonum Nees	Lauraceae	(**8, 9, 16, 154, 155, 195, 238, 293, 337, 446, 451, 452, 453**)$^{fr\text{-}EtOH}$	[16, 111, 112]
Nectandra glabrescens^{c}	Lauraceae	**9**$^{fr\text{-}EtOH}$	[113]
Nectandra miranda Sandw.	Lauraceae	(**76, 79, 137, 139, 141, 185, 294, 337, 352, 353, 392, 395, 413, 414, 415, 416, 417, 448, 449, 450**)$^{tw\text{-}PhH}$	[114]

Table 1. (Continued)

Plant species	Family	Compound	Ref.
Nectandra ssp	Lauraceae	(**342, 351, 353, 364, 397, 398, 399, 400**)$^{\text{w-PhH}}$	[16, 115]
Nicotiana tabacum L.	Solanaceae	(**94, 419**)$^{\text{s-MeOH 90\%}}$	[116]
Ochroma lagopus Swartz	Malvaceae	(**104, 105, 106**)$^{\text{w-EtOH}}$	[117]
Ocotea aciphylla (Nees) Mez.	Lauraceae	(**364, 375, 376, 377, 378**)$^{\text{w-PE}}$	[16, 118, 119]
Ocotea catharinensis Mez.	Lauraceae	(**364, 365, 366, 369, 370, 371, 372, 373**) $^{\text{bk-PE}}$	[16, 120, 121]
Ocotea macrophylla Kunth.	Lauraceae	**8**$^{\text{l-EtOH}}$, **9, 16, 154, 155, 195, 293, 337**	[16, 111, 122]
Ocotea porosa (Nees) L. Barr.	Lauraceae	(**347, 348**)$^{\text{w-PhH}}$, (**357, 358, 360, 361, 383, 384, 385, 386, 387, 388, 389, 400, 404**)$^{\text{b+w-EtOH}}$	[16, 24, 123-126]
Ocotea veraguensis Mez	Lauraceae	(**364, 402**) $^{\text{sb-PE}}$, **403** $^{\text{sb-PE and fr-EtOH}}$	[127, 128]
Otanthus maritimus L.A.	Asteraceae	**121**$^{\text{-DCM}}$	[129]
Paeonia lactiflora[a]	Ranunculaceae	(**140, 185, 193**)$^{\text{r-EW}}$	[130]
Pedicularis torta Maxim.[a]	Orobanchaceae	(**100, 102, 156, 192, 206, 211**)$^{\text{wp-EW}}$	[131]
Persea obovatifolia Kost.	Lauraceae	(**15, 82, 425, 426, 430, 443**)$^{\text{l-MeOH}}$	[132, 133]
Picea abies (L.) Karst.	Pinaceae	(**278, 313, 315, 316, 319, 320, 321, 322**) $^{\text{bk-Me}_2\text{CO}}$	[134]
Picrasma quassioides Benn.	Simaroubaceae	(**23, 27, 29, 85, 89, 91, 311**)$^{\text{st-EtOH 95\%}}$	[135, 136]
Pinus massona Lamb.	Pinaceae	(**244, 245, 247, 246, 248, 251**) $^{\text{n-EtOH}}$	[137]
Pinus thunbergii Parl.	Pinaceae	(**246, 247, 249, 251**) $^{\text{b-EW}}$	[138, 139]
Piper capense[c]	Piperaceae	(**329, 330**) $^{\text{r-CHCl}_3}$	[10]
Piper decurrens C.DC.	Piperaceae	**2**$^{\text{l-EtOH}}$	[140]
Piper regnellii[c]	Piperaceae	(**1, 41, 42, 46**)$^{\text{r-EtOAc}}$ (**41, 43, 173**)$^{\text{l-sCO}_2}$	[141-143]
Piper regnellii (Miq.) C. DC. var. *pallescens* (C. DC.) Yunck [a]	Piperaceae	(**41, 43**)$^{\text{l-EW}}$	[144]
Pittosporum glabratum Lindl.	Pittosporaceae	**212**$^{\text{r-EtOH 75\%}}$	[145]
Phyllanthus glaucus Wall.	Euphorbiaceae	(**129, 162, 181, 270, 271, 272, 273, 317**)$^{\text{wp-Me2CO}}$	[146]
Platycladus orientalis L. Franco	Cupressaceae	**133**$^{\text{l-EtOH 95\%}}$	[147]
Pleurothyrium cinereum (van der Werff)	Lauraceae	(**8, 9, 16, 154, 155, 156, 166, 195, 293, 337, 450, 451, 452, 453**)$^{\text{l-EtOH}}$	[111, 148]
Pluchea indica (L.) Less.	Asteraceae	(**31, 33, 32, 34**)$^{\text{r-MeOH}}$	[149]
Prunus domestica L.[a]	Rosaceae	**160**$^{\text{fr-EtOH 90\%}}$	[150]
Prunus tomentosa Thunb.	Rosaceae	(**9, 32, 38, 41, 42, 43, 51, 56, 81, 85, 98, 99, 101, 146, 189, 190, 193, 199, 200, 201, 219, 220, 306**) $^{\text{s-EtOH}}$	[6, 151]

Plant species	**Family**	**Compound**	**Ref.**
Rabdosia lophanthoides (Buch.-Ham.ex D.Don) Hara [a]	Lamiaceae	(**180**, **182**, **184**, **193**, **216**, **223**)$^{\text{wp-Me}_2\text{CO}}$	[152]
Rhodiola crenulata L.	Crassulaceae	(**171**, **223**, **245**, **247**, **264**)$^{\text{r-EtOH}}$	[153]
Rubus ideaus L.	Rosaceae	(**413**, **414**, **415**, **416**, **417**, **418**)$^{\text{f-EW}}$	[154]
Salix alba L.	Salicaceae	**282**$^{\text{bk-PE+MeOH}}$	[155]
Sambucus adnate[c]	Caprifoliaceae	**195**$^{\text{wp-MeOH}}$	[156]
Sambucus sieboldiana (Miq.)	Caprifoliaceae	(**128**, **140**)$^{\text{l-MeOH}}$	[157]
Schisandra sphenanthera[c]	Schisandraceae	(**161**, **203**)$^{\text{l-Me}_2\text{CO}}$	[158]
Selaginella sinensis (Desv.) Spring	Selaginellaceae	(**91**, **143**, **187**, **193**, **305**, **307**, **308**, **310**, **422**)$^{\text{wp-Me}_2\text{CO}}$	[159, 160]
Selaginella moellendorffii Hieron	Selaginellaceae	(**33**, **142**, **145**, **146**, **157**, **209**)$^{\text{wp-MeOH}}$	[161-163]
Solanum melongena L.	Solanaceae	**142**$^{\text{wp-MeOH}}$	[164]
Solanum sisymbrifolium Lam.	Solanaceae	(**92**, **115**, **116**, **146**, **160**, **168**)$^{\text{f-MeOH}}$	[165]
Tabebuia avellanedae Lorentz ex Griseb.	Bignoniaceae	**61**$^{\text{bk-bw}}$	[166]
Tabebuia roseo-alba (Ridley) Sandwith	Bignoniaceae	**192**$^{\text{wp-MeOH}}$	[167]
Taiwania cryptomerioides Hayata.	Cupressaceae	(**132**, **135**, **153**)$^{\text{l-Me}_2\text{CO}}$	[168]
Vaccinium dunalianum Wight[a]	Ericaceae	(**11**, **85**)$^{\text{bu-AW}}$	[169]
Viburnum awabuki[c]	Caprifoliaceae	**193**$^{\text{w-MeOH}}$	[170]
Virola carinata (Benth) Warburg.	Myristicaceae	(**291**, **290**)$^{\text{bk-CHCl}_3}$ (**286**, **287**, **303**, **304**)$^{\text{fr-CHCl}_3}$	[171, 172]
Vitex trifolia L.	Verbenaceae	(**193**, **306**)$^{\text{fr-BuOH}}$	[173]

[a]Herbaceous species. [b]Species reported in the original paper as *Bruguiera gymnorrhiza;* [c]the author who originally described this plant species was not provided; [d]This species currently belongs to the family Sapindaceae. [e]This species currently belongs to the family Colchicaceae; [f]This species currently belongs to the family Schisandraceae. [f]semi-synthetic derivative. bk: bark; br: branches; bu: buds; fr: fruits; la: latex; l: leaves; l+t: leaves and twigs; fp: fruit pulp; cs: core of stems; r: roots; rbk: root bark; rbe: ripe berries; rh: rhizomes; sb: stem bark; sd: seeds; st: stems; tw: trunk wood; twi: twigs; w: wood; wp: whole plant; BuOH: butanol; PhH: benzene; EtOH: ethanol; EW: ethanol/water (v/v); Hex: hexane; AW: acetone/water (v/v); EtOAc: ethyl acetate. PE: petroleum ether; MeOH: methanol; MW: methanol/water (v/v); Me_2CO: acetone; bw: boiling water; DCM: dichloromethane.

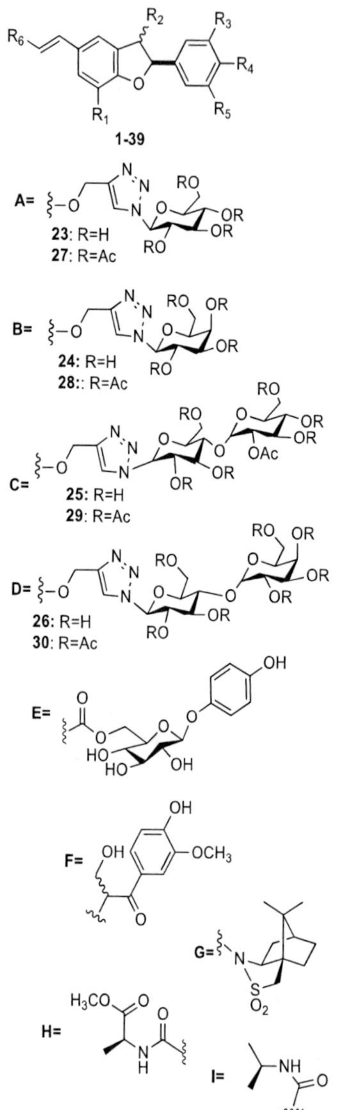

	R_1	R_2	R_3	R_4	R_5	R_6
1^{N}	H	βCH_3	H	OH	H	CH_3
2^{N}	H	αCH_3	H	OH	H	CH_3
3^{N}	H	αCH_3	H	OH	OCH_3	CH_3
4^{S}	H	βCH_3	H	OCH_3	H	CH_3
5^{N}	H	αCH_3	OCH_3	OCH_3	H	CH_3
6^{N}	OCH_3	αCH_3	OCH_3	OCH_3	OH	CH_3
7^{N}	H	αCH_3	OCH_2	OCH_2	H	CH_3
8^{S}	OCH_3	αCH_3	OCH_2	OCH_2	H	CH_3
9^{S}	OCH_3	αCH_3	H	OH	OCH_3	CH_3
10^{N}	OCH_3	αCH_3	OCH_2	OCH_2	OCH_3	CH_3
11^{N}	H	α**E**	H	OH	OH	**E**
12^{N}	OH	αCH_2OH	H	OH	OCH_3	CHO
13^{N}	OCH_3	αCH_3	OCH_3	OH	OCH_3	CH_3
14^{N}	OCH_3	αCH_3	OCH_3	OCH_3	OCH_3	CH_3
15^{NS}	OCH_3	αCH_3	OH	OH	OCH_3	CH_3
16^{S}	OCH_3	αCH_3	OCH_3	OCH_3	H	CH_3
17^{S}	OCH_3	αCH_3	OCH_3	OCH_2CCH	H	CH_3
18^{S}	OCH_3	αCH_3	OCH_3	**A**	H	CH_3
19^{S}	OCH_3	αCH_3	OCH_3	**B**	H	CH_3
20^{S}	OCH_3	αCH_3	OCH_3	**C**	H	CH_3
21^{S}	OCH_3	αCH_3	OCH_3	**D**	H	CH_3
22^{N}	OCH_3	αCH_2OH	**F**	OH	OCH_3	CHO
23^{NS}	OCH_3	αCH_2OH	H	OH	OCH_3	CH_2OH
24^{N}	OCH_3	αCH_2OH	H	OGlc	OCH_3	O-*n*-Bu
25^{N}	OCH_3	αCH_2OH	OH	H	OCH_3	CHO
26^{N}	OCH_3	αCH_2OH	OCH_3	OH	OCH_3	CH_2OH
27^{N}	OCH_3	αCH_2OH	OCH_3	OH	OCH_3	CHO
28^{N}	OCH_3	αCH_2OH	OCH_3	OCH_3	H	CH_2OGlc
29^{NS}	OCH_3	βCH_2OH	H	OH	OCH_3	CH_2OH
30^{N}	OCH_3	βCH_2OAc	H	$OGlc(OAc)_4$	OCH_3	CH_2OAc
31^{N}	OCH_3	αCH_2OGlc	H	OH	OCH_3	CHO
32^{N}	OCH_3	αCH_2OGlc	OCH_3	OH	OCH_3	CHO
33^{N}	OCH_3	αCH_2OGlc	H	OH	OCH_3	CH_2OH
34^{N}	OCH_3	αCH_2OGlc	H	OH	OCH_3	CH_2OCH_3
35^{S}	OCH_3	αCO_2H	H	OH	OCH_3	CO_2H
36^{S}	OCH_3	αCO_2CH_3	H	OH	OCH_3	CO_2CH_3
37^{S}	OCH_3	αCO-J	H	OH	OCH_3	CO-**G**
38^{NS}	OCH_3	αI	H	OH	OCH_3	**H**
39^{S}	OCH_3	αCO_2CH_3	H	OAc	OCH_3	CO_2CH_3

Scheme 5. Chemical structures of dihydrobenzofuran neolignans **1–39**. ^{S}: synthetic DBN; ^{N}: natural DBN; ^{NS}: obtained by synthesis and from natural sources, Glc: glucose.

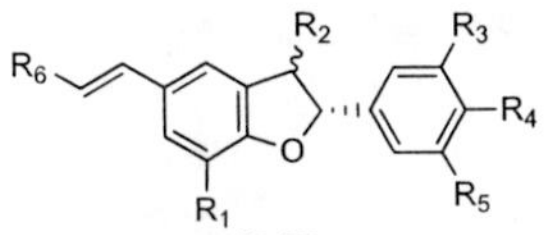

40-83

	R_1	R_2	R_3	R_4	R_5	R_6
40^{N}	H	βCH_3	H	H	H	CH_3
41^{N}	H	βCH_3	H	OH	H	CH_3
42^{N}	H	βCH_3	H	OH	OCH_3	CH_3
43^{N}	H	βCH_3	H	OCH_3	H	CH_3
44^{N}	H	αCH_3	H	OH	H	CH_3
45^{N}	H	βCH_3	H	OCH_3	OCH_3	CH_3
46^{N}	H	βCH_3	H	OCH_2	OCH_2	CH_3
47^{N}	H	αCH_3	OCH_3	OCH_3	H	CH_3
48^{N}	H	αCH_3	OCH_2	OCH_2	H	CH_3
49^{NS}	H	βCH_2OH	H	OH	H	CH_2OH
50^{NS}	H	βCH_2OH	H	OH	OCH_3	CH_2OH
51^{NS}	H	βCO_2CH_3	H	OH	H	CO_2CH_3
52^{S}	H	βCO_2CH_3	H	OH	OH	CO_2CH_3
53^{S}	H	βCO_2CH_3	H	OCH_3	H	CO_2CH_3
54^{S}	H	βCO_2CH_3	H	OAc	H	CO_2CH_3
55^{S}	H	βCO_2Et	H	OH	OCH_3	CO_2Et
56^{S}	H	βCO_2-*n*Bu	H	OH	H	CO_2-nBu
57^{S}	H	βCO_2-*n*Bu	H	OH	OH	CO_2-nBu
58^{S}	H	βCO-agmatine	H	OH	H	CO-agmatine
59^{S}	H	βCO-agmatine	H	OAc	H	CO-agmatine
60^{N}	OH	βCH_3	H	OH	OCH_3	CH_3
61^{N}	OH	βCH_2OH	H	OCH_3	OCH_3	CO_2H
62^{S}	OH	βCO_2CH_3	H	OH	H	CO_2CH_3
63^{S}	OH	βCO_2CH_3	H	OH	OH	CO_2CH_3
64^{S}	OH	βCO_2CH_3	H	OH	H	CO_2-nBu
65^{S}	OH	βCO_2-*n*Bu	H	OH	H	CO_2CH_3
66^{S}	OH	βCO_2-*n*Bu	H	OH	H	CO_2-nBu
67^{S}	OH	βCO_2-*n*Bu	H	OH	OH	CO_2-nBu
68^{S}	OH	βCO_2Et	H	OH	OH	CO_2Et
69^{S}	OH	βCO_2-*i*Pr	H	OH	OH	CO_2-iPr
70^{S}	OH	βCO_2-*t*Bu	H	OH	OH	CO_2-tBu
71^{N}	OCH_3	αCH_3	OCH_2	OCH_2	H	CH_3
72^{N}	OCH_3	βCH_3	H	OH	H	CH_3
73^{N}	OCH_3	βCH_3	H	OH	H	$(CH_2)_2OH$
74^{NS}	OCH_3	βCH_3	H	OH	OCH_3	CH_3
75^{N}	OCH_3	βCH_3	OCH_3	OH	OCH_3	CH_3
76^{N}	OCH_3	βCH_3	OCH_2	OCH_2	H	CH_3
77^{N}	OCH_3	βCH_3	OCH_3	OCH_3	H	CH_3
78^{N}	OCH_3	βCH_3	OCH_3	OCH_3	OH	CH_3
79^{N}	OCH_3	βCH_3	OCH_3	OCH_3	OCH_3	CH_3
80^{N}	OCH_3	βCH_2OH	H	OCH_3	OCH_3	CH_2OH
81^{N}	OCH_3	βCH_2OH	H	OGlc	OCH_3	CH_2OH
82^{N}	OCH_3	βCH_2OH	OH	OH	OCH_3	CH_2OH
83^{N}	OCH_3	βCH_2OH	OCH_3	OH	OCH_3	CH_2OH

Scheme 6. Chemical structures of dihydrobenzofuran neolignans **40–83**. ^{S}: synthetic DBN; ^{N}: natural DBN; ^{NS}: obtained by synthesis and from natural sources; Glc: glucose.

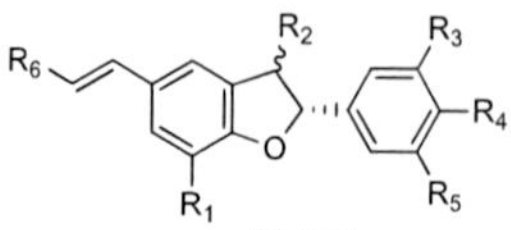

84-127

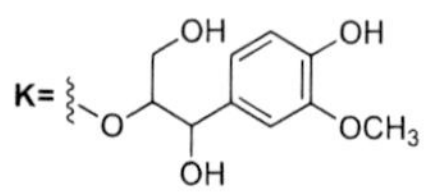

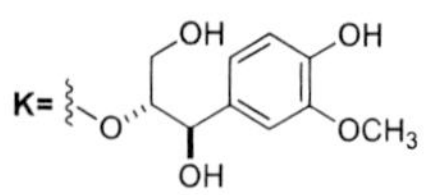

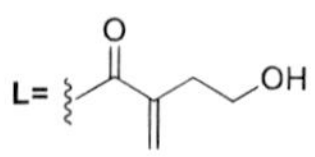

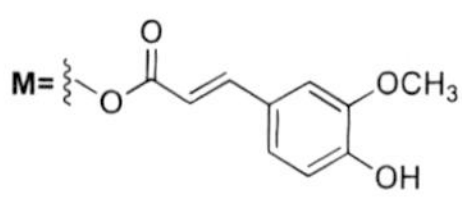

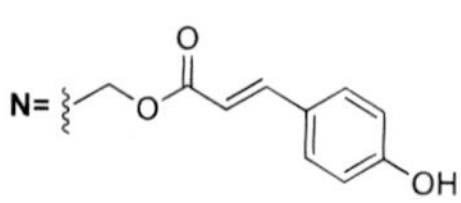

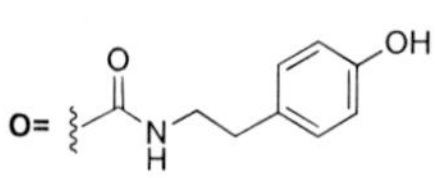

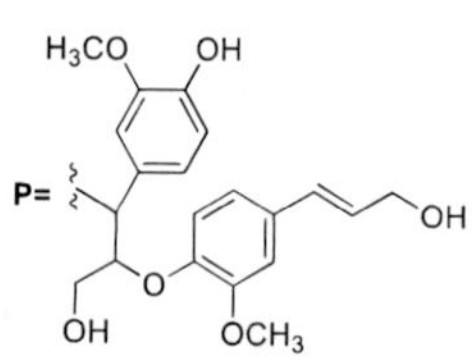

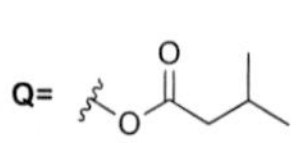

	R_1	R_2	R_3	R_4	R_5	R_6
84^{S}	OCH_3	βCH_2OH	H	OH	H	CH_2OH
85^{N}	OCH_3	βCH_2OH	H	OH	OCH_3	CH_2OH
86^{N}	OCH_3	βCH_2OH	OCH_3	O-**J**	OCH_3	CH_2OH
87^{N}	OCH_3	βCH_2OH	OCH_3	O-**K**	OCH_3	CH_2OH
88^{S}	OCH_3	CH_2OH	H	O-**P**	OCH_3	CH_2OH
89^{NS}	OCH_3	βCH_2OH	OCH_3	OH	OCH_3	CH_2OH
90^{N}	OCH_3	βCH_2OH	OCH_3	OCH_3	OCH_3	CH_2OH
91^{N}	OCH_3	βCH_2OH	H	OH	OCH_3	CHO
92^{N}	OCH_3	βCH_2OH	H	OH	OCH_3	CO_2H
93^{N}	OCH_3	βCH_2OH	H	OH	OCH_3	CH_2OGlc-**L**
94^{N}	OCH_3	βCH_2OH	OCH_3	OCH_3	OCH_3	CH_2OGlc
95^{N}	OCH_3	βCH_2OH	OCH_3	OCH_3	OCH_3	CH_2OGlc-**L**
96^{N}	OCH_3	βCH_2OH	H	OH	OCH_3	**O**
97^{N}	OCH_3	βCH_2OH	H	OCH_3	OCH_3	CHO
98^{N}	OCH_3	βCH_2OEt	OCH_3	OH	H	CH_2OH
99^{N}	OCH_3	βCH_2OGlc	H	OH	OCH_3	CH_2OH
100^{N}	OCH_3	βCH_2OGlc	H	OH	OCH_3	CH_2CHO
101^{N}	OCH_3	βCH_2OGlc-**L**	H	OH	OCH_3	CH_2OH
102^{N}	OCH_3	βCH_2OAc	OCH_3	OCH_3	OCH_3	CH_2OGlc
103^{NS}	OCH_3	βCH_2OAc	OCH_3	OCH_3	OCH_3	CH_2OGlc-**L**
104^{N}	OCH_3	βCH_2-**M**	H	OH	OCH_3	CH_2-**M**
105^{N}	OCH_3	βCH_2-**M**	OCH_3	OH	OCH_3	CH_2-**M**
106^{N}	OCH_3	βCH_2-**M**	H	OH	OCH_3	CH_2-**N**
107^{S}	OCH_3	βCO_2-**G**	H	OH	OCH_3	CO_2-**G**
108^{N}	OCH_3	β**O**	H	OH	OCH_3	**O**
109^{S}	OCH_3	β**I**	H	OH	OCH_3	**H**
110^{S}	OCH_3	βCO_2Ac	H	OAc	OCH_3	CO_2Ac
111^{S}	OCH_3	CO_2CH_3	H	OH	OCH_3	CO_2CH_3
112^{S}	OCH_3	CO_2CH_3	H	OAc	OCH_3	CO_2CH_3
113^{S}	OCH_3	βCO_2CH_3	H	O-**R**	OCH_3	OCH_3
114^{S}	OCH_3	βCO_2CH_3	H	OH	OCH_3	OCH_3
115^{N}	OCH_3	βCO_2CH_3	H	OAc	OCH_3	CH_2OAc
116^{NS}	OCH_3	βCO_2Et	H	OH	OCH_3	CO_2Et
117^{S}	OCH_3	βCO-tyramine	H	OH	OCH_3	CO-tyramine
118^{S}	OCH_3	βCO-agmatine	H	OH	H	CO-agmatine
119^{S}	OCH_3	βCO-agmatine	H	OAc	H	CO-agmatine
120^{N}	OCH_3	α$CO_2C_{18}H_{37}$	OH	H	OCH_3	$CO_2C_{18}H_{37}$
121^{N}	OCH_3	βCH_2OCO-*i*-Bu	H	OH	OCH_3	CHO
122^{N}	OCH_3	β$CH_2CO_2CH_3$	H	OH	OCH_3	CH_2OAc
123^{N}	OCH_3	β$CH_2CO_2CH_3$	OCH_3	OH	H	CHO
124^{N}	OCH_3	β$CH_2CO_2CH_3$	H	OH	OCH_3	CHO
125^{N}	OCH_3	β**Q**	OCH_3	OH	H	CHO
126^{N}	OCH_3	β**Q**	OCH_3	OH	H	CO_2H
127^{N}	OCH_3	β**Q**	OCH_3	OH	H	CO_2CH_3

Scheme 7. Chemical structures of dihydrobenzofuran neolignans **84–127**. ^{S}: synthetic DBN; ^{N}: natural DBN; ^{NS}: obtained by synthesis and from natural sources; Glc: glucose.

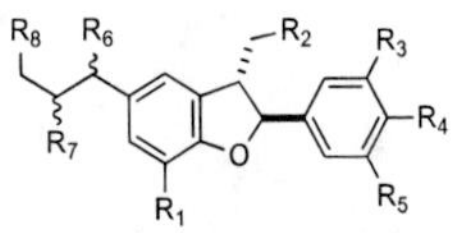

128-172

R=
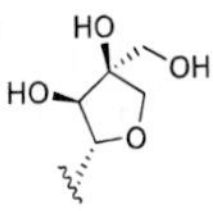

	R_1	R_2	R_3	R_4	R_5	R_6	R_7	R_8
128^{N}	H	OH	H	OGlc	OCH_3	H	H	OH
129^{N}	OH	OH	H	OH	OCH_3	H	H	OH
130^{S}	OH	OH	H	OCH_3	OCH_3	H	H	OH
131^{N}	OH	OH	H	OXyl	OCH_3	H	H	OH
132^{N}	OH	OH	H	OCH_3	OH	H	H	OH
133^{N}	OH	OH	OCH_3	OH	H	$O(CH_2)_3CH_3$	OH	OH
134^{N}	OH	ORhm	H	OH	OCH_3	H	H	OH
135^{N}	OH	OAc	H	OCH_3	OAc	H	H	OAc
136^{N}	OCH_3	H	H	OH	OCH_3	βOH	αOH	H
137^{S}	OCH_3	H	OH	CH_3	CH_3	OH	OH	H
138^{N}	OCH_3	H	OCH_3	OH	H	OH	OH	H
139^{N}	OCH_3	H	OCH_3	OCH_3	H	OH	OH	H
140^{N}	OCH_3	H	OCH_3	OGlc	H	H	H	OH
141^{S}	OCH_3	H	OCH_2	OCH_2	H	CH_3	OCH_3	H
142^{N}	OCH_3	OH	OCH_3	OH	OCH_3	H	H	OH
143^{N}	OCH_3	OH	OCH_3	OH	H	H	H	CO_2H
144^{N}	OCH_3	OH	H	OH	OCH_3	H	H	$(=O)OCH_3$
145^{N}	OCH_3	OH	OCH_3	OH	OCH_3	H	H	$(=O)OCH_3$
146^{N}	OCH_3	OH	H	OH	OCH_3	H	H	OH
147^{N}	OCH_3	OH	H	OH	OCH_3	OH	OH	OH
148^{N}	OCH_3	OH	H	OH	OCH_3	βOH	βOH	OH
149^{N}	OCH_3	OH	H	OH	OCH_3	αOH	αOH	OH
150^{N}	OCH_3	OH	H	OH	OCH_3	βOH	αOH	OH
151^{N}	OCH_3	OH	H	OH	OCH_3	αOH	βOH	OH
152^{N}	OCH_3	OH	H	OCH_3	OCH_3	H	H	OH
153^{N}	OCH_3	OH	H	OCH_2	CH_2O	H	H	OH
154^{N}	OCH_3	OH	H	OCH_2	CH_2O	OH	H	OH
155^{N}	OCH_3	OH	H	OCH_2	CH_2O	OCH_3	H	OH
156^{N}	OCH_3	OH	H	OGlc	OCH_3	H	H	OH
157^{N}	OCH_3	OH	OCH_3	OGlc	OCH_3	H	H	OH
158^{N}	OCH_3	OH	OH	H	OCH_3	=O	H	OH
159^{N}	OCH_3	OH	H	ORhm	OCH_3	H	H	OH
160^{N}	OCH_3	OGlc	H	OH	OCH_3	H	H	OH
161^{N}	OCH_3	O-**R**	H	OH	OCH_3	H	H	OH
162^{N}	OCH_3	Oxyl	H	OH	OCH_3	H	H	OH
163^{N}	OCH_3	Oxyl	H	OH	OCH_3	H	H	OGlc
164^{N}	OCH_3	Oxyl	H	OH	OCH_3	H	H	ORhm
165^{N}	OCH_3	OH	OCH_3	OH	H	H	H	ORhm
166^{N}	OCH_3	ORhm	H	OH	OCH_3	H	H	OGlc
167^{N}	OCH_3	Shikimoyl	H	OH	OCH_3	H	H	ORhm
168^{N}	OCH_3	OAc	H	OAc	OCH_3	OAc	OAc	OAc
169^{N}	OCH_3	OAc	H	OAc	OCH_3	H	H	OAc
170^{N}	OCH_3	OAc	H	OH	OCH_3	OAc	OAc	OAc
171^{N}	OGlc	OH	H	OH	OH	H	H	OH
172^{N}	OGlc	OH	H	OH	OCH_3	H	H	OH

Scheme 8. Chemical structures of dihydrobenzofuran neolignans **128–172**. ^{S}: synthetic DBN; ^{N}: natural DBN; ^{NS}: obtained by synthesis and from natural sources; Glc: glusoce; Rhm: rhamnose; Xyl: xylose.

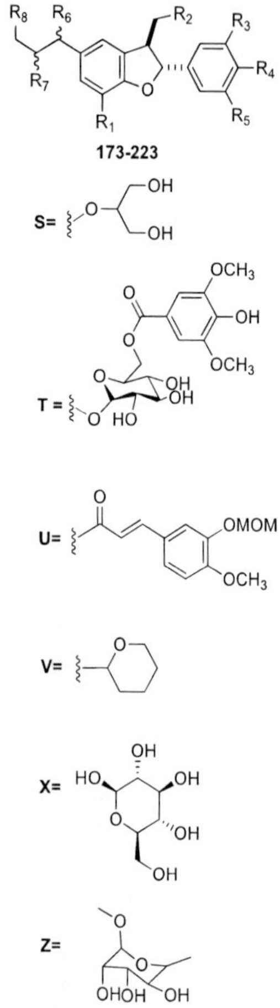

	R_1	R_2	R_3	R_4	R_5	R_6	R_7	R_8
173^{N}	H	H	H	OH	H	H	H	H
174^{N}	H	H	H	OH	H	H	H	OH
175^{N}	H	H	H	OH	OCH_3	H	H	OH
176^{N}	H	H	H	OH	H	H	βOH	H
177^{S}	H	OH	H	OH	OCH_3	H	H	OH
178^{S}	H	=O(OCH_3)	H	OCH_3	H	NH_2	H	=O(OCH_3)
179^{S}	H	=O(OCH_3)	H	OCH_3	H	H	NH_2	=O(OCH_3)
180^{N}	OH	OH	OCH_3	OH	H	H	H	OH
181^{N}	OH	OH	OCH_3	OH	OCH_3	H	H	OH
182^{N}	OH	OH	OCH_3	OGlc	H	H	H	OH
183^{N}	OH	OH	OCH_3	**S**	H	H	H	OH
184^{N}	OH	**T**	OCH_3	OH	H	H	H	OH
185^{N}	OCH_3	H	OCH_3	OGlc	H	H	H	OH
186^{N}	OCH_3	H	OCH_3	OH	H	H	=O	H
187^{N}	OCH_3	βOH	H	OH	OCH_3	H	H	=O(OCH_3)
188^{N}	OCH_3	βOH	H	OH	OCH_3	H	H	=O(OH)
189^{N}	OCH_3	OH	OCH_3	**S**	OCH_3	H	H	OH
190^{N}	OCH_3	OH	H	OH	OCH_3	H	αOH	OH
191^{N}	OCH_3	OH	OH	H	OCH_3	H	H	OH
192^{N}	OCH_3	OH	H	OGlc	OCH_3	H	H	OH
193^{S}	OCH_3	OH	H	OH	OCH_3	H	H	OH
194^{N}	OCH_3	OH	H	OH	OCH_3	H	H	**M**
195^{S}	OCH_3	**M**	H	OH	OCH_3	H	H	**M**
196^{N}	OCH_3	OH	H	OH	OCH_3	αOCH_3	βOH	OH
197^{N}	OCH_3	OH	H	OH	OCH_3	βOH	αOC_2H_5	OH
198^{N}	OCH_3	OH	H	OH	OCH_3	OEt	H	OH
199^{N}	OCH_3	OH	H	OH	OCH_3	=O	H	OH
200^{N}	OCH_3	OH	H	OH	OCH_3	=O	αOH	OH
201^{N}	OCH_3	OH	H	OH	OCH_3	=O	βOH	OH
202^{S}	OCH_3	OH	H	OCH_3	OCH_3	H	H	OH
203^{N}	OCH_3	OH	H	OCH_3	ORhm	H	H	OH
204^{N}	OCH_3	OH	H	ORhm	OCH_3	H	H	OH
205^{N}	OCH_3	OH	OH	H	OCH_3	=O	OH	OH
206^{S}	OCH_3	OH	OCH_3	OH	H	H	H	OH
207^{N}	OCH_3	OH	OCH_3	OH	H	H	H	=O(OCH_3)
208^{N}	OCH_3	OH	OCH_3	**X**	H	H	H	CH_2OH
209^{N}	OCH_3	OH	OCH_3	OH	OCH_3	H	H	OH
210^{N}	OCH_3	OH	OCH_3	OH	OCH_3	H	H	**Z**
211^{N}	OCH_3	OH	OCH_3	OH	OCH_3	H	H	OGlc
212^{N}	OCH_3	OH	OCH_3	**T**	OCH_3	H	H	OH
213^{S}	OCH_3	OH	H	O-**V**	OCH_3	H	H	OH
214^{S}	OCH_3	OH	OH	OH	H	H	H	OH
215^{N}	OCH_3	OH	OCH_3	OH	OCH_3	H	H	=O(OCH_3)
216^{N}	OCH_3	OGlc	H	OH	OCH_3	H	H	OH
217^{N}	OCH_3	OXyl	H	OH	OCH_3	H	H	OH
218^{N}	OCH_3	OGlc	H	OH	OCH_3	H	H	=O(OCH_3)
219^{N}	OCH_3	OGlc	OCH_3	**S**	OCH_3	H	H	OH
220^{N}	OCH_3	OGlc	OCH_3	OH	OCH_3	H	H	OH
221^{N}	OCH_3	O-β-Xyl	OCH_3	OH	OCH_3	H	H	OH
222^{S}	OCH_3	O-**U**	H	O-**X**	OCH_3	H	H	**U**
223^{N}	OGlc	OH	OCH_3	OH	H	H	H	OH

Scheme 9. Chemical structures of dihydrobenzofuran neolignans **173–223**. ^{S}: synthetic DBN; ^{N}: natural DBN; ^{NS}: obtained by synthesis and from natural sources; Glc: glucose; Rhm: rhamnose; Xyl: xylose.

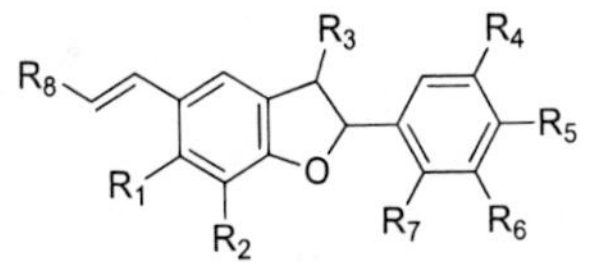

224-238

	R_1	R_2	R_3	R_4	R_5	R_6	R_7	R_8	Aryl
224^N	H	H	βCH_3	H	OH	H	H	H	
225^N	H	H	CH_3	H	OCH_3	H	H	CH_3	
226^{NS}	H	H	CH_3	H	OH	OCH_3	H	CH_3	
227^N	H	H	CH_3	H	OH	H	H	CH_3	
228^N	H	H	CH_3	H	OCH_3	H	H	CH_3	
229^N	H	OCH_3	CH_3	OCH_2	OCH_2	H	H	CH_3	
230^N	H	OCH_3	βCH_3	H	OH	OCH_3	H	CHO	
231^N	H	OCH_3	CH_2OH	H	OH	OCH_3	H	CH_2OH	
232^N	H	OCH_3	CH_2OH	H	Glc	OCH_3	H	CH_2OH	
233^N	H	OCH_3	CH_2OH	CH_2OH	H	OCH_3	OCH_3	CH_3	α
234^N	H	OCH_3	βCH_3	CH_2OH	H	OCH_3	OCH_3	CH_3	α
235^N	H	OCH_3	CH_2OH	OCH_3	O-**R'**	OCH_3	OCH_3	CH_2OH	
236^N	H	OCH_3	CH_2OH	OCH_3	O-**R'**	OCH_3	OCH_3	CHO	
237^S	H	OCH_3	CO_2CH_3	H	OCH_3	OCH_3	OCH_3	CO_2CH_3	
238^N	OH	H	αCH_3	CH_2O	CH_2O	H	H	Et	β

235, 236: R'=

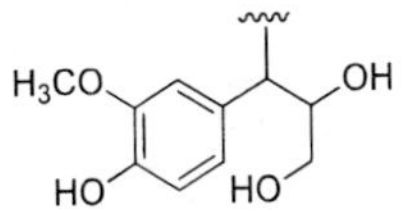

Scheme 10. Chemical structures of dihydrobenzofuran neolignans **224–238**. S: synthetic DBN; N: natural DBN; NS: obtained by synthesis and from natural sources.

Schemes 5 to 20 illustrate natural and synthetic DBN chemical structures that have been reported in the literature up to 2019. Table 1 depicts their occurrence in different plant species. Most of the phytochemical studies resulting in DBN detection (isolation and/or identification) have been carried out with the leaves of each species, mostly from the leaf extracts obtained in methanol and dichloromethane. The most frequent structural characteristic in natural DBNs is the unsaturated side chain without oxygen and the saturated C_3 side chain bearing a terminal hydroxyl group. On the other hand, nitrogen presence and oxygen absence in the side chain are rare.

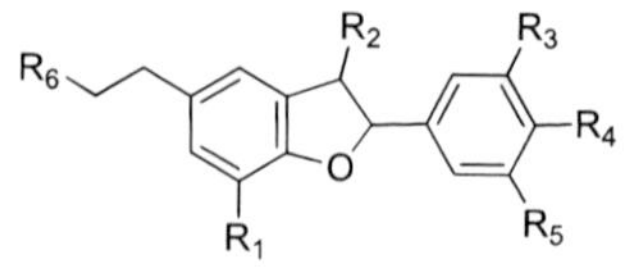

239-265

	R_1	R_2	R_3	R_4	R_5	R_6
239^{N}	H	H	H	OCH_3	H	CH_2OH
240^{N}	H	H	H	OH	OCH_3	CH_2OH
241^{N}	H	CH_3	H	OH	H	$COCH_3$
242^{N}	H	CH_3	H	OH	H	OH
243d^{S}	H	CO_2CH_3	H	OH	H	CO_2CH_3
244^{N}	OH	CH_2OH	H	OH	OCH_3	CH_2OH
245^{N}	OH	CH_2OH	H	OGlc	OCH_3	CH_2OH
246^{N}	OH	CH_2OH	H	ORhm	OCH_3	CH_2OH
247^{N}	OH	CH_2ORhm	H	OH	OCH_3	CH_2OH
248^{N}	OCH_3	CH_2OH	H	OCH_3	OCH_3	CH_2OH
249^{N}	OCH_3	CH_2OH	H	OGlc	OCH_3	CH_2OH
250^{N}	OCH_3	CH_2OH	H	OGlc	OCH_3	COOH
251^{N}	OCH_3	CH_2OH	H	ORhm	OCH_3	CH_2OH
252^{N}	OCH_3	CH_2OH	H	**R'**	OCH_3	CH_2OH
253^{N}	OCH_3	CH_2OH	OCH_3	**R'**	OCH_3	CH_2OH
254^{N}	OCH_3	CH_2OH	OCH_3	**R'**	OCH_3	CH_2OH
255^{S}	OCH_3	$CHCH_2$	OCH_3	OPiv	H	CH_2OPiv
256a^{S}	OCH_3	$CHCH_2$	OH	OPiv	H	CH_2OPiv
257^{N}	OCH_3	CH_2OXyl	H	OH	OCH_3	CH_2OH
258^{S}	OCH_3	COH	OCH_3	OPiv	H	CH_2OPiv
259^{S}	OCH_3	CO_2CH_3	H	OPiv	OCH_3	CH_2OPiv
260^{S}	OCH_3	CO_2CH_3	H	OPiv	OPiv	CH_2OPiv
261^{S}	OCH_3	CO_2CH_3	OCH_3	OCH_3	OCH_3	CH_2OPiv
262^{S}	OCH_3	CO_2CH_3	H	OCH_3	OCH_3	CO_2CH_3
263^{S}	OCH_3	CO_2CH_3	H	OH	OCH_3	CO_2CH_3
264^{N}	OGlc	CH_2OH	H	ORhm	OCH_3	CH_2OH
265^{S}	ORhm	CH_2OH	H	OH	OCH_3	CH_2OH

252, 253: R'= 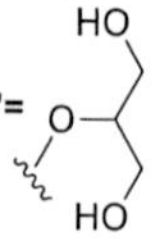

254: R'= 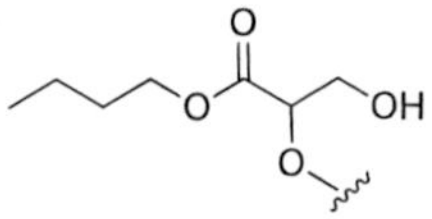

Scheme 11. Chemical structures of dihydrobenzofuran neolignans **239–265**. ^{S}: synthetic DBN; ^{N}: natural DBN; ^{NS}: obtained by synthesis and from natural sources, Glc: glucose; Rhm: rhamnose; Xyl: xylose; Piv: pivaloyl.

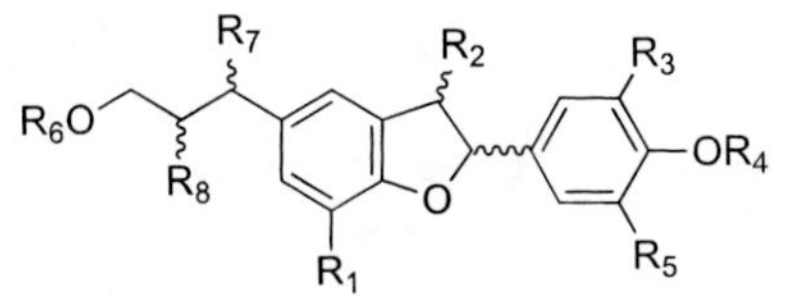

266-278

	R_1	R_2	R_3	R_4	R_5	R_5	R_5	R_8	Aryl
266[N]	OCH_3	α-CH_2OH	H	R'	OCH_3	H	H	Ac	α
267[N]	OCH_3	α-CH_2OH	OCH_3	R'	OCH_3	H	H	CH_3	α
268[N]	OCH_3	α-CH_2OH	H	R'	OCH_3	H	H	Ac	α
269[N]	OCH_3	α-CH_2OH	OCH_3	R'	OCH_3	H	H	CH_3	α
270[N]	OH	β-CH_2OH	OCH_3	R'	OCH_3	H	H	H	α
271[N]	OH	α-CH_2OH	OCH_3	R'	OCH_3	H	H	H	β
272[N]	OH	β-CH_2OH	OCH_3	R'	OCH_3	H	H	H	β
273[N]	OH	α-CH_2OH	OCH_3	R'	OCH_3	H	H	H	α
274[N]	OCH_3	α-CH_2OH	OCH_3	R'	OCH_3	H	H	H	α
275[N]	OCH_3	α-CH_2OH	H	R'	OCH_3	H	H	H	α
276[S]	OCH_3	α-CH_2OH	H	H	OCH_3	OCH_3	OH	OH	α
277[N]	OCH_3	α-CH_2OH	H	H	OCH_3	OEt	OH	OH	α
278[N]	OGlc	CH_2OH	H	H	OCH_3	H	H	H	α

264,265: R=

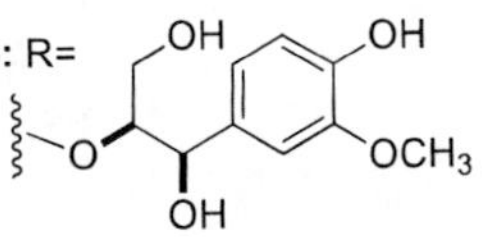

266,267: R=

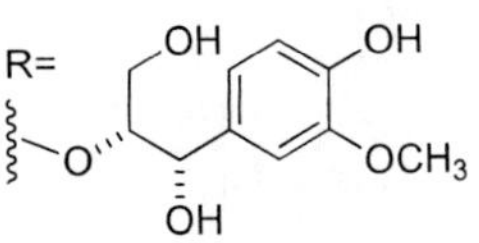

268,269: R=

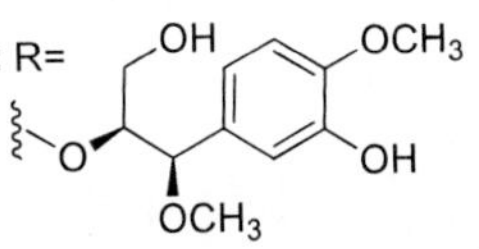

270,271: R=

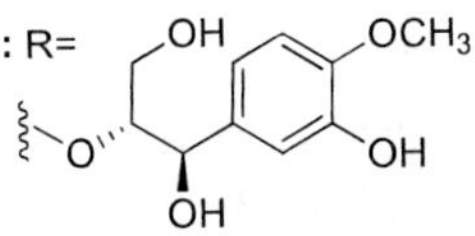

272,273: R'=

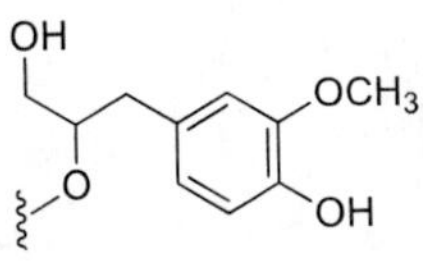

Scheme 12. Chemical structures of dihydrobenzofuran neolignans **266–278**. [S]: synthetic DBN; [N]: natural DBN; [NS]: obtained by synthesis and from natural sources.

Notwithstanding the development of new separation techniques over recent years, the DBN yields remain low. This could explain, at least in principle, the relatively low number of biological studies on DBNs as compared to the amount of investigations into their structural diversity, as will be further discussed in this chapter.

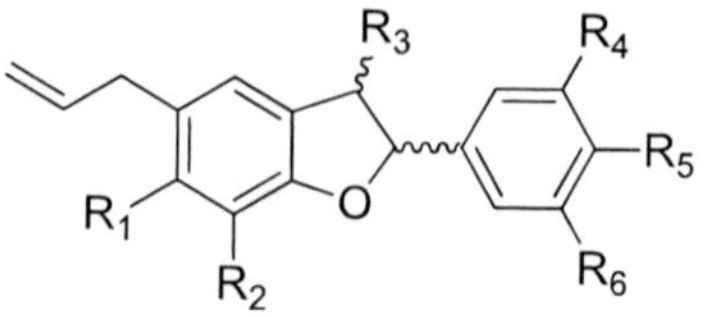

279-304

	R_1	R_2	R_3	R_4	R_5	R_6	Aryl
279[N]	H	H	αCH_3	H	OH	OCH_3	α
280[N]	H	H	αCH_3	H	OH	H	α
281[N]	H	H	αCH_3	H	OCH_3	H	α
282[N]	H	H	CH_3	H	OCH_3	H	
283[N]	H	H	CH_3	H	OH	OCH_3	
284[N]	H	H	CH_3	H	OH	H	
285[N]	H	OCH_3	CH_3	H	OH	H	
286[N]	H	OCH_3	CH_3	OCH_3	OH	H	
287[N]	H	OCH_3	CH_3	OCH_3	OCH_3	H	
288[N]	H	OCH_3	αCH_2OH	H	OH	H	α
289[N]	H	OCH_3	αCH_2OH	H	OH	OCH_3	α
290[N]	H	OCH_3	αCH_2OH	H	OCH_3	OCH_3	α
291[N]	H	OCH_3	αCH_3	H	OCH_3	OCH_3	α
292[N]	OH	H	αCH_3	OCH_3	OCH_3	OCH_3	β
293[N]	OH	H	αCH_3	OCH_2	OCH_2	H	β
294[N]	OH	H	βCH_3	H	OCH_2	OCH_2	α
295[N]	OH	H	βCH_3	OCH_3	OCH_2	OCH_2	β
296[N]	OH	H	βCH_3	OCH_3	OCH_3	OCH_3	α
297[N]	OH	H	βCH_3	OCH_3	OCH_2	OCH_2	α
298[N]	OH	H	βCH_3	OCH_3	OCH_3	OCH_3	α
299[N]	OH	H	βCH_3	OCH_3	OCH_3	OCH_3	β
300[N]	OH	H	CH_3	OCH_2	OCH_2	H	β
301[N]	OH	H	CH_3	OCH_3	OCH_3	H	β
302[N]	OH	H	CH_3	OCH_3	OCH_3	OCH_3	β
303[N]	OH	OCH_3	CH_3	OCH_3	OH	H	
304[N]	OH	OCH_3	CH_2OH	OCH_3	OCH_3	H	

Scheme 13. Chemical structures of dihydrobenzofuran neolignans **279–304**. [S]: synthetic DBN; [N]: natural DBN; [NS]: obtained by synthesis and from natural sources.

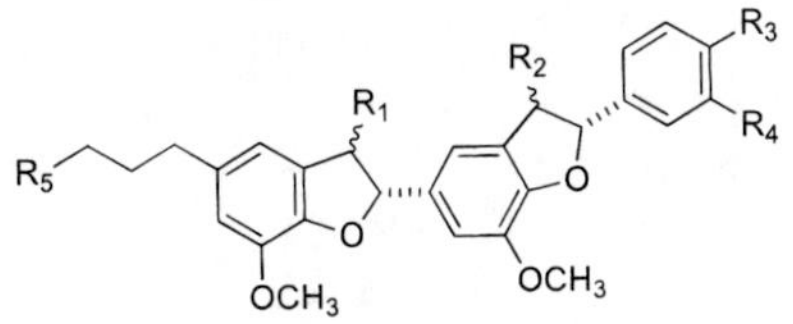

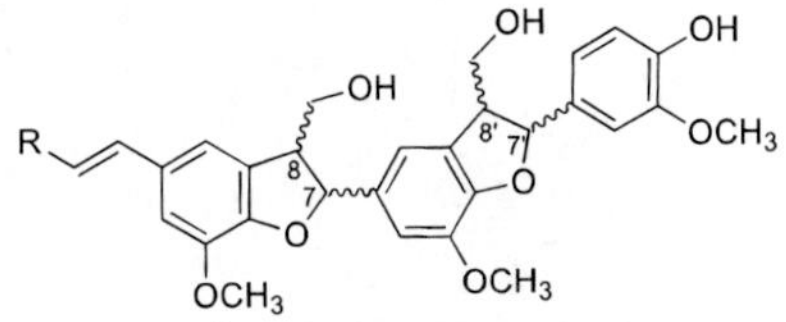

310^{N}: 7*R*,8*S*,7'*S*,8'*R*, R=CH_2OH
311^{N}: 7*S*,8*R*,7'*S*,8'*S*, R=CHO

	R_1	R_2	R_3	R_4	R_5
305^{N}	βCH_2OH	βCH_2OH	β-O-Glc	H	OH
306^{N}	βCH_2OH	βCH_2OH	OH	OCH_3	OH
307^{N}	βCH_2OH	βCH_2OH	OH	OCH_3	OH
308^{N}	βCH_2OH	βCH_2OH	OH	OCH_3	(=O)OCH_3
309^{N}	αCH_2OH	αCH_2OH	OH	OCH_3	(=O)OH

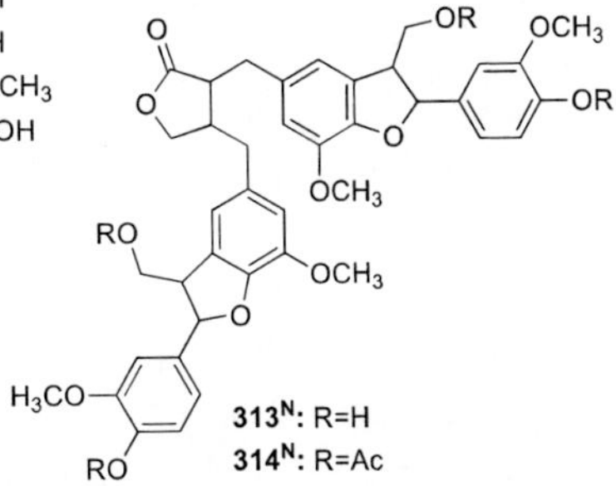

313^{N}: R=H
314^{N}: R=Ac

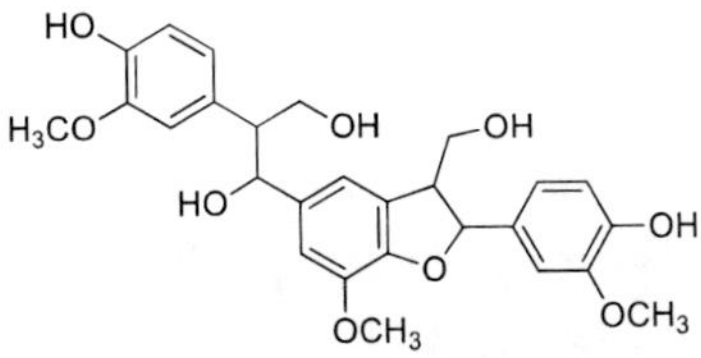

312^{N}

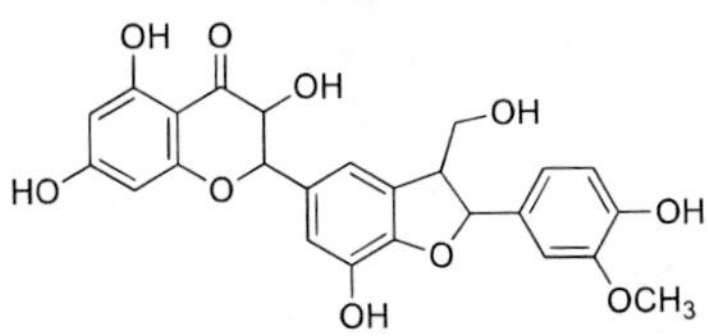

315^{N}

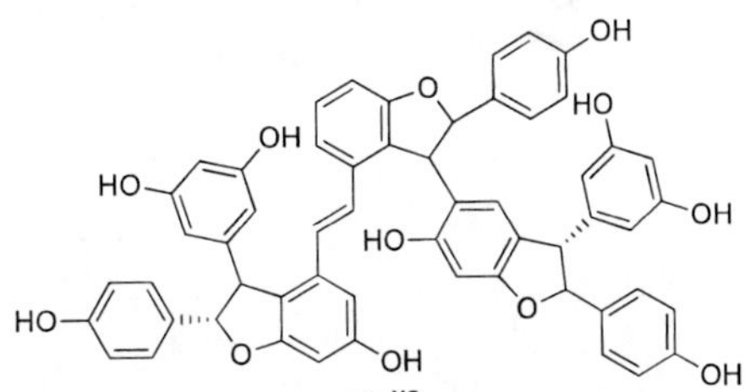

316^{NS}

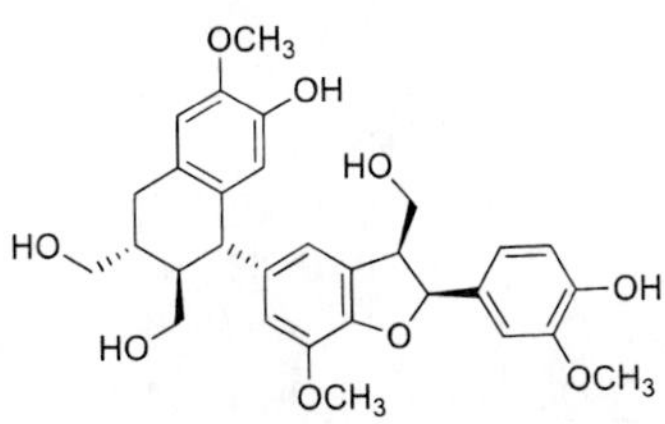

317^{N}

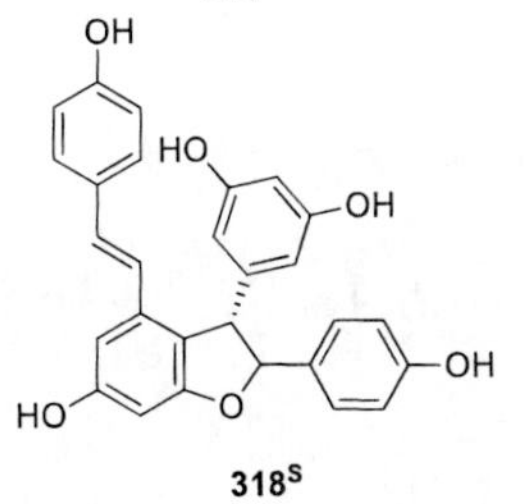

318^{S}

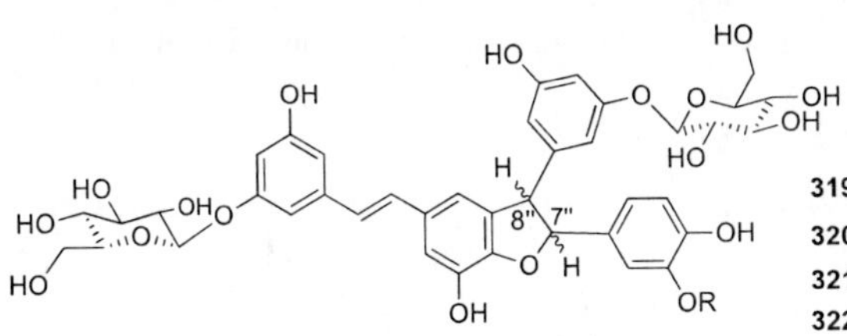

319^{N}: R = H (7"*R**, 8"*R**)
320^{N}: R = H (7"*S**, 8"*S**)
321^{N}: R = CH_3 (7"*R**, 8"*R**)
322^{N}: R = CH_3 (7"*S**, 8"*S**)

Scheme 14. Chemical structures of dihydrobenzofuran neolignans **305–322**. ^{S}: synthetic DBN; ^{N}: natural DBN; ^{NS}: obtained by synthesis and from natural sources.

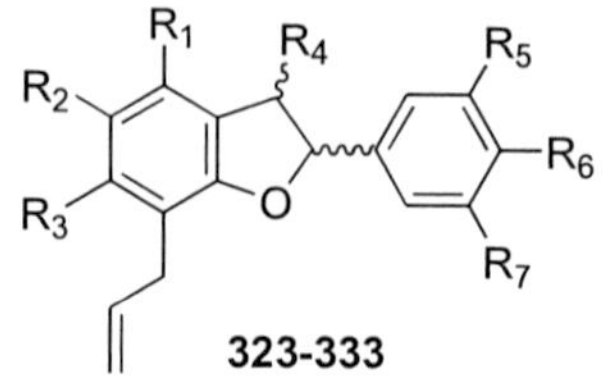

323-333

	R_1	R_2	R_3	R_4	R_5	R_6	R_7	Aryl
323[S]	H	OCH_3	H	αCH_3	H	OH	H	
324[N]	H	OH	H	βCH_3	OCH_2	OCH_2	OCH_3	β
325[N]	H	OH	H	βCH_3	OCH_2	OCH_2	OCH_3	α
326[NS]	H	OCH_3	OH	βCH_3	OCH_2	OCH_2	H	β
327[N]	H	OCH_3	OH	βCH_3	OCH_2	OCH_2	H	α
328[N]	H	OCH_3	OH	CH_3	H	OCH_2	OCH_2	
329[N]	OCH_2	OCH_2	H	αCH_3	H	OH	OCH_3	β
330[N]	OCH_2	OCH_2	H	αCH_3	H	OCH_2	OCH_2	β
331[S]	H	OCH_3	OCH_3	αCH_3	H	OCH_2	OCH_2	α
332[S]	H	OCH_3	OCH_3	αCH_3	H	OCH_3	OCH_3	α
333[S]	Cl	OCH_3	OCH_3	αCH_3	H	OCH_3	OCH_3	α

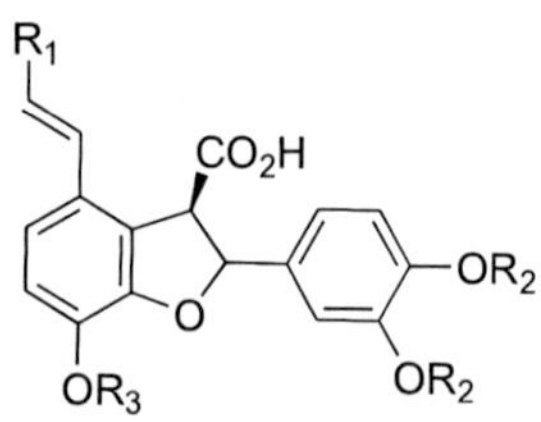

	R_1	R_2	R_3
334[S]	$CO_2CH_2CCl_3$	TIPS	MOM
335[S]	$CO_2CH_2CCl_3$	TIPS	H
336[S]	CO_2H	H	H

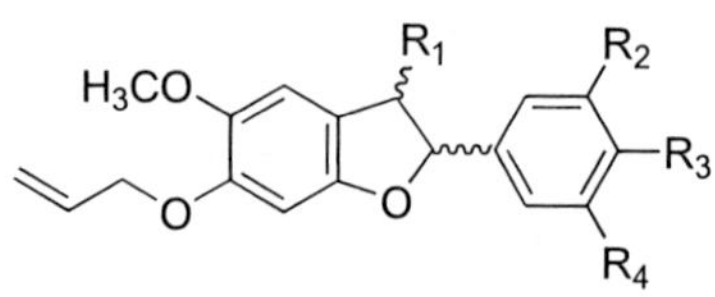

337-341

	R_1	R_2	R_3	R_4	Aryl
337[N]	αCH_3	OCH_3	OCH_3	OCH_3	β
338[N]	βCH_3	H	OCH_2	OCH_2	α
339[N]	βCH_3	OCH_3	OCH_2	OCH_2	α
340[N]	βCH_3	OCH_3	OCH_2	OCH_2	β
341[N]	βCH_3	OCH_3	OCH_3	OCH_3	β

Scheme 15. Chemical structures of dihydrobenzofuran neolignans **323–341**. [S]: synthetic DBN; [N]: natural DBN; [NS]: obtained by synthesis and from natural sources.

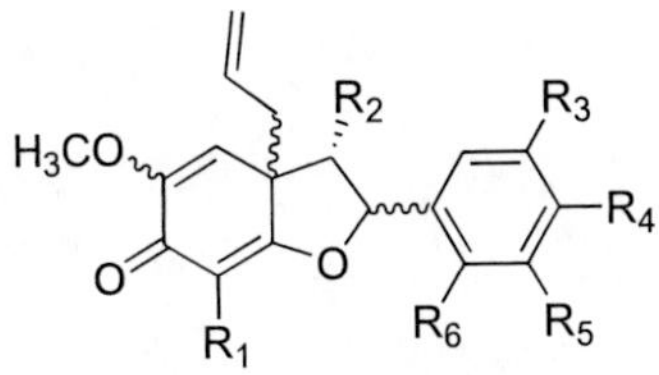

342-369

	R_1	R_2	R_3	R_4	R_5	R_6	OCH_3	Aryl	Allyl
342[N]	H	αCH_3	H	OCH_2	OCH_2	H		α	β
343[N]	H	αCH_3	H	OCH_2	OCH_2	H		β	α
344[N]	H	αCH_3	H	OCH_2	OCH_2	H		α	α
345[N]	H	αCH_3	CH_2OH	H	OCH_3	OCH_3		β	α
346[N]	H	αCH_3	OCH_2	OCH_2	OCH_3	H		α	β
347[N]	H	αCH_3	OCH_2	OCH_2	H	H		β	α
348[N]	H	αCH_3	OCH_2	OCH_2	H	H		α	β
349[N]	H	αCH_3	OCH_2	OCH_2	H	H	α	β	β
350[N]	H	αCH_3	OCH_3	OCH_2	OCH_2	H		β	α
351[N]	H	αCH_3	OCH_3	OCH_3	H	H		β	α
352[N]	H	αCH_3	OCH_3	OCH_3	OCH_3	H		β	α
353[N]	H	αCH_3	OCH_3	OCH_3	OCH_3	H		β	β
354[N]	H	αCH_3	OCH_3	OCH_2	OCH_2	H		α	
355[N]	H	CH_3	CH_2O	CH_2O	H	H		β	α
356[N]	H	αCH_3	OCH_2	OCH_2	OCH_3	H	α	β	β
357[N]	H	αCH_3	OCH_2	OCH_2	H	H	α	α	β
358[N]	H	αCH_3	OCH_3	OCH_3	H	H	α	α	β
359[N]	H	αCH_3	OCH_2	OCH_2	H	H			
360[N]	H	αCH_3	OCH_2	OCH_2	H	H		α	β
361[N]	H	βCH_3	OCH_2	OCH_2	H	H	α	α	β
362[N]	H	βCH_3	OCH_3	OCH_3	OCH_3	H	α	α	β
363[N]	OCH_3	CH_3	OCH_2	OCH_2	H	H			
364[N]	OCH_3	αCH_3	OCH_2	OCH_2	H	H		β	α
365[N]	OCH_3	αCH_3	OCH_2	OCH_2	H	H	α	β	β
366[N]	OCH_3	αCH_3	OCH_3	OCH_3	H	H	α	α	β
367[N]	OCH_3	βCH_3	OCH_2	OCH_2	OCH_3	H	α	α	β
368[N]	OCH_3	βCH_3	OCH_2	OCH_2	OCH_3	H	α	β	α
369[N]	OCH_3	βCH_3	OCH_2	OCH_2	H	H	α	α	β

Scheme 16. Chemical structures of dihydrobenzofuran neolignans **342–369**. [S]: synthetic DBN; [N]: natural DBN; [NS]: obtained by synthesis and from natural sources.

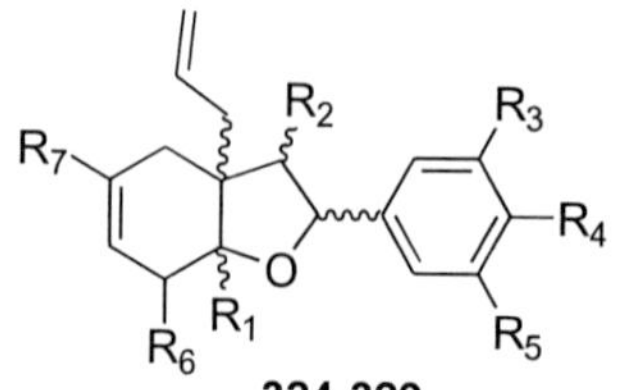

324-329

	R_1	R_2	R_3	R_4	R_5	R_6	R_7	Aryl	Allyl
370^N^	βOH	βCH_3	OCH_2	OCH_2	H	=O	H	β	β
371^N^	βOH	βCH_3	OCH_3	OCH_3	H	=O	H	α	β
372^N^	αOH	βCH_3	OCH_2	OCH_2	OCH_3	=O	H	α	β
373^N^	αOH	βCH_3	OCH_2	OCH_2	H	=O	H	α	β
374^N^	βOH	βCH_3	OCH_2	OCH_2	H	=O	H	α	α
375^N^	βOH	βCH_3	OCH_2	OCH_2	H	=O	H	β	α
336^N^	βOH	αCH_3	OCH_2	OCH_2	H	OCH_3	OCH_3	β	α
377^N^	βOH	αCH_3	OCH_2	OCH_2	H	OH	H	β	α
378^N^	βOCH_3	αCH_3	OCH_2	OCH_2	H	OH	H	β	α

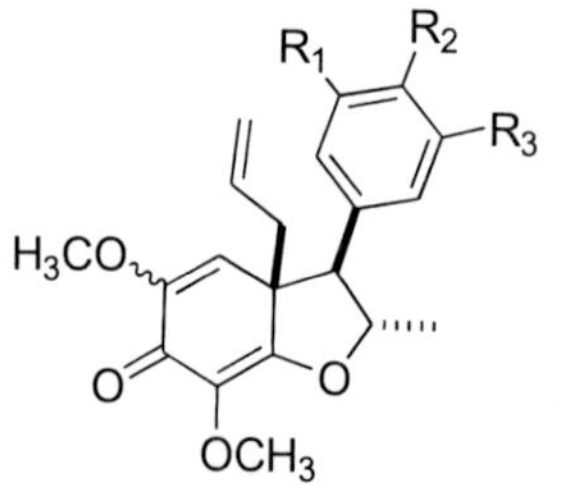

379-382

	R_1	R_2	R_3	OCH_3
379^N^	H	OCH_3	OCH_3	β
380^N^	H	OCH_2	OCH_2	β
381^N^	OCH_2	OCH_2	OCH_3	
382^N^	OCH_3	OCH_3	OCH_3	

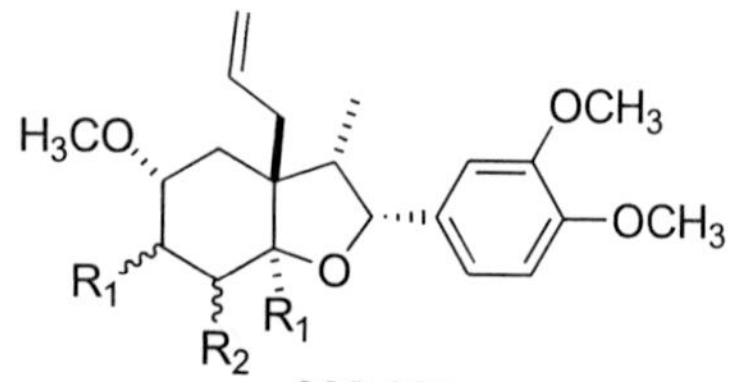

383-388

	R_1	R_2	R_3
383^N^	αOH	H	OH
384^N^	βOH	H	OH
385^N^	αOH	H	OCH_3
386^N^	αOH	αOH	OH
387^N^	βOH	αOH	OH
388^N^	β–O⁻		OH

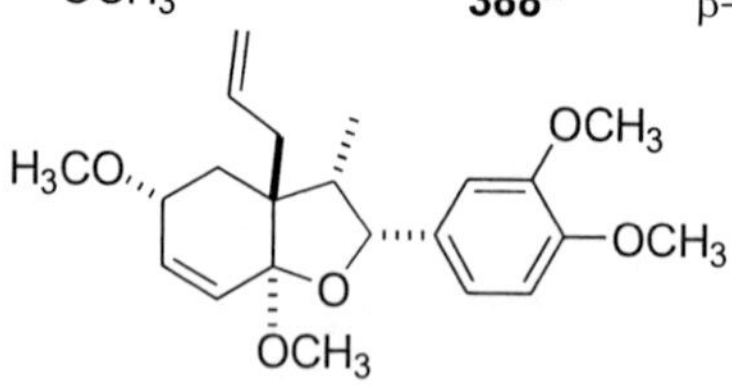

389^N^

Scheme 17. Chemical structures of dihydrobenzofuran neolignans **370–389.** ^S^: synthetic DBN; ^N^: natural DBN; ^NS^: obtained by synthesis and from natural sources.

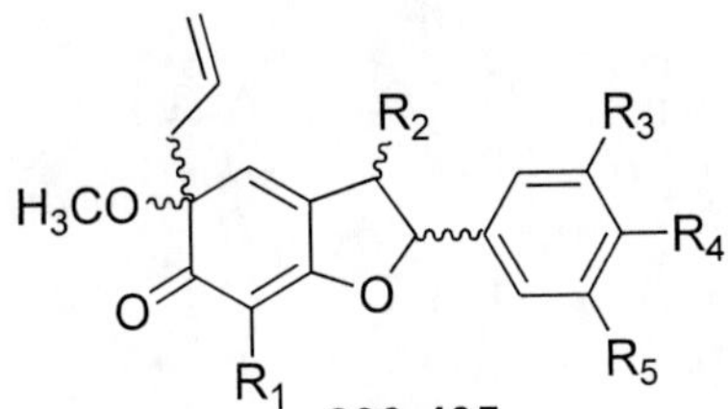

390-405

	R_1	R_2	R_3	R_4	R_5	OCH_3	Aryl	Allyl
390^{N}	H	βCH_3	H	OCH_2	OCH_2		α	β
391^{N}	H	βCH_3	OCH_3	OCH_2	OCH_2		α	
392^{N}	H	αCH_3	OCH_3	OCH_3	OCH_3		β	β
393^{N}	H	βCH_3	OCH_3	OCH_2	OCH_2		β	α
394^{N}	H	βCH_3	OCH_3	OCH_3	OCH_3		β	α
395^{N}	H	αCH_3	OCH_3	OCH_3	OCH_3		α	β
396^{N}	H	αCH_3	OCH_3	OCH_3	OCH_3	α	β	β
397^{N}	H	αCH_3	OCH_2	OCH_2	H	β	β	
398^{N}	H	βCH_3	OCH_2	OCH_2	H	α	α	
399^{N}	H	αCH_3	OCH_2	OCH_2	H		β	β
400^{N}	H	αCH_3	OCH_3	OH	H		β	α
401^{N}	H	αCH_3	OCH_2	OCH_2	H		β	α
402^{N}	H	αCH_3	OCH_2	OCH_2	OCH_3		β	β
403^{N}	OCH_3	αCH_3	OCH_2	OCH_2	OCH_3		β	β
404^{N}	H	αCH_3	OCH_2	OCH_2	H		α	β
405^{N}	H	αCH_3	OCH_2	OCH_2	OCH_3		α	β

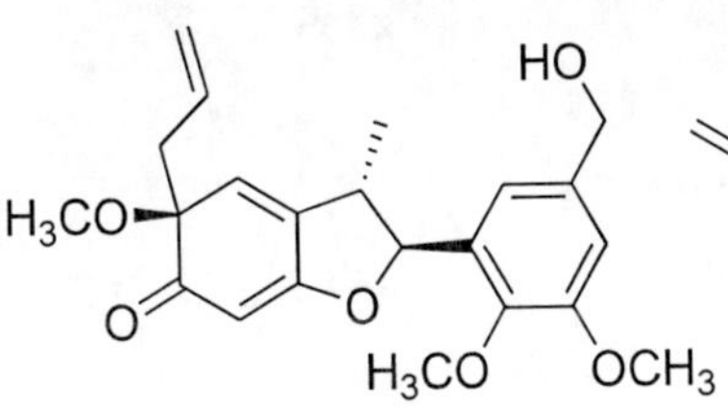

406^{N}

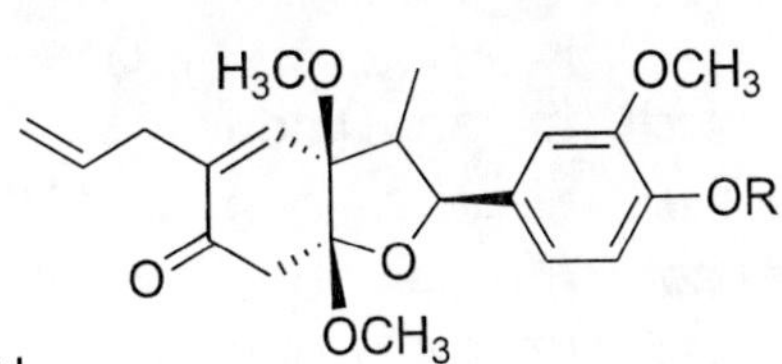

407^{N}: R= H
408^{N}: R=CH_3

Scheme 18. Chemical structures of dihydrobenzofuran neolignans **390–408.** ^{S}: synthetic DBN; ^{N}: natural DBN; ^{NS}: obtained by synthesis and from natural sources.

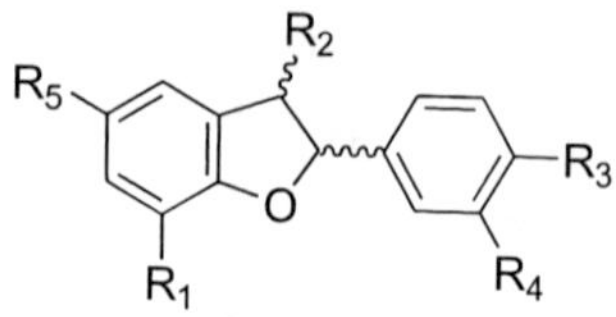

409-424

	R_1	R_2	R_3	R_4	R_5	Aryl
409[N]	H	α-CH_3	OH	OH	CHO	β
410[N]	H	α-CH_3	H	OH	CHO	β
411[N]	H	α-CH_3	H	OH	$CH_2CHOHCH_3$	β
412[N]	H	α-CH_3	H	OH	CH_2COCH_3	β
413[N]	H	α-CH_2OH	OH	H	OH	β
414[N]	H	β-CH_2OH	OH	H	OH	α
415[N]	H	α-CH_2OH	OH	H	CH_2OH	β
416[N]	H	β-CH_2OH	OH	H	CH_2OH	α
417[N]	OCH_3	α-CH_2OH	OH	H	CH_2OH	β
418[N]	OCH_3	β-CH_2OH	OH	H	CH_2OH	α
419[N]	OCH_3	β-CH_2OH	OH	OCH_3	CHO	α
420[N]	OCH_3	βCH_2OH	OCH_3	OH	$COCH_3$	α
421[N]	OCH_3	βCH_2OH	OCH_3	OGlc	$COCH_3$	α
422[N]	OCH_3	αCH_2OH	OH	OCH_3	CO_2H	β
423[N]	OCH_3	αCH_2OH	OH	OCH_3	CO_2CO_2Et	β
424[N]	OCH_3	βCH_2OH	OH	OCH_3	CH_2CO_2Et	α

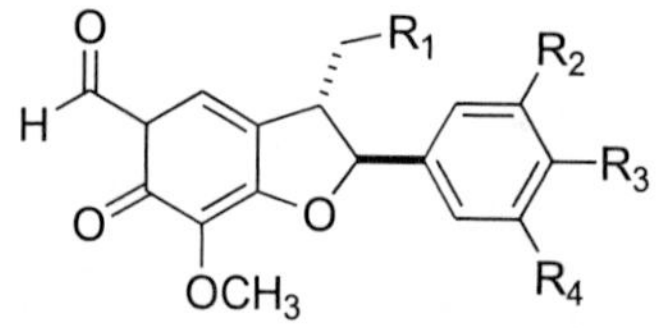

425-429

	R_1	R_2	R_3	R_4
425[N]	OH	OCH_2	OCH_2	H
426[N]	H	OCH_3	OCH_3	OH
427[N]	H	OCH_3	OH	H
428[N]	H	OH	OCH_3	H
429[N]	H	OCH_2	OCH_2	H

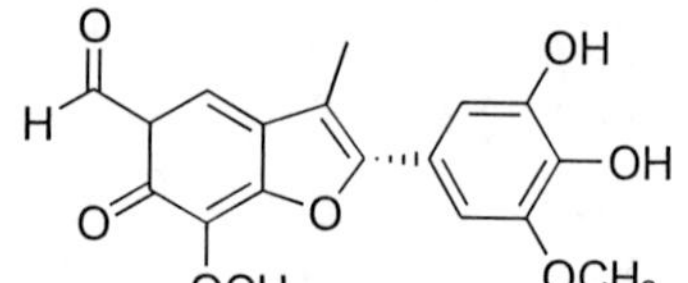

430[N]

Scheme 19. Chemical structures of dihydrobenzofuran neolignans **409–430.** [S]: synthetic DBN; [N]: natural DBN; [NS]: obtained by synthesis and from natural sources.

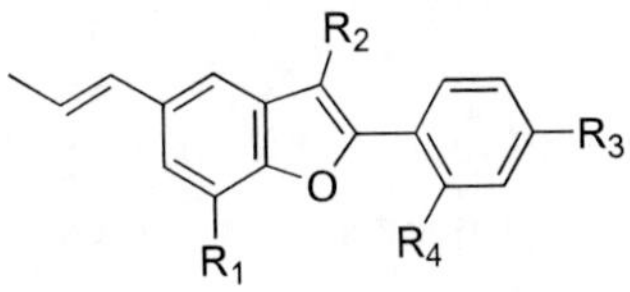

431-439

	R_1	R_2	R_3	R_4
431^{N}	H	H	OH	OH
432^{N}	H	H	OCH_3	OH
433^{N}	H	H	OH	OCH_3
434^{N}	H	H	OH	OH
435^{N}	OCH_3	OCH_3	OH	H
436^{N}	H	CH_3	OH	H
437^{N}	OCH_3	CH_3	OH	H
438^{N}	OCH_3	CHO	OH	H
439^{N}	H	CHO	OH	H

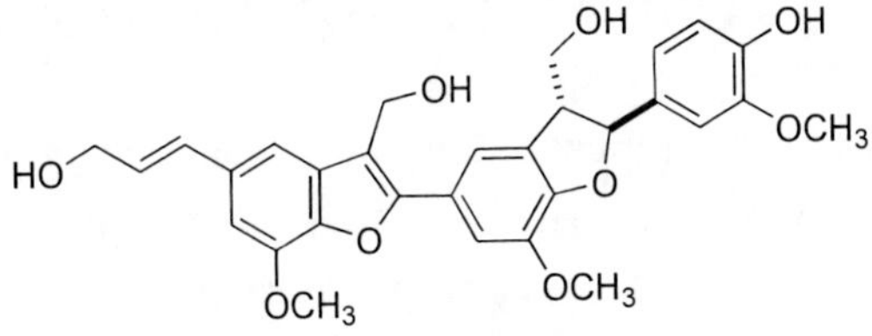

440^{N}

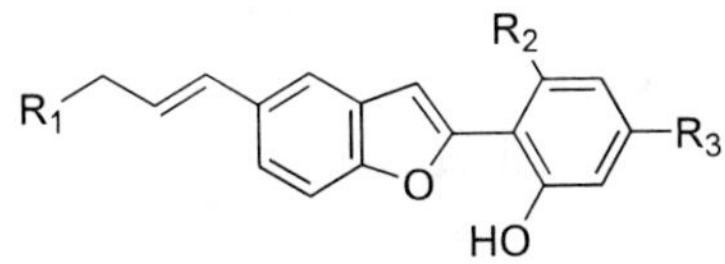

	R_1	R_2	R_3
441^{N}	H	OH	OCH3
442^{N}	OH	H	OH

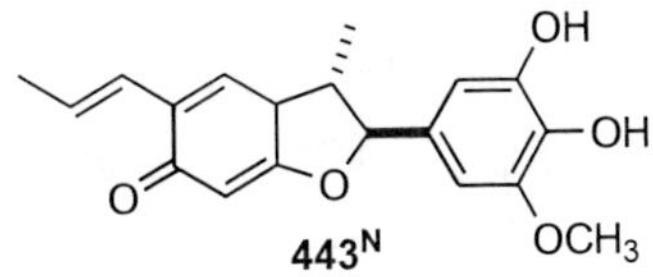

443^{N}

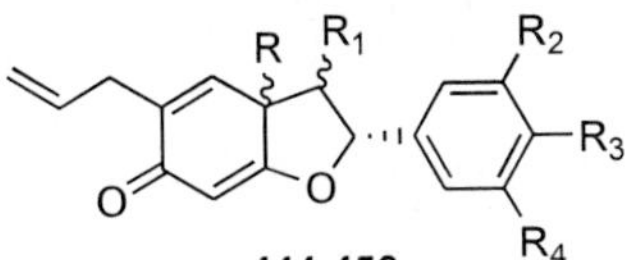

444-453

	R	R_1	R_2	R_3	R_4	Aryl
444^{N}	αOCH_3	CH_3	H	OCH_2	OCH_2	β
445^{N}	αOCH_3	CH_3	H	OCH_3	OCH_3	β
446^{N}	αOCH_3	βCH_3	H	OCH_3	OCH_3	α
447^{N}	αOCH_3	βCH_3	OCH_3	OCH_3	OCH_3	α
448^{N}	αOCH_3	CH_3	OCH_3	OCH_3	OCH_3	α
449^{N}	βOCH_3	CH_3	OCH_3	OCH_3	OCH_3	α
450^{N}	αOCH_3	CH_3	OCH_3	OCH_3	OCH_3	β
451^{N}	αOCH_3	βCH_3	H	OCH_2	OCH_2	β
452^{N}	βOCH_3	βCH_3	H	OCH_3	OCH_3	α
453^{N}	αOCH_3	βCH_3	H	OCH_3	OCH_3	β

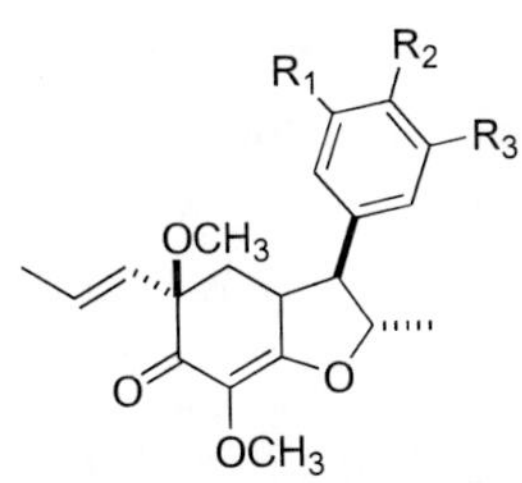

	R_1	R_2	R_3
454^{N}	OCH_2	OCH_2	OCH_3
455^{N}	OCH_3	OCH_3	OCH_3

Scheme 20. Chemical structures of dihydrobenzofuran neolignans **431–453.** ^{S}: synthetic DBN; ^{N}: natural DBN; ^{NS}: obtained by synthesis and from natural sources.

DIHYDROBENZOFURAN NEOLIGNAN SYNTHESIS

Dihydrobenzofuran neolignans display some remarkable biological activities, as will be discussed in this chapter. However, the yields of DBNs isolated from natural sources are usually low. For instance, (+)-conocarpan (**1**) and (–)-*epi*-conocarpan (**41**) have been isolated from *Conocarpus erectus* heartwood and *Piper regnellii* dried roots in 4.7 x 10^{-3} % and 2 x 10^{-2} % (m/m) yield, respectively [48, 141]. Moreover, the difficulties involved in the isolation process have led Gottlieb to describe it as "tedious" and time-consuming [14]. In this scenario, diverse DBNs have been obtained by total synthesis or structural modification with a view to exploring the biological properties of this class of compounds. For example, (±)-3',4-di-*O*-methylcedrusin (**202**), isolated from a blood-red latex produced by various American South *Crotton* species, known as "dragon's blood" and shown to have anti-cancer and anti-inflammatory effects, has been successfully synthesized by *trans*-feruloyl methyl ester oxidative coupling and further structural modifications [174].

In general, the synthetic methods that have been reported to build the DBN skeleton core can be divided into two main groups. The first group includes methodologies in which the C7-O4' bond is formed first, usually by an S_N2-type mechanism, followed by C9-C5' bond formation and consequent ring closure. The second step commonly employs Pd, Ir, Ru, or Rh compounds. For instance, Natori and co-workers have developed an alternative method to obtain (+)-conocarpan (**1**) and (-)-*epi*-conocarpan (**41**) by total synthesis in higher yields. The method involves an S_N2-type reaction, followed by a Danheiser modification of Regitz's diazo transfer reaction, which prepares the molecule for furan ring closure through C–H insertion catalyzed by a chiral dirhodium (II) complex (Scheme 21). Next, reduction with DIBAL-H, tosylation, Suzuki-Miyaura coupling, reduction with $LiEt_3BH$, and protective group removal affords **41** (global yield = 25.7%). Finally, the authors prepared (+)-conocarpan (**1**) for epimerization by employing Na_2CO_3/MeOH (global yielding = 18.3%) [175].

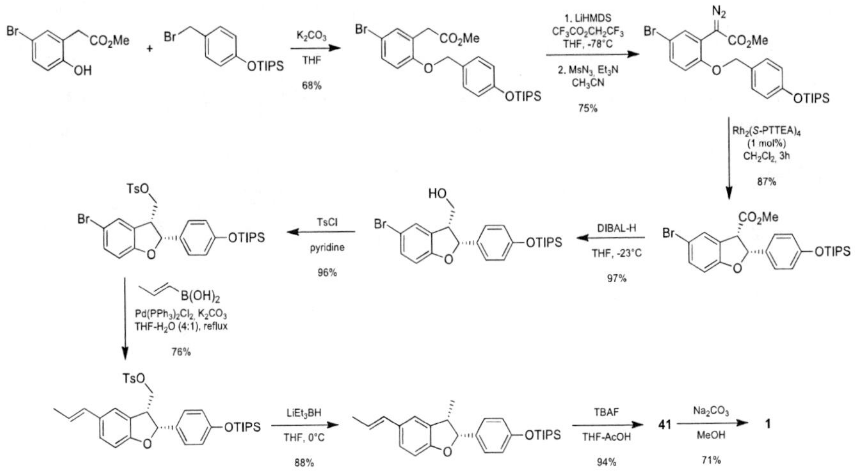

Scheme 21. Synthetic method reported by Natori and co-workers to obtain (+)-conocarpan (**1**) and (-)-*epi*-conocarpan (**41**) [175].

In the second main group of synthetic methodologies to produce DBNs, the C8-C5' bond is first formed by acylation reactions or metal-catalyzed reactions (e.g., Rh ou Pd), followed by C7-O4' bond formation and consequent five-membered ring closure through S_N2-type reactions. For instance, Chen and co-workers have developed an asymmetric synthesis of (+)-conocarpan (**1**) based on the formation of a stereocenter by enantioselective reduction. First, 2-fluoro-5-bromophenyl acetic acid is converted into its corresponding acid chloride, followed by Friedel-Crafts reaction with anisole (Scheme 22). The most important strategy of this stereoselective synthesis is to obtain a prochiral ketone by methylating the α-carbonyl position. Even though this step does not provide the process with the necessary stereochemistry because a mixture of enantiomers arises, using a chiral catalyst $RuCl_2$[(*S*)-dm-segphos][(S)-diapen] in the following step in the presence of base and H_2 (100 psi) promotes enantioselective formation of stereocenters with defined relative configuration. The furane ring is formed via a base-catalyzed S_NAr, which is followed by phenol ether demethylation. Finally, a Suzuki coupling reaction between the substrate and (*E*)-propenylboronic acid gives the DBN side chain[176].

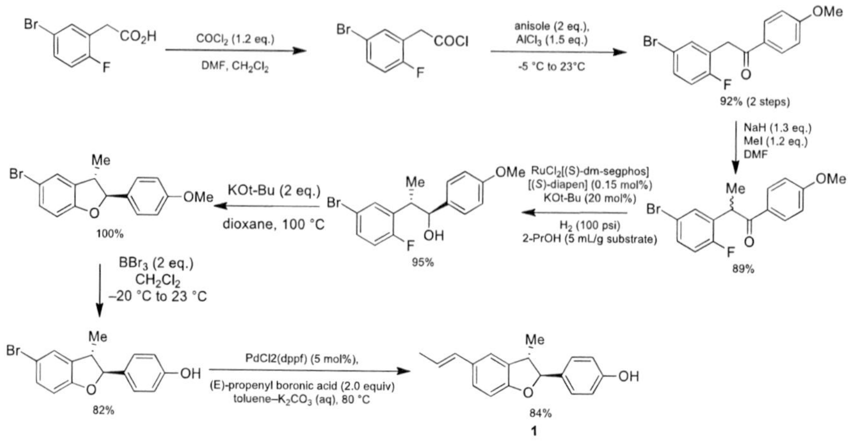

Scheme 22. Route for (+)-conocarpan (**1**) asymmetric synthesis (global yield: 55.6%).

García-Muñoz and co-workers have designed a method for the total synthesis of dihydrodehydroconiferyl alcohol (**193**), 3-*O*-demethyl dihydrodehydroconiferyl alcohol (**206**), and a derivative of these DBNs (**202**). The strategy consists in protecting the hydroxy group in the eugenol structure, followed by Claisen rearrangement and further S_N2 reaction between the deprotonated protected eugenol and a benzyl iodide (Scheme 5). Next, double bond oxidation by O_3 (ozonolysis) increases the benzylic H acidity for reaction with azide at low temperatures. The diazo group is introduced at that specific position, albeit in low yields. Ring closure is achieved in the presence of the chiral catalyst $Rh_2[(S)\text{-DOSP}]_4$ in anhydrous toluene at 0 °C, giving mostly a mixture of *cis*/*trans* diasteroisomers and showing discrete unexpected enantiomeric excess in all cases [177].

Some different metal complexes have been used as catalyst for C-C bond formation during DBN synthesis; e.g., Ir complexes, which have recently been employed by Hemelaere and co-workers. The synthesis starts from eugenol hydroboration, followed by *O*-allylation and further thermal Claisen rearrangement in *N,N*-dimethylaniline (Scheme 24). In the next two steps, the hydroxyl aliphatic and phenolic groups are selectively protected by reaction with pivaloyl chloride and *tert*-butyldimethylsilyl chloride, respectively, in good yields. Formation of an *E* double bond is driven by a first-generation Grubbs catalyst, which is followed by one-pot isomerization/allylboration catalyzed by the Ir complex catalyst ($IrH_2(thf)_2$-$(PPh_2Me)_2]PF_6$), affording *anti*-homoallylic alcohols in moderate yields and good enantiomeric excess (48 and 55%, respectively; 98% *anti*). Cyclization to produce the benzofuran ring is achieved by intramolecular Mitsunobu-type reaction after the phenolic hydroxyl group is deprotected. This enantioselective synthesis is one of the possible routes to synthesize (-)-dihydrodehydrodiconiferyl alcohol (**89**) [178].

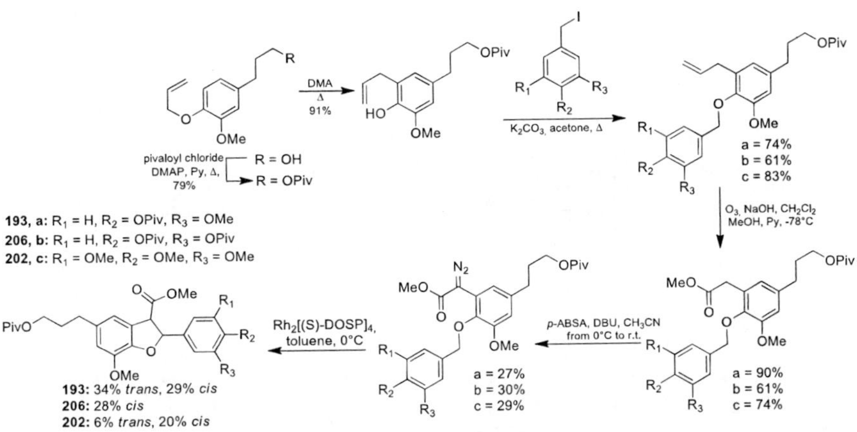

Scheme 23. Enantioselective synthesis of the DBNs **193**, **206**, and **202** catalyzed by a Rh complex (global yield of the diasteroisomer mixture: 8.1%, 2.3%, and 3.3%, respectively).

Although methodologies that use chiral metal catalysts provide DBNs in good yields, the number of steps and the diversity of reagents make their accomplishment difficult [176]. For this reason, oxidative coupling of phenylpropanoid (C_6C_3) units is by far the most frequently applied methodology. This reaction mimics natural DBN biosynthesis (Scheme 1) and affords their structural core in a single step under mild reaction conditions [174, 179]. The oxidation step can be performed both enzymatically and chemically [180], and the oxidant determines how the coupling occurs. For example, starting from the same phenylpropanoids, the use of silver oxide (Ag_2O) generates the dihydrobenzofuran skeleton, whereas the use of manganese oxide (MnO_2) gives lignans as the main products [179]. The yields of each DBN obtained by oxidative coupling vary and depend on the temperature, the concentration of phenylpropanoids and oxidants, and the solvent [179]. Indeed, the organic solvent employed in the synthesis affects the regioselectivity [181] and the reaction rate [182].

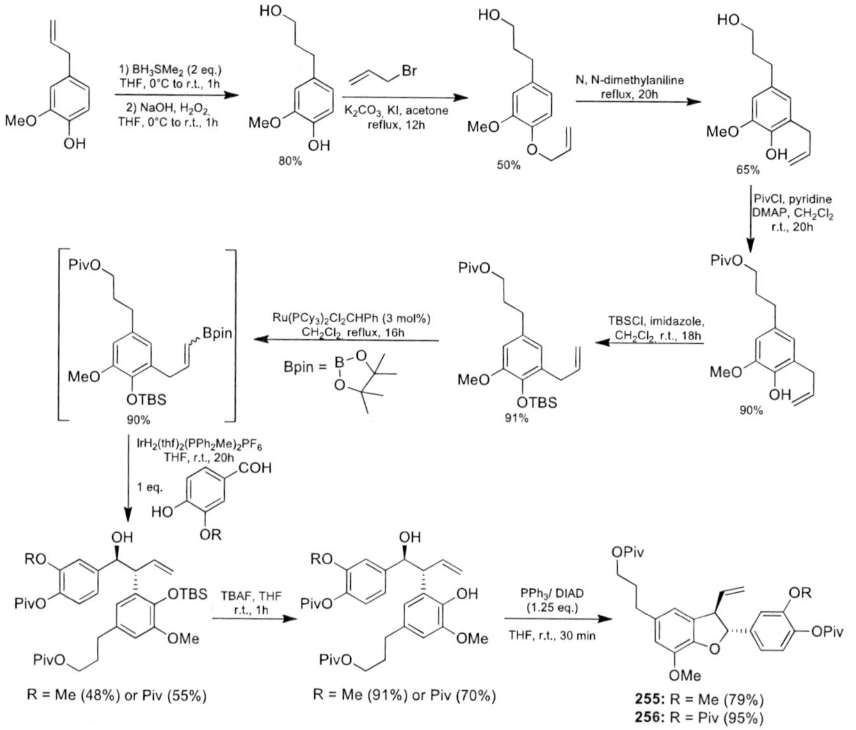

Scheme 24. Synthetic route to obtain the dihydrodiconiferyl alcohol (**89**) precursors **255** and **256** [178].

Most studies based on oxidative coupling of phenylpropanoids to synthesize DBNs have used Ag_2O as oxidant (Table 2). According to the most accepted mechanism, this coupling is initiated by O–H bond homolysis, as depicted in Scheme 2, to generate a resonance-stabilized phenoxy radical. Daquino and co-workers have proposed that the Mn(II) and Ag(I) ions can form complexes with these phenoxy radicals, hence affecting the spin density at C5/C5’ and C8/C8’, as calculated by semiempirical and quantum mechanical calculations at the AM and DFT levels of theory [179]. In the case of Mn(II) complexes, the spin density at C5 is reduced to almost zero, whilst the spin density at C8 is strongly enhanced, thereby increasing the probability of reaction at C8–C8', to produce lignans. On the other hand, formation of Ag(I) complexes does not increase the spin densities at C5 and C8 significantly. Additionally, complexation with Ag(I) constrains the approach between phenylpropanoids in a head-to-head fashion, to avoid repulsion between two positively charged regions (e.g., C5–C5'). Consequently, C5'-C8 coupling to produce DBNs is preferred when Ag_2O is the oxidant [179].

More recently, Althagafy and co-workers have offered additional insights into the role played by Ag_2O in coupling reactions. The authors have investigated the reaction mechanism of the coupling of Taxifolin with coniferyl alcohol to produce flavonolignans of biological interest, such as Silybin and Isosilybin (A and B, in both cases). Besides the desired products, the DBN dehydrodiconiferyl alcohol (**89**) is also obtained. A radical intermediate also results from coniferyl alcohol oxidation, agreeing with previous reports by Daquino and co-workers [179]. In addition, the authors have proposed that the silver salts formed within the solution act as single-electron oxidants and may coordinate the phenol moiety to the coupling position, hence stabilizing the radical intermediates [183].

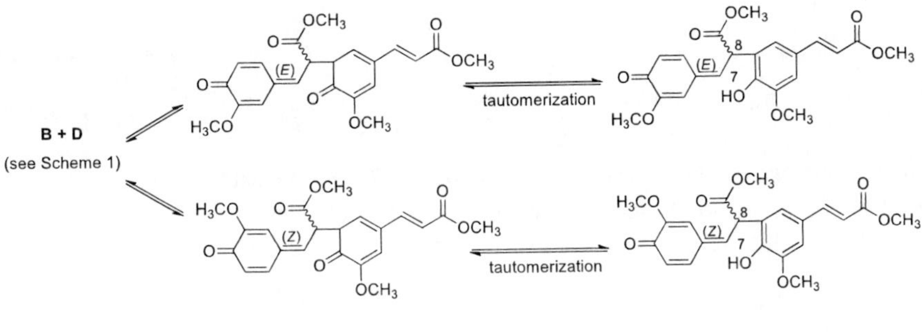

Scheme 25. Isomeric (*E*) and (*Z*) quinomethides formed by oxidative coupling of methyl ferulate.

Regarding stereoselectivity, oxidative coupling reactions of phenylpropanoids promoted by Ag_2O usually lead to diasteroselective formation of *trans*-DBN racemic mixtures [12, 180, 184, 185]. However, enantioselectivity can also be achieved when phenylpropanoids derivatized with a chiral auxiliary are employed [180]. To account for the *trans*-diasteroselectivity, Orlandi and co-workers have investigated the stereochemistry of the nucleophilic attack from the phenolic oxygen to the quinonoid double bond of the isomeric (*E*) and (*Z*) quinomethides derived from methyl ferulate (Scheme 25). These quinomethides emerge under thermodynamic control–the most stable quinomethide (*Z*) predominates in the equilibrium–and the faces of the quinomethide double bond are diasteroisomeric (Scheme 26). For the *anti* conformation, attack of the phenolic hydroxyl group to the quinomethide double bond occurs from the *Si* face, while the two *syn* conformations give nucleophilic attack from the *Re* face [180].

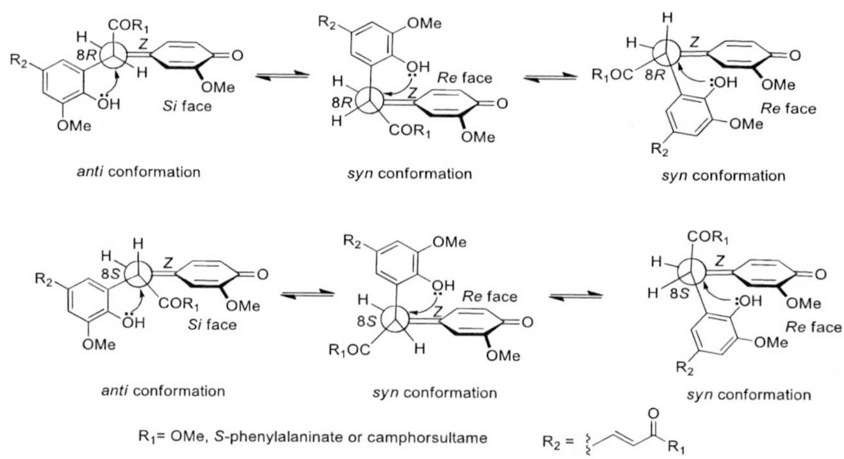

Scheme 26. Nucleophilic attack of the phenolic hydroxyl oxygen to the diasterofaces of the double bond of different (*Z*)-quinomethide intermediate conformers (adapted from ref. [180].

Conformational analysis of the rotation around the C7–C8 bond by semiempirical PM3 calculations of the MM2 optimized geometries have revealed that all the conformations leading to a *trans*-DBN are nearly 5 kcal/mol more stable than the conformations leading to the *cis*-DBN. In the presence of a chiral auxiliary (e. g., *S*-phenylalaninate or camphorsultame), an equilibrium mixture of diasteroisomeric quinomethides of different stabilities is expected. Furthermore, the presence of the chiral auxiliary also culminates in a difference of 2.0 kcal/mol among the *anti* conformations generating (7*S*,8*S*)-*trans*-DBN and the *anti* conformations generating the (7*R*,8*R*)-*trans*-DBN, causing enantioselection in the final product. Nevertheless, as in the case of most metal-mediated reactions that are used to synthesize DBNs, stereoselectivity is not observed when asymmetric aspects are not intrinsic to the catalyst structure.

Table 2 presents a literature survey of phenylpropanoid oxidative coupling for the synthesis of the DBNs shown in Schemes 5 to 20. The studies generally focus on determining the best conditions to achieve the DBNs in high yields

Methods based on enzymes, mainly horseradish peroxidase (HRP) and laccase, have also been employed to promote phenylpropanoid oxidative coupling [189]. Chioccara and co-workers have used the system HRP/H_2O_2 to synthesize the DBN from isoeugenol, methylferulate, and conyferyl alcohol by employing a citrate-phosphate buffer (pH = 3.0–7.4, with an optimized pH = 3.0) in methanol at 20 °C. *trans* β-5 dimmer is the predominant product, but traces of the mixture of β-O-4 diastereoisomers arise in the crude mixture [181]. Nonetheless, different "optimum" pH values have been used, depending on the oxidant system. For example, Yoshihara and co-workers have employed H_2O_2 (0.022%) as oxidant in phosphate buffer at a optimum pH value of 7.38 at 37°C [186], to achieve a yield of 23% for dimethyl dehydrodiferulate, which is lower as compared to the HRP/H_2O_2 system, which affords 30% yield for the same substrate. In other cases, with other substrates, 5–70% yields have been obtained with the HRP/H_2O_2 oxidant system by modifying the pH, the solvent, or even the temperature [180, 187]. More recently, some studies have described the use of the HRP/H_2O_2 system directly from coconut water (*Cocos nucifera* L.) as a natural source of HRP to produce (±)-licarin A (**8**) and other analogues starting from isoeugenol, ferulic acid, and correlated phenylpropanoids.

Table 2. Oxidant agents used for the synthesis of dihydrobenzofuran neolignan by phenylpropanoid oxidative coupling

Dihydrobenzofuran neolignan	Oxidant	Reference
9, 22, 36	H_2O_2, HRP	[181]
36, 49, 51, 58, 59, 74, 84, 89, 110, 117, 119	H_2O_2, HRP	[186]
116	H_2O_2, HRP	[187]
29, 35, 36, 37, 38, 107, 109	H_2O_2, HRP	[180]
9, 16, 17, 18, 19, 20, 21	Ag_2O	[188]
23, 29, 88	Ag_2O H_2O_2, HRP Laccase	[189]
316, 318	Ag_2O, AgOAc, Ag_2CO_3, $AgNO_3$, $Mn(OAc)_3$, CuOAc, $Cu(OAc)_2$ or $K_3Fe(CN)_6$	[190]
63, 67, 202	Ag_2O	[179]
50, 55	Ag_2O	[191]
111, 112	$FeCl_3$ or Ag_2O	[192]
51, 56, 57, 62, 63, 64, 65, 66, 67, 68, 69, 70, 130, 177, 193	Ag_2O	[185]
116	Ag_2O	[193]
113, 114, 195, 222, 276	Ag_2O	[194]
36, 51	Ag_2O	[12]
36, 39, 51, 54	Ag_2O	[195]
36, 39, 51, 54, 237, 243, 262, 263	Ag_2O	[196]
36, 39, 51, 54, 63, 237, 243, 262, 263	Ag_2O	[197]
36, 39, 51, 54, 63, 237, 243, 262, 263	Ag_2O	[198]
8	HRP, H_2O_2	[199-201]
89	Laccase	[202]
83, 89	Lacccase	[203]
35, 51, 202	Lacccase	[204]
326, 327, 328	Iodobenzene diacetate, Iodobenzene bis(trifluoroacetate)	[205]
51, 53, 178, 179	$K_3Fe(CN)_6$	[206]
9	O_2	[182]
4, 144	$Ru(bpz)_3(PF_6)_2$, hν	[207]
8	*	[208]
259, 260, 261	*	[177]
1, 4	*	[176]
206, 214, 255, 256	*	[178]
334, 335, 336	*Rh (II)	[209]
89, 114	*	[210]
226	*	[211]
2, 4	*	[212]

* Total synthesis. [a]HRP: Horseradish peroxidase; [b]TPP: tetraphenylporphyrin.

Yields range from 12 to 85%, depending on the H_2O_2 quantity and the coconut water volume [199]. The correlated methodology has already been patented and examined by other research groups [200, 201].

Another alternative enzymatic methodology to synthesize DBNs via phenylpropanoid oxidative coupling is based on laccases, which can oxidize the catechol derivatives (like phenylpropanoids) to produce radical intermediates [202, 203]. In some cases, the use of laccases as oxidant involves a simple reaction procedure that applies mild conditions and affords DBNs in yields ranging from 8 to 32% (the substrate nature and the laccase source are crucial). However, obtaining the enzymes and isolating the compounds by elaborate and time-consuming chromatographic methods are difficulties of these processes [203]. In recent studies, Basini and co-workers have employed commercial *Trametes versicolor* laccase (TvL) for methyl ferulate oxidative coupling in the presence of a biphasic system containing ethyl acetate and acetate buffer (pH = 4.7) at room temperature for 20 h. The methodology affords the DBN dimethyl dehydrodiferulate in 24% yield, which is comparable to the results achieved with the HRP/H_2O_2 system [204].

Dihydrobenzofuran Neolignan Biological Activities

3',4-Di-*O*-methylcedrusin (**202**) displays antibacterial [213], antioxidant [68], imunomudulatory, and anti-inflammatory [55] actions among others. The wound healing capacity of "sangre de drago" has been associated with the presence of **202**, which stimulates fibroblasts and collagen formation [55]. Because of these biological activities, the synthesis of 3',4-di-*O*-methylcedrusin (**202**) from phenylpropanoids has also been reported [179, 214].

Most studies on the DBN biological activities have been carried out with compounds isolated from natural sources. There has been a remarkable interest in their anti-inflammatory, antioxidant, cytotoxic, and antibiotic properties. However, other potential DBN effects have also been explored over the last twenty years, as shown in Figure 5.

The most common biological activities attributed to DBNs are discussed in the following topics.

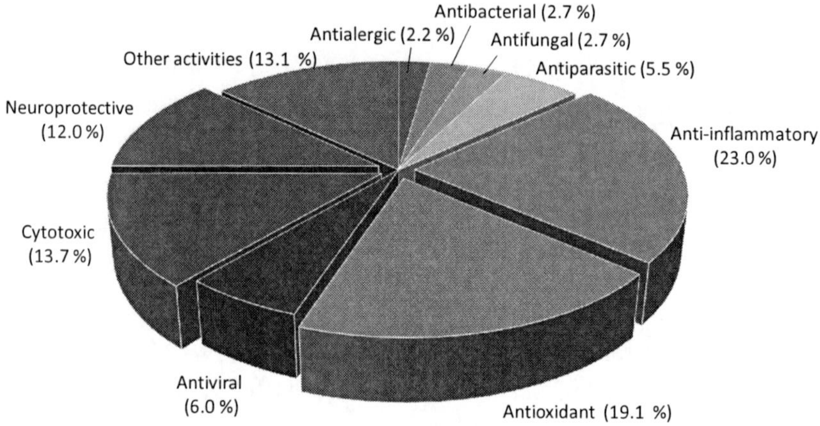

Figure 5. Main biological activities reported for natural and synthetic dihydrobenzofuran neolignans.

Anti-Inflammatory Activity

Inflammation is a multi-factorial process that results from the overproduction of pro-inflammatory cytokines and other inflammatory mediators, which are generated by enzymes such as nitric oxide synthase and cyclooxygenase [215]. DBNs have been extensively investigated for their anti-inflammatory potential (Figure 5). Some interesting results have been reported by Cho and co-workers on the anti-inflammatory effects of five DBNs called Woorenosides (**93**, **94**, **95**, **102**, and **103**), isolated from the ethanolic extract of *Coptis japonica* (Ranunculaceae) rhizomes and leaves. The tested compounds suppress necrosis factor and nitric oxide production [49]. This same suppression effect has also been observed in lymphocyte proliferation assays. Moreover, the authors have demonstrated that these compounds inhibit the mitogen response, which has been associated with their participation in anti-inflammatory process regulation.

Cistus laurifolius leaves have traditionally been employed to treat inflammatory disorders in Turkey. To determine which compounds underlie this bioactivity, some chemical compounds have been isolated on the basis of prostaglandin inhibition assays. Prostaglandins are well known for mediating inflammatory reactions, and compounds affecting their receptors may reduce inflammation, pain, and fever, among other symptoms. Chronic inflammatory conditions have also been found to enhance predisposition to developing cancer, so some prostaglandin inhibitors have been considered to prevent and to treat cancer. Compound **165** at 30 μg/mL exerts a prostaglandin inhibitory effect and has been isolated as a 2:1 mixture of two diastereomers. The absolute stereochemistry of the major compound has been determined as 7*S*,8*R* [46].

(+)-Conocarpan (**1**) and the lipophilic extract of *Krameria lappacea* roots have been evaluated in terms of their putative anti-inflammatory activities. In a mouse ear *in vivo* model, the dichloromethane extract (ID_{50} 77 μg/cm^2) and compound **1** (ID_{50} 82 μmol/cm^2) have been shown to exhibit topical antiedematous properties that are comparable to the properties of indomethacin (ID_{50} 0.29 μmol/cm^2). (+)-Conocarpan (**1**) has also been studied for its time-dependent edema development and leukocyte infiltration up to 48 h after croton oil-induced dermatitis. Compound **1** has an activity profile that resembles the hydrocortisone profile: it significantly reduces the edema (25-61%) in mice after 48 h. Its long-lasting anti-inflammatory effect peaks after 6 h (61%) and reduces the edema by 41%. The authors have also reported significantly lower leukocite infiltration, which ranges from 27 to 44% inhibition. These results are equivalent to the results obtained with the positive controls indomethacin and hydrocortisone (percentage of reductions 25–49% and 35–56%, respectively) [82].

DBN **191** isolated from the aerial parts of *Leontopodium leontopodioides* has been evaluated for its effects on nitric oxide production inhibition in lipopolysaccharide-activated macrophages. Compound **191** exhibits inhibitory activity with IC_{50} value of 35.80 μM as compared to the IC_{50} value of 40.45 μM for N-monomethyl-L-arginine, used as positive control [87].

Allergy can be defined as an exaggerated response from the immune system to bodily contact with certain foreign substances and is classified into five main types. Type I allergy is the most common allergic reaction; its clinical manifestations are caused by the release of proinflammatory mediators, such as histamine, leukotrienes, and prostaglandins. The ethanolic extract of *Pinus thunbergii* needles suppresses the antigen-mediated degranulation of rat basophilic leukemia (RBL-2H3) cells, which is used as a reliable parameter to predict possible antiallergic activities. Some DBNs have been isolated from this ethanolic extract by bioassay-guided fractionation. The inhibitory effects of the isolated constituents on the release of β-hexosaminidase from RBL-2H3 cells has also been examined. Compounds icariside E4 (**251**), massonianoside A (**247**), isomassonianoside B (**246**), and dihydrodehydrodiconiferyl alcohol-4-*O*-β-D-glucopyranoside (**249**) exhibit interesting inhibitory activity with IC_{50} values of 66.2 ± 0.5, 62.9 ± 1.5, 92.8 ± 1.4, and 52.3 ± 0.9 μM, respectively, as compare to the IC_{50} value of 78.5 ± 0.5 μM obtained for ketotifen, employed as positive control [138].

Antioxidant and Cytotoxic Activities

Normal metabolic processes that use molecular oxygen (e.g., mitochondrial oxidative metabolism, inflammation, and so on) are endogenous sources of reactive intermediates that are generally considered as toxic by-products [216]. Antioxidant activity is currently understood as the ability of a molecule to reduce free radical formation and scavenge reactive oxygen species (ROS), inhibiting and/or repairing injuries caused by biomolecule oxidation and degradation [217].

The second most frequent activity reported for DBN is the antioxidant action (Figure 5), which is typically displayed by compounds bearing a phenolic moiety. Bioactivity-guided fractionation of a methanol-soluble extract of *Euterpe oleracea* ("açai") fruits using a hydroxyl radical scavenging assay has led to the isolation of DBNs **206** and **209**. These compounds present potent antioxidant effects (IC_{50} values ranging between 0.6 and 3.7 μg/mL) in the *in vitro* hydroxyl radical scavenging assay as compared to the positive control quercetin with IC_{50} value of 0.31 μg/mL [64].

Liu and coworkers have investigated the antioxidant and anti-inflammatory actions of compounds **85**, **146**, **189**, **190**, **193**, **199**, **200**, **201**, **219**, and **220** isolated from the ethanolic extract of *Prunus tomentosa* seeds; more specifically, six DBNs named prunustosanans and prunustosanansides. Compounds **146**, **193**, and **220** provide the highest activity against the DPPH radical and ABTS, respectively. Moreover, compound **85** exhibited much stronger inhibitory effect on nitric oxide (NO) production (IC_{50} = 4.4 ± 0.5 μM) in murine microglia BV-2 as compared to minocycline (positive control, IC_{50} = 19.7 ± 1.5 μM)[151].

Oxidative stress plays a significant role in many human diseases, including cancer and neurodegenerative disorders [218]. In this way, compounds that display antioxidant and cytotoxic properties, like DBNs, are particularly useful to treat/eliminate cancer. The pinnatifidanins **91**, **158**, **193**, **196**, **197**, **198**, and **205** have been isolated from the ethanolic extract of *Crataegus pinnatifida* seeds. The antioxidant and cytotoxic properties of these DBNs have been assessed. Compound **91** is more cytotoxic than the positive control (5-fluorouracyl) against the HT-1080 cancer cell line (IC_{50} = 8.86 and 35.62 μM, respectively). The antioxidant effects of the isolated DBNs have been evaluated by the DPPH and ATBS assays. All the compounds (IC_{50} > 90 μg/mL) are less active than the positive control vitamin E (IC_{50} = 9.96 μg/mL) in the DPPH assay. On the other hand, in the ATBS assay, most DBNs show significant activity of IC_{50} < 15 μg/mL, which is comparable to the activity of the positive control vitamin E (IC_{50} = 9.30 μg/mL). The antioxidant activity is believed to be correlated to the number of free phenolic hydroxyl groups, and the antioxidant activity may dramatically decrease in the absence or during blockage of the hydroxyl groups by the methyl group [52].

Boehmenan H (**194**) isolated from the *Hibiscus cannabinus* variety *Salvador* L. (Malvaceae) is strongly cytotoxic against the HeLa (human carcinoma of the cervix) cell line (IC_{50} = 2.6 μg/mL; IC_{50} for positive control = 0.014 μg/mL). Comparison with the results obtained for other isolated compounds have suggested that dihydrofuran system loss and the presence of the feruloyl moiety significantly increase the activity against the cell lines [68]. Cytotoxicity-guided fractionation of *Helicteres hirsuta* stems has also allowed boehmenan H (**194**) to be isolated, identified, and evaluated against a small panel of cancer cell lines. Compound **194** exhibits weak cytotoxic activity [65].

The DBNs named illiciumlignans, compounds **266**, **267**, **268**, **269**, and **309**, have been isolated from *Illicium wardii* branches and leaves. The compounds present moderate cytotoxic activity against four human cancer cell lines [75].

Antiparasitic Activity

The DBN antiprotozoal activities have also been demonstrated. Van Miert and co-workers have investigated the antiprotozoal and the cytotoxic potential of a series of synthetic DBNs (**241**, **247**, **248**, **249**, **251**, **252**, **253**, **254**, and **288**) obtained by oxidative coupling of caffeic esthers promoted by Ag_2O [185]. Further acetylation, oxidation, deprotection, methylation, and reduction reactions have been carried out to afford good structural diversity. None of the synthesized DBNs described in the work exhibit promising cytotoxicity against the human tumor cell line panel. The authors have also described antiprotozoal assays against *Leishmania donovani* (axenic amastigotes and infected macrophages), chloroquine-resistant *Plasmodium falciparum* (strain K1), and *Trypanosoma brucei rhodesiense*. In the assays against axenic amastigotes, compound **251** has provided the best results with an IC_{50} value of 0.12 μg/mL (IC_{50} positive control = 0.25 μg/mL). Its activity has been confirmed in an antileishmanial assay accomplished in infected macrophages. In the latter assay, compound **251** has been the most active compound of all the tested compounds, showing an IC_{50} of 0.19 μg/mL (positive control with IC_{50} = 0.12 μg/mL). Apparently, in the case of this dihydrobenzofuran, introducing a larger ester group decreases the cytotoxicity and enhances the antileishmanial activity – the antileishmanial activity of compound **288** is almost two orders of magnitude lower as compared to compound **251**. In the assays against *P. falciparum* and *T. brucei rhodesiense*, DBN **288** provides poor results as compared to the positive control [185].

Dias and coworkers have tested a series of synthetic DBNs, namely compounds **51**, **36**, **54**, **39**, **262**, **237**, **263**, and **243** against *Schistosoma mansoni* worms *in vitro* [196]. None of the DBNs display significant schistosomicidal effects at 12.5, 25.0, or 50 μM. However, at concentrations of 100 or 200 μM, most of the compounds change the worm motor activity to some degree. The authors have also inferred that acetylation of the phenolic hydroxyl group can increase worm mortality, mainly after 24 h of incubation. Furthermore, hydrogenation at C7'-C8' or methylation of the phenolic hydroxyl group has been reported to decrease the DBN schistosomicidal effects.

Antibiotic Activity

The present section contains examples of DBNs with antibiotic activities, including antibacterial, antifungal, and antiviral actions. Most of the biological activities have been reported for isolated or natural DBNs, but both synthetic and natural compounds deserve to be highlighted as potential antibiotic agents. Although we have selected only a few examples in this chapter, DBNs can be used as building blocks to obtain new antibiotic structures, as shown below.

The DBN named callislignan A (**60**) has been isolated from *Callistemon lanceolatus* leaves, and its antibacterial activity against two strains of *Staphylococcus aureus* has been tested. One of the *S. aureus* strains is sensitive to many commercially available antibiotics, whereas the other strain is resistant to methicillin and other β-lactam antibiotics. DBN **60** displays antibacterial activity against both *S. aureus* strains, with Minimum Inhibitory Concentration (MIC) values of 200 μg/mL for the sensitive strain and 64 μg/mL for the resistant strain. Vancomycin, a standard antibiotic, affords a MIC value of 1 μg/mL [40].

The supercritical CO_2 extract from *Piper regnellii* var. *pallescens* leaves has been purified by chromatographic methods, yielding the DBN conocarpan (**1**) and its derivatives **4** and **173**. These compounds have been screened for their anti-*Mycobacterium tuberculosis* effects. Conocarpan (**1**) has an IC_{50} value of 97.6 μg/mL and selectivity index (SI) of 6.5. Although this MIC value is higher than the value of the reference drug isoniazid (MIC 0.03 μg/mL), SI is used to estimate the therapeutic window of a drug and to identify drug candidates for further studies. An SI value greater than 1 indicates that a compound is more selective for microorganisms [143]. *Piper regnellii* var. *pallescens* is a medicinal plant that has been traditionally used to treat infectious diseases in Brazil. The extracts obtained from its leaves along with the isolated DBN conocarpan (**1**) have been investigated for their antifungal activities against the yeasts *Candida albicans*, *C. krusei*, *C. parapsilosis*, and *C. tropicalis.* Conocarpan (**1**) at concentrations of 6.3 to 12.5 μg/mL (MIC values) is active against the yeasts. The positive control nystatin displays MIC values of 1–8 μg/mL. Next, conocarpan (**1**) methylation at C-4 affords *O*-methylconocarpan, which is inactive against all the tested yeasts. The antifungal activity of compound **1** can be related to the presence of the phenolic hydroxyl at C-4 [144].

Fukui and coworkers have evaluated the antimicrobial activity of a series of synthetic DBNs (compounds **51**, **36**, **54**, **39**) in therms of their MIC against bucal pathogens that cause caries, such as *Streptococcus salivarius*, *S. sanguinis*, *S. mitis*, *S.mutans, S. sobrinus, Enterococcus faecalis*, and *Lactobacillus casei*. In addition, the authors have tested these compounds against dermathomycosis and aspergilosis agents, including *Aspergillus brasiliensis*, *A. fumigatus*, *Rihizopus stolonifer,* and *Mucor rouxii* [195]. Compound **36** displays moderate activity against *S. mitis* and *L. casei*, whereas compounds **51** and **39** have moderate activity against *S. mitis* at the same MIC values (MIC = 400 μg/mL). Compound **54** has moderate activity against *A. brasiliensis* (MIC = 125 μg/mL) and *M. rouxii* (MIC = 250 μg/mL). The cytotoxyc effects of these compounds have been tested on V79 cells, and all these DBNs demonstrate some level of cytotoxicity. Besides that, compounds **51**, **36**, and **39** decrease the viability of V79 cells to 75–79%, whereas compound **54** at a concentration of 100 μM reduces V79 cell viability to 27%. Compound **54** presents the lowest IC_{50} value (67.97 μM) among the tested compounds after 24 h of incubation. The authors have proposed that the presence of methyl ester groups at C9 and C9' decrease the lipophilicity of the tested compounds. On the other hand, the presence of an acetoxyl group in componud **54** and the absence of a methoxyl group at C3 and C3' increase the lipophylicity and explain the antifungal effect of this compound as compared to the other tested DBNs. The same explanation has been used to account for the high cytotoxicity of compound **54** to V79 cells.

Sawasdee and co-workers have isolated DBNs from the methanolic extract of *Miliusa mollis* leaves (**241**, **279**, **280**, **281**, and **288**). The inhibitory effects of the isolated DBNs on the herpes simplex virus (HSV) types 1 and 2 have been examinated. Compound **280** exerts weak activity against HSV-1 and HSV-2 (IC_{50} = 155.3 and 222.0 μM, respectively) as compared to the positive control acyclovir (IC_{50} = 1.9 and 2.1 μM, respectively) [107].

For the first time, Yang and co-workers have obtained the DBN herpepropenal (**23**) from *Herpestospermum caudigeranum* seeds and have assayed the inhibitory effects of this isolated compound on hepatitis B virus (HBV DNA) and the replication or expression of HBsAg and HBeAg in HepG2.2.15 cells, which is considered a good strategy to screen anti-HBV drugs *in vitro*. The experiments have revealed that compound **23** at 25, 50, 100, and 200 μg/mL exhibits 14.31%, 26.12%, 35.39%, and 42.34% inhibitory effects on the replication rates (IR) of HBV DNA, respectively. These results demonstrate that herpepropenal is effective against hepatitis B virus in a dose-dependent manner. Compound **23** behaves similarly to the positive control at the same concentration. In summary, herpepropenal is an anti-HBV drug candidate [67].

Some analysis on the DBN structure-activity relationship (QSAR) have also been reported. The presence of OH at C-4 seems to be important for the antimicrobial activity [143, 213]. When the antiprotozoal and cytotoxic actions are compared, it is evident that: (i) the presence of the phenolic OH at C-3' in the benzofuran moiety does not change either of the activities; (ii) compounds that contain the hydroxyl group instead of an H atom at C-3 in the pending phenyl ring display the highest cytotoxic activity, while this structural feature does not affect the antiprotozoal action; (iii) bulkier and more lipophilic alkyl moieties (ethyl; *n*- and *i*-propyl; and *n*-, *i*-, and *tert*-butyl) at C-9 decrease the cytotoxicity.

Other Activities

Besides the bioactivities highlighted below, the neuroprotective activity against glutamate-induced cell death displayed by DBNs **2**, **4**, **5**, **7**, **8**, **43**, **44**, **45**, **46**, **47**, and **48,** isolated from the ethanolic extract of *Aristolochia fordiana* stems [33], and the protective effects of DBNs **413**, **414**, **417**, and **418**, isolated from *Rubus idaeus* [154], against H_2O_2-induced neurotoxicity deserve to be mentioned. Moreover, the synthetic DBNs **18**, **19**, **20**, and **21** exhibit moderate activity on α-glucosidase inhibition, which is useful for diabetes profilaxy and treatment [188].

Resistance to conventional drugs and chemicals must be overcome in applications of new biologically active compounds. DBNs have been synthesized for this purpose. For example, Baldin and co-workers have studied synthetic DBN derivatives as insecticidal agents. The derivatives **36**, **39**, **51**, **54**, **63**, **237**, **243**, **262**, and **263** have been assayed against second instar *Bemisia tabaci*. The tested compounds have remarkable insecticidal potential, especially compounds **54** and **237**, which act as potential repellents and oviposition deterrents for adult whiteflies after 24 and 48 h. In addition, most of the tested compounds kill 100% of *B. tabaci*, so they are potentially efficient for future applications in the pest management field [197].

DBNs have been exploited against Alzheimer's disease. Liu and co-workers have isolated compounds **140**, **185**, and **193** from *Paeonia lactiflora* and tested their inhibitory effects on Aβ1-42 aggregation through ThT-based fluorometric assay, on the basis of the β-amyloid aggregation on the brain hypothesis as a factual event of this pathogenesis [130]. According to the authors, compound **193** exhibits optimal aggregation inhibition (80.1% at 20 μM) as compared to the positive control. On the basis of these results, compound **193** is a promising agent against Alzheimers's disease, in contrast to docking studies reported by the authors in the same paper.

Conclusion

This chapter has provided an overview of DBN occurrence, synthesis, and biological activities. Unfortunately, despite the wide DBN structural diversity, the nomenclature used over the last four decades to refer to these compounds is not uniform, which still represents a challenge in terms of literature review. Although the anti-inflammatory, antitumor, and antiparasitic activities of some DBNs have been highlighted and their structural diversity has been discussed herein, studies on the biological activities of this class of compounds are still limited due to difficulties in isolating DBNs. In this scenario, only a small percentage of the currently known DBNs have been investigated for their biological activities. On the other hand, hyphenated techniques have boosted natural DBN isolation, whereas the use of enzymatic methods or metallic complexes as a catalyst in organic reactions for C-C bond formation have increased the variety of synthetic DBNs. However, knowledge of the biological activities of DBNs remains a long way to be paved.

References

[1] Moss, G. P. (2000). "Nomenclature of lignans and neolignans". *Pure and Applied Chemistry* 72:1493-1523.

[2] Gottlieb, O. R. (1978). "Neolignans". In *Fortschritte der Chemie Organischer Naturstoffe/Progress in the Chemistry of Organic*

Natural Products, edited by Herz, W., Grisebach, H. and Kirby, G. W., 1-72. Vienna: Springer Vienna.

[3] Haworth, R. D. (1942). "The chemistry of the lignan group of natural products". *Journal of the Chemical Society (Resumed)* 448-456.

[4] Gottlieb, O. R. and Yoshida, M. 1984. "Lignoides com atenção especial à química das neolignanas". *Química Nova* 250-273.

[5] Dewick, P. M. (2002). *Medicinal natural products: a biosynthetic approach.* London: John Wiley & Sons.

[6] Liu, Q.-B., Huang, X.-X., Yan, X.-J., Bai, M., Yu, L.-H., Hu, C., Zhu, T., Li, L.-Z. and Song, S.-J. (2014). "Neolignans from the seeds of *Prunus tomentosa* (Rosaceae) and their chemotaxonomic interest". *Biochemical Systematics and Ecology* 55:236-240.

[7] Barceló, A. R. and Pomar, F. (2002). "Plant peroxidases: versatile catalysts in the synthesis of bioactive natural products". *Studies in Natural Products Chemistry* 27:735-791.

[8] Yoshida, M. (1991). "Chemical transformations of neolignans". *Memórias do Instituto Oswaldo Cruz* 86:39-42.

[9] Zhuang, P.-Y., Zhang, G.-J., Wang, X.-J., Zhang, Y., Yu, S.-S., Ma, S.-G., Liu, Y.-B., Qu, J., Li, Y., Xu, S., Lü, H.-N., Chen, X., Li, L., Si, Y.-K. and Zhang, D. (2013). "Prenylated C_6–C_3 compounds from the roots of *Illicium henryi*". *Phytochemistry* 86:176-183.

[10] Green, T. P., Galinis, D. L. and Wiemer, D. F. (1991). "Three neolignans from the roots of *Piper capense*". *Phytochemistry* 30:1649-1652.

[11] Li, S. L., Iliefski, T., Lundquist, K. and Wallis, A. F. A. (1997). "Reassignment of relative stereochemistry and C-7 and C-8 in arylcoumaran neolignans". *Phytochemistry* 46:929-934.

[12] Medeiros, T. C. T., Dias, H. J., Silva, E. O., Fukui, M. J., Soares, A. C. F., Kar, T., Heleno, V. C. G., Donate, P. M., Parreira, R. L. T. and Crotti, A. E. M. (2016). "Detailed ^{1}H and ^{13}C NMR spectral data assignment for two dihydrobenzofuran neolignans". *Journal of the Brazilian Chemical Society* 27:136-143.

[13] Iwashima, T. 2000. "The structure and distribution of the flavonoids in plants". *Journal of Plant Research* 113:287-299.

[14] Gottlieb, O. R. (1972). "Chemosystematics of the Lauraceae". *Phytochemistry* 11:1537-1570.

[15] Andrade, C. H. S., Filho, R. B. and Gottlieb, O. R. (1980). "Neolignans from *Aniba ferrea*". *Phytochemistry* 19:1191–1194.

[16] Li, Y., Xie, S., Ying, J., Wei, W. and Gao, K. (2018). "Chemical structures of lignans and neolignans isolated from Lauraceae". *Molecules* 23:1-18.

[17] Alvarenga, M. A. D., Brocksom, U., Oscar, C. C. and Magalhães, M. T. (1977). "Neolignans from *Aniba burchellii*". 16:1797–1799.

[18] Lima, O. A., Gottlieb, O. R. and Magalhães, M. T. (1972). "Burchellin, a neolignan from *Aniba burchellii*". *Phytochemistry* 11:2031-2037.

[19] Bülow, M. V. v., Franca, N. C., Gottlieb, O. R. and Puentes Suarez, A. M. (1973). "Guianin: a neolignan from *Aniba guianensis*". *Phytochemistry* 12:1805-1808.

[20] Fernandes, J. B., Gottlieb, O. R. and Maia, J. G. S. 1976. "Neolignans from an *Aniba species*". *Phytochemistry* 15:1033-1036.

[21] Aiba, C. J., Corrêa, R. G. C. and Gottlieb, O. R. (1973). "Natural occurrence of Erdtman's dehydrodiisoeugenol". *Phytochemistry* 12:1163-1164.

[22] Juan, C. M. V., Maia, J. S. and Yoshida, M. (1980). "Neolignans from an *Aniba species*". *Phytochemistry* 19:474–476.

[23] Gotlieb, O. R., Mourão, J. C., Yoshida, M., Mascarenhas, Y. P., Rodrigues, M., Rosenstein, R. D. and Tomita, K. (1977). "Absolute configuration of the benzofuranoid neolignans". *Phytochemistry* 16:1003-1006.

[24] Dias, D. A., Yoshida, M. and Gottlieb, O. R. (1986). "Further neolignans from *Ocotea porosa*". *Phytochemistry* 25:2613–2616.

[25] Trevisan, L. M. V. and Yoshida, M. (1984). "Hexahydrobenzofuranoid neolignans from an *Aniba species*". *Phytochemistry* 23:661–665.

[26] Barbosa-Filho, J. M., Yoshida, M., Gottlieb, O. R., Barbosa, R. d. C. S. B. C., Giesbrecht, A. M. and Young, M. C. M. (1978). "Benzoyl esters and amides, styrylpyrones and neolignans from the fruits of *Aniba riparia*". 26:2615–2617.

[27] Aiba, C. J., Alvarenga, M. A. D., Oscar, C. C., Giesbrecht, A. M. and Pagliosa, F. M. (1977). "Benzofuranoid neolignans from *Aniba simulans*". *Phytochemistry* 16:741–743.

[28] Gottlieb, O. R., Da Silva, M. L. and Ferreira, Z. S. (1975). "Neolignans from *Aniba terminalis*". *Phytochemistry* 14:1825-1827.

[29] Wu, Y., Liu, C., Li, H.-F., Sun, J.-B., Li, Y.-Y., Gu, W., Wang, D.-Y., Liu, J.-G. and Hu, Y.-L. (2014). "A novel neolignan glycoside from *Aquilaria sinensis*". *Biochemical Systematics and Ecology* 55:41-45.

[30] Ichihara, A., Kanai, S., Nakamura, Y. and Sakamura, S. (1978). "Structures of lappaol F and H, dilignans from *Arctium lappa* L." *Tetrahedron Letters* 5:3035-3038.

[31] Zhou, G.-S., Ma, X.-L., Zeng, K.-W., Tu, P.-F. and Jiang, Y. (2013). "Inhibitory activity of chemical constituents from *Arenaria serpyllifolia* on nitric oxide production". *Planta Medica* 79:687-692.

[32] França, V. C., Agra, M. F., Barbosa-Filho, J. M., da Cunha, E. V. L. and da Silva, M. S. (2003). "Physcion and dihydrocarinatin from *Aristolochia birostris*". *Biochemical Systematics and Ecology* 31:1341-1343.

[33] Tang, G.-H., Chen, Z.-W., Lin, T.-T., Tan, M., Gao, X.-Y., Bao, J.-M., Cheng, Z.-B., Sun, Z.-H., Huang, G. and Yin, S. (2015). "Neolignans from *Aristolochia fordiana* prevent oxidative stress-induced neuronal death through maintaining the Nrf2/HO-1 pathway in HT22 cells". *Journal of Natural Products* 78:1894-1903.

[34] Nascimento, I. R. and Lopes, L. M. X. (1999). "2,3-Dihydrobenzofuran neolignans from *Aristolochia pubescens*". *Phytochemistry* 52:345-350.

[35] Wu, J., Zhang, X.-H., Zhang, S.-W. and Xuan, L.-J. (2008). "Three novel compounds from the flowers of *Bombax malabaricum*". *Helvetica Chimica Acta* 91:136–143.

[36] Odonbayar, B., Murata, T., Suganuma, K., Ishikawa, Y., Buyankhishig, B., Batkhuu, J. and Sasaki, K. J. (2019). "Acylated lignans isolated from *Brachanthemum gobicum* and their trypanocidal activity". *Journal of Natural Products* 86:774−784.

[37] Li, Y.-P., Dong, L.-B., Chen, D.-Z., Li, H.-M., Zhong, J.-D., Li, F., Liu, X., Wang, B. and Li, R.-T. (2013). "Two new dihydrobenzofuran-type neolignans from *Breynia fruticosa*". *Phytochemistry Letters* 6:281-285.

[38] Shang, S. and Long, S. (2008). "Brugnanin, a new *syn*-2,3-dihydrobenzofuran neolignan dioate from the mangrove *Bruguiera gymnorrhiza*". *Chemistry of Natural Compounds* 44:186-189.
[39] Houghton, P. J. (1985). "Lignans and neolignans from *Buddleja davidii*". *Phytochemitry* 24:819-826.
[40] Rattanaburi, S., Mahabusarakam, W., Phongpaichit, S. and Carroll, A. R. (2012). "Neolignans from *Callistemon lanceolatus*". *Phytochemistry Letters* 5:18-21.
[41] Phatchana, R., Thongsri, Y., Somwaeng, R., Piboonpol, K. and Yenjai, C. (2015). "Canangalias A and B from the stem bark of *Cananga latifolia*". *Phytochemistry Letters* 13:147-151.
[42] Anh, N. H., Ripperger, H., Porzel, A., Van Sung, T. and Adam, G. (1997). "Neolignans from *Caryodaphnopsis baviensis*". *Phytochemistry* 46:569 -571.
[43] Agrawal, P. K., Agarwal, S. K. and Rastogi, R. P. (1980). "A new neolignan and other phenolic constituents from *Cedrus deodara*". *Phytochemistry* 19:1260-1261.
[44] Valcic, S., Montenegro, G. and Timmermann, B. N. (1998). "Lignans from *Chilean Propolis*". *Journal of Natural Products* 61:771-775.
[45] Zhang, S., Huang, Y., Li, Y., Wang, Y. and He, X. (2019). "Anti-neuroinflammatory and antioxidant phenylpropanoids from Chinese olive". *Food Chemistry* 286:421-427.
[46] Sadhu, S. K., Okuyama, E., Fujimoto, H., Ishibashi, M. and Yesilada, E. (2006). "Prostaglandin inhibitory and antioxidant components of *Cistus laurifolius*, a Turkish medicinal plant". *Journal of Ethnopharmacology* 108:371-378.
[47] Seidel, V., Bailleul, F. and Waterman, P. G. (2000). "Novel oligorhamnosides from the stem bark of *Cleistopholis glauca*". *Journal of Natural Products* 63:6-11.
[48] Hayashi, T. and Thomson, R. H. (1975). "New lignans in *Conocarpus erectus*". *Phytochemistry* 14:1085-1087.
[49] Cho, J. Y., Baik, K. U., Yoo, E. S., Yoshikawa, K. and Park, M. H. (2000). "*In vitro* antiinflammatory effects of neolignan woorenosides from the rhizomes of *Coptis japonica*". *Journal of Natural Products* 63:1205-1209.

[50] Yoshikawa, K., Kinoshita, H., Kan, Y. and Arihara, S. (1995). "Neolignans and phenylpropanoids from the rhizomes of *Coptis japonica* var. *dissecta*". *Chemical and Pharmaceutical Bulletin* 43:578-581.

[51] Guoa, R., Lva, T., Hana, R. Y., Linb, B., Yaoa, G. D., Wangc, X. B., Huanga, X. X. and Songa, S. J. (2019). "Chiral resolution and neuroprotective activities of enantiomeric dihydrobenzofuran neolignans from the fruit of *Crataegus pinnatifida*". *Bioorganic Chemistry* 85:469-474.

[52] Huang, X.-X., Zhou, C.-C., Li, L.-Z., Peng, Y., Lou, L.-L., Liu, S., Li, D.-M., Ikejima, T. and Song, S.-J. (2013). "Cytotoxic and antioxidant dihydrobenzofuran neolignans from the seeds of *Crataegus pinnatifida*". *Fitoterapia* 91:217-223.

[53] Pieters, L. A. C., Berghe, D. A. V. and Vlietinck, A. J. (1990). "A dihydrobenzofuran lignan from *Croton erythrochilus*". *Phytochemistry* 29:348-349.

[54] Pieters, L., de Bruyne, T., Claeys, M., Vlietinck, A., Calomme, M. and Berghe, D. V. (1993). "Isolation of a dihydrobenzofuran lignan from South American Dragon's Blood (*Croton* spp.) as an inhibitor of cell proliferation". *Journal of Natural Products* 56:899-906.

[55] Pieters, L., De Bruyne, T., Van Poel, B., Vingerhoets, R., Totté, J., Berghe, D. V. and Vlietinck, A. (1995). "*In vivo* wound healing activity of Dragon's Blood (*Croton* spp.), a traditional South American drug, and its constituents". *Phytomedicine* 2:17-22.

[56] Guo, R., Luo, M., Long, C. L., Li, M. L., Ouyang, Z. Q., Zhou, Y. P., Wang, Y. H., Li, X. Y. and Shi, Y. N. (2008). "Two new lignans from *Dipteronia dyeriana*". *Chinese Chemical Letters* 19:1215-1217.

[57] Cho, N., Yang, H., Kim, J. W., Kim, Y. C. and Sung, S. H. (2014). "Chemical constituents isolated from *Disporum viridescens* leaves and their inhibitory effect on nitric oxide production in BV2 microglial cells". *Bioorganic & Medicinal Chemistry Letters* 24:5675-5678.

[58] Rudiyansyah and Garson, M. J. (2006). "Secondary metabolites from the wood bark of *Durio zibethinus* and *Durio kutejensis*". *Journal of Natural Products* 69:1218-1221.

[59] Miyase, T., Ueno, A., Takizawa, N., Kobayashi, H. and Ogutchi, H. (1989). "Ionone and lignan glycosides from *Epimedium diphyllum*". *Phytochemistry* 28:3483-3485.

[60] Deyama, T., Ikawa, T., Kitagawa, S. and Nishibe, S. (1987). "The constituents of *Eucommia ulmoides* Oliv. V. isolation of dihydroxydehydrodiconiferyl alcohol isomers and phenolic compounds". *Chemical and Pharmaceutical Bulletin* 35:1785-1789.

[61] Song, C.-W., Wang, S.-M., Zhou, L.-L., Hou, F.-F., Wang, K.-J., Han, Q.-B., Li, N. and Cheng, Y.-X. (2011). "Isolation and identification of compounds responsible for antioxidant capacity of *Euryale ferox* seeds". *Journal of Agricultural and Food Chemistry* 59:1199-1204.

[62] Kim, N.-H., Yang, M. H., Heo, J.-D., Sung, S. H. and Jeong, E. J. (2016). "Dihydrobenzofuran neolignans isolated from *Euonymus alatus* leaves and twigs attenuated inflammatory responses in the activated RAW264.7 macrophage cells". *Natural Product Sciences* 22:53-59.

[63] Salama, O., Chaudhuri, R. K. and Sticher, O. (1981). "A lignan glucoside from *Euphrasia rostkoviana*". *Phytochemistry* 20:2603-2604.

[64] Chin, Y.-W., Chai, H.-B., Keller, W. J. and Kinghorn, D. A. (2008). "Lignans and other constituents of the fruits of *Euterpe oleracea* (Açaí) with antioxidant and cytoprotective activities". *Journal of Agricultural and Food Chemistry* 56:7759-7764.

[65] Chin, Y.-W., Jones, W. P., Rachman, I., Riswan, S., Kardono, L. B. S., Chai, H.-B., Farnsworth, N. R., Cordell, G. A., Swanson, S. M., Cassady, J. M. and Kinghorn, D. A. (2006). "Cytotoxic lignans from the stems of *Helicteres hirsuta* collected in Indonesia". *Phytotherapy Research* 20:62-65.

[66] Gao, Y., Liu, Y., Wang, Z.-g. and Zhang, H.-l. (2014). "Chemical constituents of *Heracleum dissectum* and their cytotoxic activity". *Phytochemistry Letters* 10:276-280.

[67] Yang, F., Zhang, H.-J., Zhang, Y.-Y., Chen, W.-S., Yuan, H.-L. and Lin, H.-W. (2010). "A Hepatitis B virus inhibitory neolignan from *Herpetospermum caudigerum*". *Chemical and Pharmaceutical Bulletin* 58:402-404.

[68] Moujir, L., Seca, A. M. L., Silva, A. M. S., Lópes, M. R., Padilla, N., Cavaleiro, J. A. S. and Neto, C. P. (2007). "Cytotoxic activity of lignans from *Hibiscus cannabinus*". *Fitoterapia* 78:385-387.

[69] Seca, A. M., Silva, A. M., Silvestre, A. J., Cavaleiro, J. A., Domingues, F. M. and Pascoal-Neto, C. (2001). "Lignanamides and other phenolic constituents from the bark of kenaf (*Hibiscus cannabinus*)". *Phytochemistry.* 58:1219-1223.

[70] Seca, A. M., Silva, A. M., Silvestre, A. J., Cavaleiro, J. A., Domingues, F. M. and Pascoal-Neto, C. (2001). "Phenolic constituents from the core of Kenaf (*Hibiscus cannabinus*)". *Phytochemistry* 56:759-767.

[71] Ma, C.-Y., Liu, W. K. and Che, C.-T. (2002). "Lignanamides and nonalkaloidal components of *Hyoscyamus niger* seeds". *Journal of Natural Products* 65:206-209.

[72] Kouno, I., Yanagida, Y., Shimono, S., Shintomi, M., Ito, Y. and Chun-Shu, Y. (1993). "Neolignans and a phenylpropanoid glucoside from *Illicium difengpi*". *Phytochemistry* 32:1573-1577.

[73] Xiang, W.-J., Ma, L. and Hu, L.-H. (2010). "Neolignans and flavonoids from the root bark of *Illicium henryi*". *Fitoterapia* 81:1228-1231.

[74] Tang, W.-Z., Liu, Y.-B., Yu, S.-S., Qu, J. and Su, D.-M. (2007). "New sesquiterpene lactone and neolignan glycosides with antioxidant and anti-inflammatory activities from the fruits of *Illicium oligandrum*". *Planta Medica* 53:484-490.

[75] Yea, F.-M., Yang-guo Xie, Y.-g., Ren, J., Ye, J., Guo, Y.-G., Yan, S.-K., Jin, H.-Z. and Zhang, W.-D. (2016). "Six new dihydrobenzofuran lignans from the branches and leaves of *Illicium wardii* and their cytotoxic activities". *Phytochemistry Letters* 17:263-269.

[76] De Marino, S., Cattaneo, F., Festa, C., Zollo, F., Iaccio, A., Ammendola, R., Incollingo, F. and Iorizzi, M. (2011). "Imbricatolic acid from *Juniperus communis* L. prevents cell cycle progression in CaLu-6 cells". *Planta Medica* 77:1822-1828.

[77] Nakanishi, T., Iida, N., Inatomi, Y., Murata, H., Inada, A., Murata, J., Lang, F. A., Iinuma, M. and Tanaka, T. (2004). "Neolignan and flavonoid glycosides in *Juniperus communis* var. *depressa*". *Phytochemistry* 65:207-213.

[78] Achenbach, H., Grob, J., Dominguez, X. A., Cano, G., Star, J. V., Brussolo, L. D. C., Muñoz, G., Salgado, F. and López, L. (1987). "Lignans neolignans and norneolignans from *Krameria cystisoides*". *Phytochemistry* 26:1159-1166.
[79] Dominguez, X. A., Sanchez, H. V., Espinoza, G. C. B., Verde, J. S., Achenbach, H. and Utz, W. (1990). "Lignans and nor-neolignans from *Krameria interior*". *Phytochemistry* 29:2651-2653.
[80] Achenbach, H., Utz, W., Usubillaga, A. and Rodriguez, H. A. (1991). "Lignans from *Krameria ixina*". *Phytochemistry* 30:3753-3757.
[81] Achenbach, H., Grob, J., Bauereib, P., Dominguez, X. A., Vega, H. S., Star, J. V. and Romboldt, C. (1989). "Neolignans and nor-neolignans from *Krameria lanceolata*". *Phytochemistry* 28:1959-1962.
[82] Baumgartner, L., Sosa, S., Atanasov, A. G., Bodensieck, A., Fakhrudin, N., Bauer, J., Favero, G. D., Ponti, C., Heiss, E. H., Schwaiger, S., Ladurner, A., Widowitz, U., Loggia, R. D., Rollinger, J. M., Werz, O., Bauer, R., Dirsch, V. M., Tubaro, A. and Stuppner, H. (2011). "Lignan derivatives from *Krameria lappacea* roots inhibit acute inflammation in vivo and pro-inflammatory mediators in vitro". *Journal of Natural Products* 74:1779-1786.
[83] Achenbach, H., Utz, W., Lozano, B., Guajardo, E. M. T. and Moreno, S. (1996). "Lignans and neolignans from *Krameria parvifolia*". *Phytochemistry* 43:1093-1095.
[84] Miki, K., Takehara, T., Sasaya, T. and Sakakibara, A. (1980). "Lignans of *Larix leptolepis*". *Phytochemistry* 19:449-453.
[85] Yang, L.-j., Chen, W., Luo, Y.-P., Li, G.-P., Yang, X.-D. and Li, L. (2009). "Lignans and ketonic compounds from *Litsea chinpingensis* (Lauraceae)". *Biochemical Systematics and Ecology* 37:696–698.
[86] Tun, N. T., Hu, D.-B., Xia, M.-Y., Zhang, D.-D., Yang, J., Oo, T. N., Wang, Y.-H. and Yang, X.-F. (2019). "Chemical constituents from ethanoic extracts of the aerial parts of *Leea aequata* L., a traditional folk medicine of Myanmar". *Natural Products and Bioprospecting* 9:243-249.
[87] Li, X., Luo, J.-G., Wang, X.-B., Luo, J., Wang, J.-S. and Kong, L.-Y. (2012). "Phenolics from *Leontopodium leontopodioides* inhibiting nitric oxide production". *Fitoterapia* 83:883-887.

[88] Aiba, C. J., Gottlieb, O. R., Maia, J. G. S., Pagliosa, F. M. and Yoshida, M. (1978). "Benzofuranoid neolignans from *Licaria armeniaca*". *Phytochemistry* 17:2038-2039.

[89] Barbosa-Filho, J. M., Yoshida, M. and Gottlieb, O. R. (1987). "Neolignans from the fruits of *Licaria armeniaca*". *Phytochemistry* 26:319-321.

[90] Giesbrecht, A. M., Franca, N. C., Gottlieb, O. R. and Rocha, A. I. D. (1974). "The neolignans of *Licaria canella*". *Phytochemistry* 13:2285–2293

[91] Ferreira, Z. S., Roque, N. C., Gottlieb, O. R. and Gottlieb, H. E. (1982). "An unusual porosin type neolignan from *Licaria chrysophylla*". *Phytochemistry* 21:2756–2758.

[92] Lopes, M. N., Silva, M. S. D., Barbosa, J. M. A., Ferreira, Z. S., Yoshida, M. and Gottlieb, O. R. (1986). "Unusual benzofuranoid neolignans from *Licaria chrysophylla*". *Phytochemistry* 25:2609–2612.

[93] Suh, W. S., Kim, K. H., Kim, H. K., Choi, S. U. and Lee, K. R. (2015). "Three new lignan derivatives from *Lindera glauca* (Siebold et Zucc.) Blume". *Helvetica Chimica Acta* 98:1087-1094.

[94] Ahmad, W., Zahira, A., Nadeem, M., Garros, L., Drouet, S., Renouard, S., Doussot, J., Giglioli-Guivarc'h, N., Hanob, C. and Abbasi, B. H. (2019). "Enhanced production of lignans and neolignans in chitosan-treated flax (*Linum usitatissimum* L.) cell cultures". *Process Biochemistry* 79:155-165.

[95] Zahira, A., Ahmada, W., Nadeema, M., Giglioli-Guivarc'h, N., Hano, C. and Abbasia, B. H. (2018). "In vitro cultures of *Linum usitatissimum* L.: Synergistic effects of mineral nutrients and photoperiod regimes on growth and biosynthesis of lignans and neolignans". *Journal of Photochemistry and Photobiology B: Biology* 187:141-150.

[96] Hung, T. M., Lee, J. S., Chuong, N. N., Kim, J. A., Oh, S. H., Woo, M. H., Choi, J. S. and Min, B. S. (2015). "Kinetics and molecular docking studies of cholinesterase inhibitors derived from water layer of *Lycopodiella cernua* (L.) Pic. Serm. (II)". *Chemico-Biological Interactions* 240:74-82.

[97] Tsai, I.-L., Chen, J.-H., Duh, C.-Y. and Chen, I.-S. (2000). "Cytotoxic neolignans from the stem wood of *Machilus obovatifolia*". *Planta Medica* 66:403-407.

[98] Phan, M. G., Phan, T. S., Matsunami, K. and Otsuka, H. (2006). "New neolignans and lignans from vietnamese medicinal plant *Machilus odoratissima* NEES". *Chemical and Pharmaceutical Bulletin* 54:380-383.

[99] Shimomura, H., Sashida, Y. and Oohara, M. (1987). "Lignans from *Machilus thunbergii*". *Phytochemistry* 26:1513–1515.

[100] Yu, Y. U., Kang, S. Y., Park, H. Y., Sung, S. H., Lee, E. J., Kim, S. Y. and Kim, Y. C. (2000). "Antioxidant lignans from *Machilus thunbergii* protect CCl4–injured primary cultures of rat hepatocytes". *Journal of Pharmacy and Pharmacology* 52:1163–1169.

[101] Ma, C. J., Sung, S. H. and Kim, Y. C. 2004. "Neuroprotective lignans from the bark of *Machilus thunbergii*". *Planta Medica* 70:79–80.

[102] Li, G., Lee, C.-S., Woo, M.-H., Lee, S.-H., Chang, H. W. and Son, J.-K. (2004). "Lignans from the bark of *Machilus thunbergii* and their DNA topoisomerases I and II inhibition and cytotoxicity". *Biological and Pharmaceutical Bulletin* 24:1147-1150.

[103] Iida, T. and Ito, K. (1983). "Four phenolic neolignans from *Magnolia liliflora*". *Phytochemistry* 22:763-766.

[104] Jiang, Z.-H., Liu, Y.-P., Huang, Z.-H., Wang, T.-T., Feng, X.-Y., Yue, H., Guo, W. and Fu, Y.-H. (2017). "Cytotoxic dihydrobenzofuran neolignans from *Mappianthus iodoies*". *Bioorganic Chemistry* 17:260-264.

[105] Wang, L., Li, F., Yang, C.-Y., Khan, A.-A., Liu, X. and Wang, M.-K. (2014). "Neolignans, lignans and glycoside from the fruits of *Melia toosendan*". *Fitoterapia* 99:92-98.

[106] Sawasdee, K., Chaowasku, T., Lipipun, V., Dufat, T.-H., Michel, S. and Likhitwitayawuid, K. (2013). "New neolignans and a lignan from *Miliusa fragrans*, and their anti-herpetic and cytotoxic activities". *Tetrahedron Letters* 54:4259-4263.

[107] Sawasdee, K., Chaowasku, T., Lipipun, V., Dufat, T.-H., Michel, S. and Likhitwitayawuid, K. (2013). "Neolignans from leaves of *Miliusa mollis*". *Fitoterapia* 85:49-56.

[108] Jaidee, W., Maneerat, W., Andersen, R. J., Patrick, B. O., Pyne, S. J. and Laphookhieo, S. (2018). "Antioxidant neolignans from the twigs and leaves of *Mitrephora wangii* HU". *Fitoterapia* 130:219-224.

[109] Francis, S. K., James, B., Varughese, S. and Nair, M. S. (2019). "Phytochemical investigation on *Myristica fragrans* stem bark". *Natural Product Research* 33:1204-1208.

[110] Chiu, S., Wang, T., Belski, M. and Abourashed, E. A. (2016). "HPLC-guided isolation, purification and characterization of phenylpropanoid and phenolic constituents of Nutmeg Kernel (*Myristica fragrans*)". *Natural Product Communications* 11:483–488.

[111] Barrera, E. D. C., Cuca-Suárez, L. E. and Sefkow, M. (2009). "COX, LOX and platelet aggregation inhibitory properties of Lauraceae neolignans". *Bioorganic & Medicinal Chemistry Letters* 19:6922-6925.

[112] Barrera, E. D. C. and Suárez, L. E. C. (2009). "Three new 7.3',8.5'-connected bicyclo[3.2.1]octanoids and other neolignans from leaves of *Nectandra amazonum* NEES. (Lauraceae)". *Chemical and Pharmaceutical Bulletin* 57:639-642.

[113] Barbosa-Filho, J. M., Yoshida, M. and Gottlieb, O. R. (1989). "Lignoids from *Nectandra amazonum* and *N. glabrescens*". *Phytochemistry* 28:1991-1991.

[114] Aiba, C. J., Gottlieb, O. R., Pagliosa, F. M., Yoshida, M. and Magalhães, M. T. (1977). "Neolignans from *Nectandra miranda*". *Phytochemistry* 16:745-748.

[115] Braz-Filho, R., Figliuolo, R. and Gottlieb, O. R. (1980). "Neolignans from a *Nectandra* species". *Phytochemistry* 19:659–662.

[116] Zhang, T., Shen, Q., Liu, C., Zhang, F., He, P., Wnag, K., Zhu, R., Yang, G. and Liu, Z. (2015). "A new dihydrobenzofuran neolignan from stem of flue-cured tobacco and its anti-tobacco mosaic virus activity". *Asian Journal of Chemistry* 27:2964-2966.

[117] Paula, V. F., Barbosa, L. C. A., Howarth, O. W., Demuner, A. J., Cass, Q. B. and Vieira, I. J. C. (1995). "Lignans from *Ochroma lagopus* Swartz". *Tetrahedron* 51:12453-12462.

[118] Romoff, P., Yoshida, M. and Gottlieb, O. R. 1984. "Neolignans from *Ocotea aciphylla*". *Phytochemistry* 23:2101–2104.

[119] Felicio, J. D. A., Motidome, M., Yoshida, M. and Gottlieb, O. R. (1986). "Further neolignans from *Ocotea aciphylla*". *Phytochemistry* 25:1707–1710.

[120] Haraguchi, M., Motidome, M., Yoshida, M. and Gottlieb, O. R. (1983). "Neolignans from *Ocotea catharinensis*". *Phytochemistry* 22:561–563.

[121] Ishige, M., Motidome, M., Yoshida, M. and Gottlieb, O. R. (1991). "Neolignans from *Ocotea catharinensis*". *Phytochemistry* 30:4121–4128.

[122] Barrera, E. D. C., Cuca-Suárez, L. E. and Sefkow, M. (2009). "PAF-antagonistic bicyclo[3.2.1]octanoid neolignans from leaves of *Ocotea macrophylla* Kunth. (Lauraceae)". *Phytochemistry* 70:1309-1314.

[123] Aiba, C. J., Filho, R. B. and Gottlieb, O. R. (1973). "Porosin: a neolignan from *Ocotea porosa*". *Phytochemistry* 12:413-416.

[124] Carvalho, M. G. D., Yoshida, M., Gottlieb, O. R. and Gottlieb, H. E. (1988). "Bicyclooctanoid, carinatone and megaphone type neolignans from *Ocotea porosa*". *Phytochemistry* 27:2319–2323.

[125] David, J. M., Yoshida, M. and Gottlieb, O. R. (1994). "Neolignans from bark and leaves of *Ocotea porosa*". *Phytochemistry* 36:491–499.

[126] Marques, M. O. M., Gomes, M. C. C. P., Yoshida, M. and Gottlieb, O. R. (1992). "Bicyclo [3.2.1] octanoid neolignans from *Ocotea porosa*". *Phytochemistry* 31:275–277.

[127] Khan, M. R., Gray, A. I. and Waterman, P. G. (1987). "Neolignans from stem bark of *Ocotea veraguensis*". *Phytochemistry* 26:1155–1158.

[128] Dodson, C. D., Stermitz, F. R., Oscar, C. C. and Janzen, D. H. (1987). "Neolignans from fruits of *Ocotea veraguensis*". *Phytochemistry* 26:2037–2040.

[129] Ruiu, S., Anzani, N., Orrù, A., Floris, C., Caboni, P., Maccioni, E., Distinto, S., Alcaro, S. and Cottiglia, F. (2013). "*N*-alkyl dien- and trienamides from the roots of *Otanthus maritimus* with binding affinity for opioid and cannabinoid receptors". *Bioorganic & Medicinal Chemistry* 21:7074-7082.

[130] Liu, X., Yang, M.-H., Wang, X.-B., Xie, S.-S., Li, Z.-R., Kim, D.-H., Park, J.-S. and Kong, L.-Y. (2015). "Lignans from the root of

Paeonia lactiflora and their anti-β-amyloid aggregation activities". *Fitoterapia* 103:136-142.

[131] Changzeng, W. and Zhongjian, J. (1997). "Lignan, phenylpropanoid and iridoid glycosides from *Pedicularis torta*". *Phytochemistry* 45:159-166.

[132] Tsai, I.-L., Hsieh, C.-F., Duh, C.-Y. and Chen, I.-S. (1996). "Cytotoxic neolignans from *Persea obovatifolia*". *Phytochemistry* 43:1261–1263.

[133] Tsai, I.-L., Hsieh, C.-F. and Duh, C.-Y. (1998). "Additional cytotoxic neolignans from *Persea obovatifolia*". *Phytochemistry* 48:1371-1375.

[134] Li, S.-H., Niu, X.-M., Zahn, S., Gershenzon, J., Weston, J. and Schneider, B. (2008). "Diastereomeric stilbene glucoside dimers from the bark of *Norway spruce* (*Picea abies*)". *Phytochemistry* 69:772-782.

[135] Jiao, W.-H., Gao, H., Zhao, F., He, F., Zhou, G.-X. and Yao, X.-S. (2011). "A new neolignan and a new sesterterpenoid from the stems of *Picrasma quassioides* Bennet". *Chemistry & Biodiversity* 8:1163-1169.

[136] Lou, L.-L., Yao, G.-D., Wang, J., Zhao, W.-Y., Wang, X.-B., Huang, X.-X. and Song, S.-J. (2018). "Enantiomeric neolignans from *Picrasma quassioides* exhibit distinctive cytotoxicity on hepatic carcinoma cells through ROS generation and apoptosis induction". *Bioorganic & Medicinal Chemistry Letters* 28:1263-1268.

[137] Lundgren, L. N., Shen, Z. and Theander, O. (1985). "The constituents of conifer needles, dilignol glycosides from *Pinus massoniana* Lamb". *Acta Chemica Scandinavica* 39:241-248.

[138] Hong, S. S., Jeong, W., Kim, J. K., Kwon, J. G., Lee, J. Y., Ahn, E.-K., Oh, J., Seo, D.-W. and Oh, J.-S. (2014). "Neolignan inhibitors of antigen-induced degranulation in RBL-2H3 cells from the needles of *Pinus thunbergii*". *Fitoterapia* 99:347-351.

[139] Dellus, V., Mila, I., Scalbert, A., Menard, C., Michon, V. and Herve du Penhoat, C. L. M. (1997). "Douglas-fir polyphenols and heartwood formation". *Phytochemistry* 45:1573–1578.

[140] Chauret, D. C., Bernard, C. B., Arnason, J. T., Durst, T., Krishnamurty, H. G., Sanchez-Vindas, P., Moreno, N., San Roman,

L. and Poveda, L. (1996). "Insecticidal neolignans from *Piper decurrens*". *Journal of Natural Products* 59:152-155.

[141] Benevides, P. J. C., Sartorelli, P. and Kato, M. J. (1999). "Phenylpropanoids and neolignans from *Piper regnellii*". *Phytochemistry* 52:339-343.

[142] Bernard, C. B., Krishnamurty, H. G., Chauret, D., Durst, T., Philogéne, B. J. R., Vindas, P. S., Hasbun, C., Poveda, L., Roman, L. S. and Arnason, J. T. (1995). "Insecticidal defenses of Piperaceae from the neotropics". *Journal of Chemical Ecology* 21:801–814.

[143] Scodro, R. B. L., Pires, C. T. A., Carrara, V. S., Lemos, C. O. T., Cardozo-Filho, L., Maniglia, V. A. S., Corrêa, A. G., Siqueira, V. L. D., Lonardoni, M. V. C., Cardoso, R. F. and Cortez, D. A. G. (2013). "Anti-tuberculosis neolignans from *Piper regnellii*". *Phytomedicine.* 20:600-604.

[144] Pessini, G. L., Dias Filho, B. P., Nakamura, C. V. and Cortez, D. A. G. (2005). "Antifungal activity of the extracts and neolignans from *Piper regnellii* (Miq.) C. DC. var. *pallescens* (C. DC.) Yunck". *Journal of the Brazilian Chemical Society* 16:1130-1133.

[145] Zhao, H., Nie, T., Guo, H., Li, J. and Bai, H. (2012). "Two new neolignan glycosides from *Pittosporum glabratum* Lindl". *Phytochemistry Letters* 5:240-243.

[146] Wu, Z., Lai, Y., Zhou, L., Wu, Y., Zhu, H., Hu, Z., Yang, J., Zhang, J., Wang, J., Luo, Z., Xue, Y. and Zhang, Y. (2016). "Enantiomeric lignans and neolignans from *Phyllanthus glaucus*: enantioseparation and their absolute configurations". *Scientific Reports* 6:1-13.

[147] Fan, S.-Y., Pei, Y.-H., Zeng, H.-W., Zhang, S.-D., Li, Y.-L., Li, L., Ye, J., Pan, Y.-X., Li, H.-L. and Zhang, W.-D. (2011). "Compounds from *Platycladus orientalis* and their inhibitory effects on nitric oxide and TNF-α production". *Planta Medica* 77:1623-1630.

[148] Barrera, E. D. C. and Suárez, L. E. C. (2008). "Chemical constituents from *Pleurothyrium cinereum* (van der Werff) (Lauraceae) from Colombia". *Biochemical Systematics and Ecology* 36:674–677.

[149] Uchiyama, T., Miyase, T., Ueno, A. and Usmanghani, K. (1991). "Terpene and lignan glycosides from *Pluchea indica*". *Phytochemistry* 30:655-657.

[150] Kikuzaki, H., Kayano, S.-i., Fukutsuka, N., Aoki, A., Kasamatsu, K., Yamasaki, Y., Mitani, T. and Nakatani, N. (2004). "Abscisic acid

related compounds and lignans in prunes (*Prunus domestica* L.) and their oxygen radical absorbance capacity (ORAC)". *Journal of Agricultural and Food Chemistry* 52:344-349.

[151] Liu, Q.-B., Huang, X.-X., Bai, M., Chang, X.-B., Yan, X.-J., Zhu, T., Zhao, W., Peng, Y. and Song, S.-J. (2014). "Antioxidant and anti-inflammatory active dihydrobenzofuran neolignans from the seeds of *Prunus tomentosa*". *Journal of Agricultural and Food Chemistry* 62:7796-7803.

[152] Feng, W.-S., Zang, X.-Y., Zheng, X.-K., Wang, Y.-Z., Chen, H. and Li, Z. (2010). "Two new dihydrobenzofuran lignans from *Rabdosia lophanthoides* (Buch.-Ham.ex D.Don) Hara". *Journal of Asian Natural Products Research* 12:557-561.

[153] Yang, Y.-n., Liu, Z.-z., Feng, Z.-m., Jiang, J.-s. and Zhang, P.-c. (2012). "Lignans from the root of *Rhodiola crenulata*". *Journal of Agricultural and Food Chemistry* 60:964-972.

[154] Zhou, L., Wang, J., Guo, R., Lin, B., Wang, X.-B., Huang, X.-X. and Songa, S.-J. (2018). "Discovery of dihydrobenzofuran neolignans from *Rubus ideaus* L. with enantioselective anti-Aβ1–42 aggregation activity". *Bioorganic Chemistry* 80:64-69.

[155] Du, Q., Jerz, G., Shen, L., Xiu, L. and Winterhalter, P. (2007). "Isolation and structure determination of a lignan from the bark of *Salix alba*". *Natural Product Research* 21:451–454.

[156] Sasaki, T., Li, W., Morimura, H., Li, S., Li, Q., Asada, Y. and Koike, K. (2011). "Chemical constituents from *Sambucus adnata* and their protein-tyrosine phosphatase 1B inhibitory activities". *Chemical and Pharmaceutical Bulletin* 59:1396-1399.

[157] Machida, K., Takano, M., Kakuda, R., Yaoita, Y. and Kikuchi, M. (2002). "A new lignan glycoside from the leaves of *Sambucus sieboldiana* (MIQ.) BLUME ex. GRAEBN". *Chemical and Pharmaceutical Bulletin* 50:669—671.

[158] Liang, C.-Q., Hu, J., Luo, R.-H., Shi, Y.-M., Shang, S.-Z., Gao, Z.-H., Wang, R.-R., Zheng, Y.-T., Xiong, W.-Y., Zhang, H.-B., Xiao, W.-L. and Sun, H.-D. (2013). "Six new lignans from the leaves and stems of *Schisandra sphenanthera*". *Fitoterapia* 86:171-177.

[159] Chen, H., Hao, Z.-Y., Wang, X.-L., Zheng, X.-K., Feng, W.-S. and Wang, Y.-Z. (2014). "Sinensioside A, a new sesquilignan glycoside

from *Selaginella sinensis*". *Chinese Journal of Natural Medicines* 12:148-150.

[160] Chena, X., Xua, P.-S., Zou, Z.-X., Liu, Y., Zhou, W.-H., Ren, Q., Li, D., Lia, X.-M., Xu, K.-P. and Tan, G.-S. (2019). "Sinensiols B-G, six novel neolignans from *Selaginella sinensis*". *Fitoterapia* 134:256-263.

[161] Wang, Y.-H., Sun, Q.-Y., Yang, F.-M., Long, C.-L., Zhao, F.-H., Tang, G.-H., Niu, H.-M., Huang, Q.-Q., Xu, J.-J. and Ma, L.-J. (2010). "Neolignans and caffeoyl derivatives from *Selaginella moellendorffii*". *Helvetica Chimica Acta* 93:2467-2477.

[162] Zheng, X. K., Li, K. K., Wang, Y. Z. and Feng, W. S. (2008). "A new dihydrobenzofuran lignanoside from *Selaginella moellendorffii* Hieron". *Chinese Chemical Letters* 19:79-81.

[163] Zhuo, J.-X., Wang, Y.-H., Su, X.-L., Mei, R.-Q., Yang, J., Kong, Y. and Long, C.-L. (2016). "Neolignans from *Selaginella moellendorffii*". *Natural Products and Bioprospecting* 6:161-166.

[164] Sun, J., Huo, H.-X., Zhang, J., Huang, Z., Zheng, J., Zhang, Q., Zhao, Y.-F., Li, J. and Tu, P.-F. (2015). "Phenylpropanoid amides from the roots of *Solanum melongena* L. (Solanaceae)". *Biochemical Systematics and Ecology* 58:265-269.

[165] Chakravarty, A. K., Mukhopadhyay, S., Saha, S. and Pakrashi, S. C. (1996). "A neolignan and sterols in fruits of *Solanum sisymbrifolium*". *Phytochemistry* 41:935-939.

[166] Suo, M., Isao, H., Kato, H., Takano, F. and Ohta, T. (2012). "Anti-inflammatory constituents from *Tabebuia avellanedae*". *Fitoterapia* 83:1484-1488.

[167] Ferreira-Júnior, J. C., Conserva, L. M., Lemos, R. P. L., Omena-Neta, G. C., Cavalcante-Neto, A. and Barreto, E. (2014). "Isolation of a dihydrobenzofuran lignan, icariside E4, with an antinociceptive effect from *Tabebuia roseo-alba* (Ridley) Sandwith (Bignoniaceae) bark". *Archives of Pharmacal Research* 38:950-956.

[168] Lin, W.-H., Fang, J.-M. and Cheng, Y.-S. (1999). "Lignans from *Taiwania cryptomerioides*". *Phytochemistry* 50:653-658.

[169] Zhao, P., Tanaka, T., Hirabayashi, K., Zhang, Y.-J., Yang, C.-R. and Kouno, I. (2008). "Caffeoyl arbutin and related compounds from the buds of *Vaccinium dunalianum*". *Phytochemistry* 69:3087-3094.

[170] Fukuyama, Y., Nakahara, M., Minami, H. and Kodama, M. (1996). "Two new benzofuran-type lignans from the wood of *Viburnum awabuki*". *Chemical and Pharmaceutical Bulletin* 44:1418-1420.

[171] Cavalcante, S. H., Yoshida, M. and Gottlieb, O. R. (1985). "Neolignans from *Virola carinata* fruit". *Phytochemistry* 24:1051-1055.

[172] Kawanishi, K., Uhara, Y. and Hashimoto, Y. (1982). "Neolignans of *Virola carinata* bark". *Phytochemistry* 21:2725-2728.

[173] Gu, Q., Zhang, X.-M., Zhou, J., Qiu, S.-X. and Chen, J.-J. (2008). "One new dihydrobenzofuran lignan from *Vitex trifolia*". *Journal of Asian Natural Products Research* 10:499-502.

[174] Lemière, G., Gao, M., De Groot, A., Dommisse, R., Lepoivre, J., Pieters, L. and Buss, V. (1995). "3',4-di-*O*-methylcedrusin: synthesis, resolution and absolute configuration". *Perkin Transactions* 1775-1779.

[175] Natori, Y., Tsutsui, H., Sato, N., Nakamura, S., Nambu, H., Shiro, M. and Hashimoto, S. (2009). "Asymmetric synthesis of neolignans (−)-epi-conocarpan and (+)-conocarpan via Rh(II)-catalyzed C−H insertion process and revision of the absolute Configuration of (−)-epi-conocarpan". *The Journal of Organic Chemistry* 74:4418-4421.

[176] Chen, C.-y. and Weisel, M. (2013). "Concise asymmetric synthesis of (+)-conocarpan and obtusafuran". *Synlett* 24:189-192.

[177] Garcia-Muñoz, S., Alvarez-Corral, M., Jimenez-Gonzalez, L., Lopez-Sanchez, C., Rosales, A., Muñoz-Dorado, M. and Rodriguez-Garcia, I. (2006). "Synthesis of dihydrodehydrodiconiferyl alcohol and derivatives through intramolecular C-H insertion". *Tetrahedron* 62:12182-12190.

[178] Hemelaere, R., Carreaux, F. and Carboni, B. (2015). "A diastereo-selective route to *trans*-2-aryl-2,3-dihydrobenzofurans through sequential cross-metathesis/isomerization/allylboration reactions: synthesis of bioactive neolignans". *European Journal of Organic Chemistry* 2015:2470-2481.

[179] Daquino, C., Rescifina, A., Spatafora, C. and Tringali, C. (2009). "Biomimetic synthesis of natural and "unnatural" lignans by oxidative coupling of caffeic esters". *European Journal of Organic Chemistry* 36:6289-6300.

[180] Orlandi, M., Rindone, B., Molteni, G., Rummakko, P. and Brunow, G. (2001). "Asymmetric biomimetic oxidations of phenols: the mechanism of the diastereo- and enantioselective synthesis of dehydrodiconiferyl ferulate (DDF) and dehydrodiconiferyl alcohol (DDA)". *Tetrahedron* 57:371-378.

[181] Chioccara, F., Poli, S., Rindone, B., Pilati, T., Brunow, G., Pietikäinen, P., Setälä, H. and Tsuda, T. (1993). "Regio- and diastereo-selective synthesis of dimeric lignans using oxidative coupling". *Acta Chemica Scandinavica* 47:610-616.

[182] Drago, R. S., Corden, B. B. and Barnes, C. W. (1986). "Novel cobalt(II)-catalyzed oxidative cleavage of a carbon-carbon double bond". *Journal of the American Chemical Society* 108:2453-2454.

[183] Althagafy, H. S., Meza-Aviña, M. E., Oberlies, N. H. and Croatt, M. P. (2013). "Mechanistic study of the biomimetic synthesis of flavonolignan diasteroisomers in milk thistle". *Journal of Organic Chemistry* 78:7594-7600.

[184] Davin, L. B., Wang, H.-B., Crowell, A. L., Bedgar, D. L., Martin, D. M., Sarkanen, S. and Lewis, N. G. (1997). "Stereoselective bimolecular phenoxy radical coupling by an auxiliary (dirigent) protein without an active center". *Science* 275:362-366.

[185] Van Miert, S., Dyck, S. V., Schmidt, T. J., Brun, R., Vlietinck, A., Lemière, G. and Pieters, L. (2005). "Antileishmanial activity, cytotoxicity and QSAR analysis of synthetic dihydrobenzofuran lignans and related benzofurans". *Bioorganic & Medicinal Chemistry* 13:661-669.

[186] Yoshihara, T., Yamaguchi, K. and Sakamura, S. (1983). "The relative configuration of Grossamide and Hordatines". *Agricultural and Biological Chemistry* 47:217-220.

[187] Ralph, J., Conesa, M. T. G. and Williamson, G. (1998). "Simple preparation of 8-5-coupled diferulate". *Journal of Agricultural and Food Chemistry* 46:2531-2532.

[188] Liu, S.-Y., Wang, G.-Q., Liang, Z.-Y. and Wang, Q.-A. (2013). "Synthesis of dihydrobenzofuran neoligans licarin a and dihydrocarinatin as well as related triazolylglycosides". *Chemical Research in Chinese Universities* 29:1119-1124.

[189] Quideuau, S. and Ralph, J. (1994). "A biomimetic route to lignin model compounds via silver (I) oxide oxidation: 1. synthesis of dilignols and non-cyclic benzyl aryl ethers". *Holzforshung* 48:12-22.

[190] Sako, M., Hosokawa, H., Ito, T. and Iinuma, M. (2004). "Regioselective oxidative coupling of 4-hydroxystilbenes: synthesis of resveratrol and epsilon-viniferin (*E*)-dehydrodimers". *The Journal of Organic Chemistry* 69:2598-2600.

[191] Hu, K. and Jeong, J.-H. (2006). "A convenient synthesis of an anti-helicobacter pylori agent, dehydrodiconiferyl alcohol". *Archives of Pharmacal Research* 29:563-565.

[192] Kuo, Y.-H. and Wu, C.-H. (1996). "Synthesis of 5-(3-hydroxypropyl)-7-methoxy-2-3'-methoxy-4'-hydroxyphenyl)-3-benzo[*b*]furancarbaldehyde, a novel adenosine A_1 receptor ligand from the root of *Salvia miltiorrhiza*". *Journal of Natural Products* 59:625-628.

[193] Xia, Y.-M., Li, C., Zhang, H., Lin, J. and Chai, C. (2015). "Total synthesis of a lignanamide from *Aptenia cordifolia*". *Journal of Chemical Research* 39:535-538.

[194] Xia, Y.-M., Dai, X., Liu, H. and Chai, C. (2014). "First total synthesis of Boehmenan". *Journal of Chemical Sciences* 126:813-820.

[195] Fukui, M. J., Dias, H. J., Severiano, M. E., de Souza, M. G. M., de Oliveira, P. F., Ambrósio, S. R., Martins, C. H. G., Tavares, D. C. and Crotti, A. E. M. (2018). "Antimicrobial and cytotoxic activity of dihydrobenzofuran neolignans". *ChemistrySelect* 3:1836-1839.

[196] Dias, H. J., Patrocínio, A. B., Pagotti, M. C., Fukui, M. J., Rodrigues, V., Magalhães, L. G. and Crotti, A. E. M. (2018). "Schistosomicidal activity of dihydrobenzofuran neolignans". *Chemistry & Biodiversity* 15:e1800134.

[197] Baldin, E. L. L., Dias, H. J., de Souza, C. M., Soares, M. C. E., Grundman, C. O., Santos, T. L. B. and Crotti, A. E. M. (2019). "Insecticidal and inhibitory effects of dihydrobenzofuran neolignans on *Bemisia tabaci*". *Journal of Pest Science* 92:861-869.

[198] Pieters, L., Van Dyck, S., Gao, M., Bai, R., Hamel, E., Vlietinck, A. and Lemiere, G. (1999). "Synthesis and biological evaluation of dihydrobenzofuran lignans and related compounds as potential

antitumor agents that inhibit tubulin polymerization". *Journal of Medicine Chemistry* 42:5475-5481.

[199] Nelo, R. A. (2016). "Synthesis of analogs the licarina A for research trypanocidal activit". Master Thesis, Universidade Federal do ABC.

[200] Marques, S. D. G. (2014). "Lignan synthesis by oxidative coupling of phenylpropanoids". PhD. Thesis, Universidade Federal da Paraíba.

[201] Rodrigues, L. C. and Barbosa-Filho, J. M. (2011). "*Request for Patent to INPI-Brazilian Institute of Intellectual Property*", published in Revista da Propriedade Industrial (RPI), 2011, No. 2157., PI 1101946-8.

[202] Erdtman, H. (1933). "Dehydrierungen in der Coniferylreihe. II. Dehydrodi-isoeugenol". *Justus Liebigs Annalen der Chemie* 503:283-294.

[203] Shiba, T., Xiao, L., Miyakoshi, T. and Chen, C.-L. (2000). "Oxidation of isoeugenol and coniferyl alcohol catalyzed by laccases isolated from *Rhus vernicifera* Stokes and *Pycnoporus coccineus*". *Journal of Molecular Catalysis B: Enzymatic* 10:605-615.

[204] Basini, G., Spatafora, C., Tringali, C., Bussolati, S. and Grasselli, F. (2014). "Effects of a ferulate-derived dihydrobenzofuran neolignan on angiogenesis, steroidogenesis, and redox status in a swine cell model". *Journal of Biomolecular Screening* 19:1282-1289.

[205] Wang, S., Gates, B. D. and Swenton, J. S. (1991). "A convergent route to dihydrobenzofuran neolignans via a formal 1,3-cycloaddition to oxidized phenols". *The Journal of Organic Chemistry* 56:1979-1981.

[206] Wasserman, H. H., Brunner, R. K., Buynak, J. D., Carter, C. G., Oku, T. and Robinson, R. P. (1985). "Total synthesis of (.+-.)-o-methylorantine". *Journal of the American Chemical Society* 107:519-521.

[207] Blum, T. R., Zhu, Y., Nordeen, S. A. and Yoon, T. P. (2014). "Photocatalytic synthesis of dihydrobenzofurans by oxidative [3+2] cycloaddition of phenols". *Angewandte Chemie International Edition* 53:11056-11059.

[208] Akai, S., Morita, N., Iio, K., Nakamura, Y. and Kita, Y. (2000). "Ambident effect of a *p*-sulfinyl group for the introduction of two

carbon substituents to phenol rings: a convergent synthesis of diverse benzofuran neolignans". *Organic Letters* 2:2279-2282.

[209] Ito, M., Namie, R., Krishnamurthi, J., Miyamae, H., Takeda, K., Nambu, H. and Hashimoto, S. (2014). "Asymmetric total synthesis of (–)-trans-blechnic acid via Rhodium(II)-catalyzed C–H insertion and Palladium(II)-catalyzed C–H olefination reactions". *Synlett* 25:288-292.

[210] Matsumoto, S., Asakawa, T., Hamashima, Y. and Kan, T. (2012). "A short-step asymmetric synthesis of dehydrodiconiferyl alcohol via C-H insertion reaction". *Synlett* 23:1082-1084.

[211] Coy, B. E. D., Cuca, S. L. E. and Sefkow, M. (2010). "The first diastereoselective synthesis of cinerins A-C, PAF-antagonistic macrophyllin-type bicyclo[3.2.1]octane neolignans, using a novel Pd-catalysed oxyarylation". *Organic and Biomolecular Chemistry* 8:2003-2005.

[212] Zheng, S.-L., Yu, W.-Y., Xu, M.-X. and Che, C.-M. (2003). "First synthesis of naturally occurring (±)-*epi*-conocarpan". *Tetrahedron Letters* 44:1445-1447.

[213] Moreira, D. L., Paiva, R. A., Marques, A. M., Borges, R. M., Barreto, A. L. S., Curvelo, J. A. R., Cavalcalti, J. F., Romanos, M. T. V., Soares, R. M. A. and Kaplan, M. A. C. (2016). "Bioactive neolignans from the leaves of *Piper rivinoides* Kunth (Piperaceae)". *Records of Natural Products* 10:472-484.

[214] Van Dyck, S. M. O., Lemière, G. L. F., Jonckers, T. H. M. and Dommisse, R. (2000). "Synthesis of 4-*O*-methylcedrusin. Selective protection of catechols with diphenyl carbonate". *Molecules* 5:153–161.

[215] Chen, L., Deng, H., Cui, H., Fang, J., Zuo, Z., Junliang, D., Li, Y., Wang, X. and Zhao, L. (2018). "Inflammatory responses and inflammation-associated diseases in organs". *Oncotarget* 9:7204-7218.

[216] Bhattacharyya, A., Chattopadhyay, R., Mitra, S. and Crowe, S. (2014). "Oxidative stress: an essential factor in the pathogenesis of gastrointestinal mucosal diseases". *Physiology Reviews* 94:329-354.

[217] Roginsky, V. and Lissi, E. A. (2005). "Review of methods to determine chain-breaking antioxidant activity in food". *Food Chemistry* 92:235-254.

[218] Kozarski, M., Klaus, A., Niksic, M., Vrvic, M. M., Todorovic, N., Jakovljevic, D. and Van Griensven, L. J. L. D. (2012). "Antioxidative activities and chemical characterization of polysaccharide extracts from the widely used mushrooms *Ganoderma applanatum*, *Ganoderma lucidum*, *Lentinus edodes* and *Trametes versicolor*". *Journal of Food Composition and Analysis* 26:144-153.

Biographical Sketches

Herbert J. Dias

Affiliation: Professor of Chemistry at Barão de Mauá School Professor of Chemistry at Vivenda School

Education: He took his BS in Chemistry at University of Franca (2010). He obtained his MSc degree in Sciences at University of Franca (2013) and PhD in Sciences at University of São Paulo (2018) with Dr. Antônio Eduardo Miller Crotti.

Research and Professional Experience: Dr. Herbert J. Dias has experience in Natural Products, where he developed projects employing Mass Spectrometry in the determination of fragmentation patterns to structural elucidation using electrospray ionization tandem mass spectrometry. Dr. Dias took projects in determination of the volatile compounds of essential oils, as well as the synthesis of Natural Product derivatives and determination of potential biological properties in the organic compounds.

Honors:

- Honor for completing the B.S. in Chemistry in the first place by computing all the curricular activities developed during the course, University of Franca, Brazil, 2010.

- Prize for best work in the area of Exact and Technological Sciences at the 5th Meeting of Graduate Researchers (University of Franca, Brazil, 2012).
- Honorable Mention for the work: Synthesis and detailed nuclear magnetic resonance study of some dihydrobenzofuran neolignans, USP 23rd International Symposium on Scientific and Technological Initiation Symposium (University of São Paulo, Brazil, 2015).

Publications from the Last 3 Years:

- Soares, A. C.F., Matos, P.M., Dias, H.J., Aguiar, G.P., Dos Santos, E.S., Martins, C.H.G., Veneziani, R.C.S., Ambrosio, S.R., Heleno, V.C.G. Variability of the antibacterial potential among analogue diterpenes against Gram-positive bacteria: considerations on the structure-activity relationship. *Canadian Journal of Chemistry*. v.97, p.568-575, 2019.
- Souza, J.M., Candido, A.C.B.B., Pagotti, M.C., Dias, H.J., Aguiar, G.P., Arantes, A.T., Cunha, W.R., Groppo, M., Crotti, A.E.M., Parreira, R.L.T., Bernatchez, J.A., Magalhães, L.G. In vitro evaluation of the leishmanicidal potential of selected plant-derived extracts against *Leishmania* (Leishmania) amazonensis. *International Journal of Complementary & Alternative Medicine*. v.12 (1) 36-41, 2019.
- Soares, A.C.F., Matos, P.M., Silva, K.F., Martins, C.H.G., Veneziani, R.C.S., Ambrosio, S.R., Dias, H.J., Santos, R.A., Heleno, V.C.G. Antimicrobial potential of natural and semi-synthetic *ent*-kaurane and *ent*-pimarane diterpenes against clinically isolated gram-positive multidrug-resistant bacteria. *Journal of the Brazilian Chemical Society*. v.30 (2), p. 333-341, 2019.
- Baldin, E.L.L., Dias, H.J., Souza, C.M., Soares, M.C.E., Grundman, C.O., Santos, T.L.B., Crotti, A.E.M. Insecticidal and

Inhibitory effects of dihydrobenzufuran neolignans on *Bemisia tabaci*. *Journa of Pest Science*. v. 2019, p.1053-1062.

- Dias, H.J., Baguenard, M., Crevelin, E.J., Palaretti, V., Gates, P.J., Vessecchi, R., Crotti, A.E.M. Gas-phase fragmentation reactions of protonated benzofuran- and dihydrobenzofuran-type neolignans investigated by accurate-mass electrospray ionization tandem mass spectrometry. *Journal of Mass Spectrometry*. v.54, 35-46, 2019.
- Aguiar, G.P., Crevelin, E.J., Dias, H.J., Ambrosio, S.R., Bastos, J.K., Heleno, V.C.G., Vessecchi, R., Crotti, A.E.M. Electrospray ionization tandem mass spectrometry of labdane-type acid diterpenes. *Journal of Mass Spectrometry*. v.53, p. 1086-1096, 2018.
- Dias, H.J., Patrocínio, A.B., Pagotti, M.C., Fukui, M.J., Rodrigues, V., Magalhães, L.G., Crotti, A.E.M. Schistosomicidal activity of dihydrobenzofuran neolignans. *Chemistry & Biodiversity*. v.15, p. e1800134, 2018.
- Dias, H. J.; Bento, M. V. B., Silva, E. H., Saturnino Junior, A., Oliveira, M. F., Vessecchi, R., Parreira, R. L. T.; Crotti, A. E.M. Gas-phase fragmentation reactions of protonated cocaine: new details to an old story. *Journal of Mass Spectrometry*, v. 53, p. 203-213, 2018.
- Fukui, M. J., Dias, H.J., Severiano, M.E., De Souza, M.G.M., De Oliveira, P.F., Ambrósio, S.R., Martins, C.H.G., Tavares, D.C., Crotti, A.E.M. Antimicrobial and cytotoxic activity of dihydrobenzofuran neolignans. *ChemistrySelect*, v. 3, p. 1836-1839, 2018.
- Dias, H. J.; Aquaroni, N.A.S; Pietro, R. C. L. R., Crotti, A. E.M. Antifungal activity of dihydrobenzofuran neolignans. *MOJ Proteomics & Bioinformatics*, v. 6, p. 00201, 2017.
- Dias, H. J.; Vieira, T. M., Carvalho, C. E.; Aguiar, G. P.; Wakabayashi, K. A. L.; Turatti, I.C.C., Willrich, G. B., Groppo, M.; Cunha, W. R.; Martins, C. H. G.; Crotti, A.E.M. Screening of selected plant-derived extracts for their antimicrobial activity

against oral pathogens. *International Journal of Complementary and Alternative Medicine*, v. 6, p. 00188-00192, 2017.

- Vieira, T. M., Dias, H. J.; Grundmann, C. O., Groppo, M.; Heleno, V. C. G., Martins, C. H. G.; Cunha, W. R.; Crotti, A.E.M., Silva, E. O. Chemical composition and antimicrobial activity of the essential oil of *Artemisia absinthium* Asteraceae leaves. *Journal of Essential Oil-Bearing Plants*, v. 20, p. 123-131, 2017.
- Martins, M. H. G., Fracarolli, L.; Vieira, T. M., Dias, H. J.; Cruz, M. G., Deus, C. C. H., Nicolella, H. D., Stefani, R., Rodrigues, V.; Tavares, D. C.; Magalhães, Lg; Crotti, A. E. M. Schistosomicidal effects of the essential oils of *Citrus limonia* and *C. reticulata* against *Schistosoma mansoni*. *Chemistry & Biodiversity* (Print), v. 14, p. e1600194, 2017.
- Soares, M.H., Dias, H.J., Vieira, T.M., De Souza, M.G.M., Cruz, A. F. F., Badoco, F. R., Nicolella, H. D., Cunha, W.R..; Groppo, M., Martins, C. H. G., Tavares, D. C.,Magalhães, L.G., Crotti, A.E.M. Chemical composition, antibacterial, schistosomicidal, and cytotoxic activities of the essential oil of *Dysphania ambrosioides* (L.) Mosyakin & Clemants (Chenopodiaceae). *Chemistry & Biodiversity*, v. 14, p. e1700149, 2017.
- Dias, H.J., Vieira, Tatiana M., Crevelin, Eduardo J., Donate, Paulo M., Vessecchi, Ricardo; Crotti, A. E.M. Fragmentation of 2-aroylbenzofuran derivatives by electrospray ionization tandem mass spectrometry. *Journal of Mass Spectrometry*, v. 52, p. 809-816, 2017.
- Dias, H.J., Stefani, R., Tomaz, J.C., Vessecchi, R., Crotti, A.E.M. Differentiation between 3,4- and 4,15-epoxyeudesmanolides by electrospray ionization tandem mass spectrometry. *Journal of Analytical Methods in Chemistry*, v. 2017, p. 1-9, 2017.
- Alcoba, A. E. T., Melo, D. C. De; De Andrade, P. M., Dias, H.J., Pagotti, M. C., Magalhães, L.G.; Júnior, W. G. F., Crotti, A.E.M., Miranda, M. L. D. Chemical composition and *in vitro* antileishmanial and cytotoxic activities of the essential oils of

Ocotea dispersa (Nees) Mez and *Ocotea odorifera* (Vell) Rohwer (Lauraceae). *Natural Product Research*. v.2017, p 1-4, 2017.

Eliane de Oliveira Silva

Affiliation: Organic Chemistry Department, Chemistry Institute, Federal University of Bahia, Salvador, Bahia, Brazil

Research and Professional Experience: Natural Products, Endophytic Fungi, Biotransformation

Professional Appointments: Professor, Federal University of Bahia – Chemistry Institute, Organic Chemistry Department

Honors: Young Researcher Award from Financial Support Agency (Bahia Research Foundation - FAPESB), 2017

Publications from the Last 3 Years: (2017-2019)

1. Nascimento, J. S., Silva, F. M., Magallanes-Noguera, C., Kurina-Sanz, M., Santos, E. G., Caldas, I. S., Luiz, J. H. H., Silva, E. O. Natural trypanocidal product produced by endophytic fungi through co-culturing. *Folia Microbiologica*, Epub ahead of print (2019).
2. Santos, V. H. P., Silva, Eliane O. Endophytic fungi from the Brazilian flora and their employment in biotransformation reactions. *Química Nova*, v. 42, n. 7, 784-791, 2019.
3. Catarine Santos Rodrigues, Carla; Latércia Tranches Dias, Amanda; De Oliveira Silva, Eliane. Unprecedented derivatization of ferulic acid through selective methoxylation by *Aspergillus brasiliensis* ATCC 16404. *Biocatalysis and Biotransformation*, v. 37, p. 1-5, 2019.

4. Fernandes, V. H. C., Vieira, N. B., Zanini, L. B. L., Silva, A. F., Salem, P. P. O., Soares, M. G., Nicacio, K. J., Paula, A. C. C., Virtuoso, L. S., Oliveira, T. B., Silva, Eliane O., Dias, D. F., Chagas-Paula, D.A. Fast and Efficient Method to Obtain Tagitinin F by Photocyclization of Tagitinin C. *Photochemistry and Photobiology*, Epub ahead of print (2019).
5. Sousa, I. P., Chagas-Paula, D. A., Tiossi, R. F. J., Silva, E. O., Miranda, M. A., Oliveira, R. B., Spadaro, A. C. C., Bastos, J. K., Furtado, N.A.J.C., Costa, F. B. Essential oils from *Tithonia diversifolia* display potent anti-oedematogenic effects and inhibit acid production by cariogenic bacteria. *Journal of Essential Oil Research*, v. 31, p. 43-52, 2019.
6. Bernardes, Wagner A., Silva, Eliane O., Crotti, Antônio E.M., Baldin, Edson L.L. Bioactivity of selected plant-derived essential oils against *Zabrotes subfasciatus* (Coleoptera: Bruchidae). *Journal of Stored Products Research*, v. 77, p. 16-19, 2018.
7. Nascimento, J. S., Conceição, J. C. S., Silva, E. O. Biotransformation of Coumarins by Filamentous Fungi: An Alternative Way for Achievement of Bioactive Analogs. *Mini-Reviews in Organic Chemistry*, v. 15, p. online, 2018.
8. Paludo, Camila Raquel; Da Silva-Junior, Eduardo Afonso; De Oliveira Silva, Eliane; Vessecchi, Ricardo; Peporine Lopes, Norberto; Tallarico Pupo, Mônica; Da Silva Emery, Flavio; Dos Santos Gonçalves, Natália; Alves Dos Santos, Raquel; Jacometti Cardoso Furtado, Niege Araçari. Inactivation of β-Lapachone Cytotoxicity by Filamentous Fungi that Mimic the Human Blood Metabolism. *European Journal of Drug Metabolism and Pharmacokinetics*, v. 42, p. 213-220, 2017.
9. Vieira, Tatiana M., Dias, Herbert J., Medeiros, Talita C.T., Grundmann, Carlismari O., Groppo, Milton; Heleno, Vladimir C. G., Martins, Carlos H.G., Cunha, Wilson R., Crotti, Antônio E. M., Silva, Eliane O. Chemical Composition and Antimicrobial Activity of the Essential Oil of *Artemisia absinthium* Asteraceae Leaves. *Journal of Essential Oil Bearing Plants*, v. 20, p. 1-9, 2017.

10. Goncalves, R., Ayres, V. F. S., Carvalho, C. E., Souza, M. G. M., Guimaraes, A. C., Correa, G. M., Martins, C. H. G., Takeara, R., Silva, E.O., Crotti, A. E. M. Chemical Composition and Antibacterial Activity of the Essential Oil of *Vitex agnus-castus* L. (Lamiaceae). *Anais Da Academia Brasileira De Ciencias*, v. 89, p. 2825-2832, 2017.
11. De Oliveira Silva, Eliane; Batista, Ronan. Ferulic Acid and Naturally Occurring Compounds Bearing a Feruloyl Moiety: A Review on their Structures, Occurrence, and Potential Health Benefits. *Comprehensive Reviews in Food Science and Food Safety*, v. 16, p. 580-616, 2017.

Tatiana Manzini Vieira, Msc

Affiliation: Chemistry Departament, Faculdade de Filosofia, Ciências e Letras de Ribeirão Preto – Universidade de São Paulo

Research and Professional Experience: Mass spectrometry, synthesis, organic chemistry

Professional Appointments: PhD student, Chemistry Departament, Faculdade de Filosofia, Ciências e Letras de Ribeirão Preto – Universidade de São Paulo

Honors: Financial Support Agency (FAPESP- 2019/11700-0)

Publications from the Last 3 Years:

1. Vieira, Tatiana M., Dos Santos, Isabella A., Silva, Thayná S., Martins, Carlos H. G., Crotti, Antônio E. M. Antimicrobial activity of monoketone curcuminoids against cariogenic bacteria. *Chemistry & Biodiversity*, 2018, 15, e1800216.

2. Vieira, T. M., Ambrosio, M.A. L. V., Carlos H. G., Crotti, Antônio E. M. Structure-antimicrobial activity relationships of monoketone curcuminoids. *Int J Complement Alt Med*, 2018, 11, 347-350.
3. Vieira, T. M., Dias, H. J., Grundmann, C. O., Groppo, M., Heleno, V. C. G., Martins, C. H. G., Cunha, W. R., Crotti, Antonio E. M., Silva, E. O. Chemical Composition and Antimicrobial Activity of the Essential Oil of *Artemisia absinthium* Asteraceae Leaves. *Journal of Essential Oil-Bearing Plants*, 2017, 20, 1-9.
4. Dias, Herbert J., Vieira, Tatiana M., Crevelin, Eduardo J., Donate, Paulo M., Vessecchi, Ricardo; Crotti, Antônio E.M. Fragmentation of 2-aroylbenzofuran derivatives by electrospray ionization tandem mass spectrometry. *Journal of Mass Spectrometry*, 2017, 52, 809-816.
5. Dias, H. J., Vieira, T. M., Carvalho, C. E., Aguiar, G. P., Wakabayashi, K. A. L., TURATTI, Isabel Cristina Casanova; Willirich, G. B., Groppo, M., Cunha, W. R., Martins, C. H. G., Crotti, AEM. Screening of selected plant-derived extracts for their antimicrobial activity against oral pathogens. *International Journal of Complementary and Alternative Medicine*, 2017, 6, 00188-00192.
6. Lopes, M. H. S., Dias, H. J., Vieira, T. M., Souza, M. G. M., Cruz, A. F. F., Badoco, F. R., Nicolella, H. D., Cunha, W. R., Groppo, M., Martins, C. H. G., Tavares, D. C., Magalhaes, L. G., Crotti, AEM. Chemical Composition, Antibacterial, Schistosomicidal, and Cytotoxic Activities of the Essential Oil of *Dysphania ambrosioides* (L.) Mosyakin & Clemants (Chenopodiaceae). *Chemistry & Biodiversity*, 2017, 14, e1700149-e1700149.
7. Martins, M. H. G., Fracarolli, L., Vieira, T. M., Dias, H. J., Cruz, M. G., Deus, C. C. H., Nicolella, H. D., Stefani, R., Rodrigues, V., Tavares, D. C., Magalhães, LG; Crotti, A. E. M. Schistosomicidal Effects of the Essential Oils of *Citrus limonia* and *C. reticulata* Against *Schistosoma mansoni*. *Chemistry & Biodiversity*, 2017, 14, e1600194.

8. Santiago, M. B., Moraes, T. Da S., Massuco, J. E., Silva, L. O., Lucarini, R., Da Silva, D. F., Vieira, T. M., Crotti, A. E. M., Martins, C. H. G. In vitro evaluation of essential oils for potential antibacterial effects against *Xylella fastidiosa*. *Phytopathology*, 2018, 166, 790-798.
9. Singulani, J. L., Pedroso, R. S., Ribeiro, A. B; Nicolella, H. D., Freitas, K. S., Damasceno, J. L., Vieira, T. M., Crotti, A. E. M; Tavares, D. C., Martins, C. H. G., Mendes-Giannini, Maria J. S., Pires, R. H. Geraniol and linalool anticandidal activity, genotoxic potential and embryotoxic effect on zebrafish. *Future Microbiol*, 2018, 13, 1637–1646.
10. Almeida, K. C. R., Silva, B. B., Alves, C. C. F., Vieira, T.M., Crotti, A. E. M., Souza, J. M., Martins, C. H. G., Ribeiro, A. B., Squarisi, I. S., Tavares, D. C., Bernabé, L. D. S., Magalhães, L. G., Miranda, M. L. D. Biological properties and chemical composition of essential oil from *Nectandra megapotamica* (Spreng.) Mez. leaves (Lauraceae). *Nat Prod. Res*, 2019 (Aceito para publicação).

Antônio Eduardo Miller Crotti, PhD

Affiliation: Departamento de Química, Faculdade de Fiulosofia, Ciências e Letras de Ribeirão Preto, Universidade de São Paulo

Research and Professional Experience: Natural Products, Mass Spectrometry, Organic Synthesis

Professional Appointments: Professor, Universidade de São Paulo Ribeirão Preto, SP, Brazil

Publications from the Last 3 Years: (2017-2019)

1. Durazzini, A. M. S., Machado, C. H. M., Fernandes, C. C., Willrich, G. B., Crotti, A. E. M., Candido, A. C. B. B., Magalhães,

L. G., Squarisi, I. S., Ribeiro, A. B., Tavares, D. C., Martins, C. H. G., Miranda, M. L. D. *Eugenia pyriformis* Cambess: a species of the Myrtaceae family with bioactive essential oil. *Natural Product Research* 2019 (Accepted for publication). DOI: 10.1080/14786419.2019.1669031.

2. Gonçalves, R., Ayres, V. F. S., Magalhães, L. G., Crotti, A. E. M., Corrêa, G. M., Guimarães, A. C., Takeara, R. Chemical composition and schistosomicidal activity of essential oils of two *Piper* species from the Amazon region. *Journal of Essential Oils Bearing Plant* 2019 (Accepted for publication). DOI: 10.1080/0972060X.2019.1631720.
3. Oda, F. B., Crevelin, E. J., Crotti, A. E. M., Orlando, A. B., Medeiros, A. I., Nogueira, F. A. R., Santos, A. G. Acidic and hepatic derivatives of bioactive clerodane diterpenes casearins. *J* and *O. Fitoterapia* 2019 (Accepted for publication). DOI: 10.1016/j.fitote.2019.104197.
4. Moreira, R. R. D., dos Santos, A. G., Carvalho, F. A., Perego, C. H., Crevelin, E. J., Crotti, A. E. M., Cogo, J., Cardoso, M. L. C., Nakamura, C. V. Antileishmanial activity of the essential oils of *Melampodium divaricatum* and *Casearia sylvestris* on Leishmania amazonensis. *Revista do Instituto de Medicina Tropical de São Paulo* 2019, 61, e33. DOI: 10.1590/S1678-9946201961033. [*Magazine of the Institute of Tropical Medicine of São Paulo*]
5. Silva, F. F. F., Alves, C. C. F., Santiano, B., Martins, C. H. G., Vieira, T. M., Crotti, A. E. M., Miranda, M. L. D. Chemical composition and *in vitro* antibacterial activity of essential oils from *Murraya paniculata* (L.) Jack (Rutaceae) ripe and unripe fruits against bacterial genera Mycobacterium and Streptococcus. *Brazilian Journal of Pharmaceutical Sciences* 2019 (Accepted for publication).
6. Silva, F. F. A., Alves, C. C. F., Oliveira-Filho, J. G., Vieira, T. M., Crotti, A. E. M., Miranda, M. L. D. Chemical constituents of essential oil from *Murraya paniculata* leaves and its application to in vitro biological control of the fungus *Sclerotinia sclerotiorum*.

Food Science and Technology 2019 (Accepted for publication). DOI: 10.1590/fst.20218.

7. Almeida, K. C. R., Silva, B. B., Alves, C. C. F., Vieira, T. M., Crotti, A. E. M., Souza, J. M., Martins, C. H. G., Ribeiro, A. B., Squarisi, I. S., Tavares, D. C., Bernabé, L. S., Magalhães, L. G., Miranda, M. L. D. Biological properties and chemical composition of essential oil from *Nectandra megapotamica* (Spreng.) Mez. leaves (Lauraceae). *Natural Product Research 2019.* DOI: 10.1080/14786419.2019.1608539 (Accepted for publication).
8. Cabral, R. C. S., Alves, C. C. F., Batista, H. R. F., Sousa, W. C., Abrahão, I. S., Crotti, A. E. M., Santiago, M. B., Martins, C. H. G., Miranda, M. L. D. Chemical composition of essential oils from different parts of *Protium heptaphyllum* (Aubl.) *Marchand and their in vitro antibacterial activity. Natural Product Research 2019* (Accepted for publication). DOI: 10.1080/14786 419.2018.1536659.
9. Cabral, F. D., Alves, C. C. F., Cabral, R. S. C., Willrich, G. B., Crotti, A. E. M., Miranda, M. L. D. Chemical constituents of essential oils extracted from the leaves and flowers of *Spiranthera odoratissima* A. St. Hil. (Rutaceae). *Records on Natural Products* 2019 (Accepted for publication). DOI: 10.25135/rnp.95.18.07.117
10. Souza, J. M., Candido, A. C. B. B., Pagotti, M. C., Dias, H. J., Aguiar, G. P., Arantes, Andréia T., Cunha, W. R., Groppo, M., Cunha, W. R., Crotti, A. E. M., Parreira, R. L. T., Bernatchez, J. L. Magalhães, M. G. *In vitro* evaluation of the leishmanicidal potential of selected plant-derived extracts against *Leishmania amazonensis. International Journal of Complementary and Alternative Medicine* 2019, 12, 36-41. DOI: 10.15406/ijcam. 2019.12.00446.
11. Spósito, L., Oda, F. B., Vieira, J. H., Carvalho, F. A., Ramos, M. A. S., Castro, R. C., Crevelin, E. J., dos Santos, A. G., Crotti, A. E. M., Silva, P. G., Chorilli, M., Bauab, T. M. In vitro and in vivo anti-Helicobacter pylori activity from *Casearia sylvestris* leaf

derivatives. *Journal of Ethnopharmacology* 2019, 233, 1-12. DOI: 10.1016/j.jep.2018.12.032.

12. Baldin, E. L. L., Dias, H. J., Souza, C. M., Soares, M. C. E., Grundmann, C. O., Santos, T. L. B., Crotti, A. E. M. Insecticidal and inhibitory effects of dihydrobenzofuran neolignans on *Bemisia tabaci*. *Journal of Pest Science* 2019, 92, 861-869. DOI:10.1007/s10340-018-1053-2.
13. Dias, H. J., Baguenard, M., Crevelin, E. J., Palaretti, V., Gates, P. J., Vessecchi, R., Crotti, A. E. M. Gas-phase fragmentation reactions of protonated benzofuran- and dihydrobenzofuran-type neolignans investigated by accurate-mass electrospray ionization tandem mass spectrometry. *Journal of Mass Spectrometry* 2019, 54, 35-46.DOI: 10.1002/jms.4304.
14. Vieira, T. M., Ambrósio, M. A. L. V., Martins, C. H. G., Crotti, A. E. M. Structure-antimicrobial activity relationships of monoketone curcuminoids. *International Journal of Complementary and Alternative Medicine* 2018, 11, 347-350. DOI: 10.15406/ijcam.2018.11.00423.
15. Singulani, J. L., Damasceno, J. L., Pedroso, R. S., Ribeiro, A. B., Nicolella, H. D., Frieitas, K. S., Vieira, T. M., Crotti, A. E. M., Tavares, D. C., Martins, C. H. G., Mendes-Gianini, M. J. S., Pires, R. H. Geraniol and linalool anticandidal activity, genotoxic potential, and embryotoxic effect on zebrafish. *Future Microbiology* 2018, 166, 790-798. DOI: 10.2217/fmb-2018-0200.
16. Santiago, M. B., Moraes, T. S., Massuco, J. E., Silva, L. O., Lucarini, R., Silva, D. F., Vieira, T. M., Crotti, A. E. M., Martins, C. H. G. In vitro evaluation of the essential oils for potential antibacterial effects against *Xylella fastidiosa*. *Journal of Phytophatology* 2018, 166, 790-798. DOI: 10.1111/jph.12762.
17. Aguiar, G. P., Crevelin, E. J., Dias, H. J., Ambrósio, S. R., Bastos, J. K., Heleno, V. C. G., Vessecchi, R., Crotti, A. E. M. Electrospray ionization tandem mass spectrometry of labdane-type acid diterpenes. *Journal of Mass Spectrometry* 2018, 53, 1086-1096. DOI: 10.1002/jms.4284.

18. Vieira, T. M., Santos, I. A., Silva, T. S., Martins, C. H. G., Crotti, A. E. M. Antimicrobial activity of monoketone curcuminoids against cariogenic bacteria. *Chemistry and Biodiversity* 2018, 15, e1800216. DOI: 10.1002/cbdv.201800216.
19. Dias, H. J., Patrocínio, A. B., Pagotti, M. C., Fukui, M. J., Rodrigues, V., Magalhães, L. G., Crotti, A. E. M. Schistosomicidal activity of dihydrobenzofuran neolignans. *Chemistry and Biodiversity* 2018, 15, e1800134. DOI: 10.1002/cbdv.201800134.
20. Valadares, C. F., Alves, C. C. F., Alves, J. M., Deus, I. P. B., Oliveira-Filho, J. G., Santos, T. C. L., Dias; H. J., Crotti, A. E. M., Miranda, M. L. D. Essential oils from *Piper aduncum* inflorescences and leaves: chemical composition and antifungal activity against *Sclerotinia sclerotiorum. Annals of the Brazilian Academy of Sciences* 2018, 90, 2691-2699. DOI: 10.1590/0001-3765201820180033.
21. Lemes, R. S., Alves, C. C. F., Estevam, E. B. B., Santiago, M. B., Martins, C. H. G., Santos, T. C. L., Crotti, A. E. M., Miranda, M. L. D. Chemical composition and antibacterial activity of essential oils from *Citrus aurantifolia* leaves and fruit peel against oral pathogenic bacteria. *Annals of the Brazilian Academy of Sciences* 2018, 90, 1285-1292. DOI: 10.1590/0001-3765201820170847.
22. Bernardes, W. A., Silva, E. O., Crotti, A. E. M., Baldin, E. L. L. Bioactivity of selected plant-derived essential oils against *Zabrotes subfasciatus* (Coleoptera: Bruchidae). *Journal of Stored Products Research* 2018, 77, 16-19. DOI: 10.1016/j.jspr.2018.02.007.
23. Fukui, M. J., Dias, H. J., Severiano, M. E., Souza, M. G. M., Oliveira, P. F., Ambrósio, S. R., Martins, C. H. G., Tavares, D. C., Crotti, A. E. M. Antimicrobial and cytotoxic activity of dihydrobenzofuran neolignans. *Chemistry Select* 2018, 3, 1836-1839. DOI: 10.1002/slct.201703024.
24. Dias, H. J., Bento, M. V. B., Silva, E. H., Saturnino-Júnior, A., Oliveira, M. F., Vessecchi, R., Parreira, R. L. T., Crotti, A. E. M. Gas-phase fragmentation reactions of protonated cocaine: new

details to an old story. *Journal of Mass Spectrometry*, 2018, 53, 203-213. DOI: 10.1002/jms.4053.

25. Oliveira, A. P., Coppede, J. S., Bertoni, B. W., Crotti, A. E. M., Franca, S. C., Pereira, A. M. S., Taleb-Contini, S. H. *Costus spiralis* (Jacq.) Roscoe: a novel source of flavones with alpha-glycosidase inhibitory activity. *Chemistry & Biodiversity* 2018, 14, e1700421. DOI: 10.1002/cbdv.201700421.
26. Alcoba, A. E. T., de Melo, D. C., de Andrade, P. M., Dias, H. J., Pagotti, M. C., Magalhães, L. G., Ferreira-Júnior, W. G., Crotti, A. E. M., Miranda, M. L. D. Chemical composition and in vitro antileishmanial and cytotoxic activities of the essential oils of *Ocotea dispersa* (Nees) Mez and *Ocotea odorifera* (Vell) Rohwer (Lauraceae). *Natural Product Research* 2018, 32, 2865-2868. DOI: 10.1080/14786419.2017.1385007.
27. Dias, H. J., Aquaroni, N. A. S., Pietro, R. C. L. R., Crotti, A. E. M. Antifungal activity of dihydrobenzofuran neolignans. *MOJC Proteomics & Bioinformatics,* 2017, 6, 00201. DOI: 10.15406/mojpb.2017.06.00201.
28. Dias, H. J., Vieira, T. M., Crevelin, E. J., Donate, P. M., Vessecchi, R., Crotti, A. E. M. Fragmentation of 2-aroylbenzofuran derivatives by electrospray ionization tandem mass spectrometry. *Journal of Mass Spectrometry* 2017, 52, 809-816. DOI: 10.1002/jms.4024.
29. Andrade, P. M., dos Santos, T. C. L., Pagotti, M. C., Júnior, W. G. F., Magalhães, L. G., Crotti, A. E. M., Miranda, M. L. D. Chemical composition and evaluation of the antileishmanial and cytotoxic activities of the essential oils from the leaves of *Cryptocarya aschersoniana* Mes (Lauraceae Juss.). *Annals of the Brazilian Academy of Sciences* 2017, 90, 2671-2678. DOI: 10.1590/0001-3765201820170332.
30. Gonçalves, R., Ayres, V. F. S., Carvalho, C. E., Souza, M. G. M., Guimarães, A. C., Corrêa, G. M., Martins, C. H. G., Takeara, R., Silva, E. O., Crotti, A. E. M. Chemical composition and antibacterial activity of the essential oil of *Vitex agnus-castus* L.

(Lamiaceae). *Annals of the Brazilian Academy of Sciences 2017*, 89, 2825-2832. DOI: 10.1590/0001-3765201720170428.

31. Soares, M. H., Dias, H. J., Vieira, T. M., de Souza, M. G. M., Cruz, A. F. F., Badoco, F. R., Nicolella, H. D., Cunha, W. R., Groppo, M., Martins, C. H. G., Tavares, D. C., Magalhães, L. G., Crotti. A. E. M.. Chemical composition, antibacterial, schistosomicidal, and cytotoxic activities of the essential oil of *Dysphania ambrosioides* (L.) Mosyakin & Clemants (Chenopodiaceae). *Chemistry & Biodiversity* 2017, 14, e1700149. DOI: 10.1002/cbdv.201700149.
32. Dias, H. J., Stefani, R., Tomaz, J. C., Vessecchi, R., Crotti, A. E. M. Differentiation between 3,4- and 4,15-epoxyeudesmanolides by electrospray ionization tandem mass spectrometry. *Journal of Analytical Methods in Chemistry* 2017, 2017, Article ID 7921867. DOI: 10.1155/2017/7921867.
33. Dias, H. J., Vieira, T. M., Carvalho, C. E., Aguiar, G. P., Wakabayashi, K. A. L., Turatti, I. C. C., Willrich, G. B., Groppo, M., Cunha, W. R., Martins, C. H. G., Crotti, A. E. M. Screening of selected plant-derived extracts for their antimicrobial activity against oral pathogens. *International Journal Complementary & Alternative Medicine* 2017, 6, 0188. DOI: 10.15406/ijcam.2017.06.00188.
34. Crevelin, E. J., Possato, B., Lopes, J. L. C., Lopes, N. P., Crotti, A. E. M. A precursor ion scan mode-based strategy for fast screening of polyether ionophores by copper-induced gas-phase radical fragmentation reactions. *Analytical Chemistry* 2017, 89, 3929-3936. DOI: 10.1021/acs.analchem.6b02855.
35. Carrão, D. B., Albuquerque, N. C. P., Marques, L. M. M., Crotti, A. E. M., Bolzani, V. S., Berretta, A. A., Oliveira, A. R. M. In vitro metabolism of Artepilin C by rat and human liver microsomes. *Planta Medica* 2017, 83, 737-745. DOI: 10.1055/s-0042-124359
36. Costa, J. C., Ramos, G. M., Valadão, G. M. R., Pala, G., Gallani, S. U., Kotzent, S., Crotti, A. E. M., Fracarolli, L., Silva, J. J. M.,

Pilarski, F. Copaifera duckei oleoresin as a novel alternative for treatment of monogenean infections in pacu Piaractus mesopotamicus. *Aquaculture* 2017, 471, 72-79. DOI: 10.1016/j.aquaculture.2016.11.041.

37. Martins, M. H. G., Fracarolli, L., Vieira, T. M., Dias, H. J., Cruz, M. G., Deus, C. H., Nicolella, H. D., Stefani, R., Rodrigues, V., Tavares, D. C., Magalhães, L. G., Crotti, A. E. M. Schistosomicidal effects of the essential oils of *Citrus limonia* and *C. reticulata* against *Schistosoma mansoni. Chemistry & Biodiversity* 2017, 14, e1600194. DOI: 10.1002/cbdv.201600194.
38. Vieira, T. M., Dias, H. J., Medeiros, T. C. T., Grundmann, C. O., Groppo, M., Heleno, V. C. G., Martins, C. H. G., Cunha, W. R., Crotti, A. E. M., Silva, E. O. Chemical composition and antimicrobial activity of the essential oil of *Artemisia absinthium* (Asteraceae) leaves. *Journal of Essential Oil Bearing Plants* 2017, 20, 123-131. DOI: 10.1080/0972060X.2016.1257370.
39. Silva, M. L. A., Pereira, A. C., Esperandim, V. R., Simaro, G. V., Lima, T. C., Januario, A. H., Pauletti, P. M., Rehder, V. L. G., Crevelin, E. J., Cunha, W. R., Crotti, A. E. M., Bastos, J. K. *In vitro* trypanocidal activities of *Pfaffia glomerata* root extract, its hydrolyzed fractions and pfaffic acid against *Trypanosoma cruzi* trypomastigotes. *Chemistry & Biodiversity* 2017, 14, e1600175. DOI: 10.1002/cbdv.201600175.

In: Benzofuran: Production and Applications ISBN: 978-1-53617-873-9
Editor: Murilo Souza Barros

Chapter 3

Palladium or Rhodium Catalyzed Benzofuran and Derivatives Synthesis via C-H Functionalizations

Bo Sun, PhD, Tian-Yu Yao, MD, Lu-Lu Zhang, MD, Ming-Li, PhD and Lin-Bao Zhang*, PhD
College of Chemistry and Molecular Engineering,
Qingdao University of Science and Technology, Qingdao, P. R. China

Abstract

Benzofurans and their derivatives are a class of important structural motifs owing to their wide existence in natural products, pharmaceuticals and biologically active compounds, thus inspiring extensive research for their synthesis under mild and simple reaction conditions. During the past decades, various strategies have been developed to synthesize such useful heterocyles. Among these methods, the annulation reaction represents one of the most straightforward ways to produce the target molecules. In this context, great attention has been paid to the protocols that possess the atom and step efficiency, with which multiple C-H functionalizations could be accomplish the expectations. This review is mainly focused on the palladium and rhodium catalyzed multiple C-H functionalizations of

* Corresponding Author's E-mail: zhang_linbao@126.com.

the development in the field of benzofurans and their derivatives' synthesis. The use of Ir or Ru salts will be slightly referred to at the end of the context.

Keywords: palladium, rhodium, C-H functionalization, benzofuran

Introduction

Owing to important biological activities, the synthesis of benzofurans and their derivatives has drawn great attention for several decades [1-4]. Among the various methodologies, transition-metal-catalyzed cross-coupling reactions have been used as versatile methods to construct the core of heterocyclic systems [5-7]. However, the prefunctionalized cross-coupling substrates require tedious synthetic steps as well as the undesired toxic side products and heavy metal waste are produced in the synthetic route [8]. As a consequence, C-H functionalization, as a convenient and powerful alternative to cross-coupling reactions, has been profoundly developed from the viewpoint of sustainable and enviro-friendly benign features [9-13]. Moreover, a library of viable protocols have been established to construct the particular heterocyclic core based on C-H activation [14, 15]. In this context, Pd- or Rh-catalyzed C-H bond functionalizations to construct these heterocycles will be summarized. Other findings including Ru, Ir-catalyzed reactions will be also discussed at the end of this review.

Methods

Pd-Catalyzed Benzofuran Synthesis via C-H Functionalizations

Traditionally, the direct method for the synthesis of benzofuran involves Lewis acid-catalyzed condensation reactions [16-19]. Transition-metal-catalyzed oxidative anulation of ortho-disubstituted arene precursor was also reported to synthesize heterocycle [20-23]. Recently, direct C-H bond functionalization has attracted great attention as it enables a wide range of simple starting materials without producing any unwanted waste material. We will summarize the palladium-catalyzed benzofuran synthesis via C-H functionalization in this section. Moreover, the mechanism understandings will be also discussed in this review.

Larock and his co-workers have established an efficient protocol to synthesize the substituted dibenzofurans 3 in moderate to excellent yields via the palladium-catalyzed cross-coupling of alkynes 2 and substituted aryl iodides 1 (Scheme 1) [24]. This process proceeded through the carbopalladation of the alkyne 2, especially heteroatom-directed vinylic to aryl palladium migration, and intramolecular arylation to furnish target molecules. 3-Iodophenyl phenyl ether derivatives 1 reacted with internal alkyne 2 in the presence of Pd(OAc)2 as catalyst to get dibenzofurans 3. They felt that an increase in the electron density of the aromatic ring should facilitate vinylic to aryl palladium migration in this transformation.

The obtained product patterns could be explained by their proposed mechanism. First, the oxidative addition of palladium catalyst with aryl iodide 1 followed by an intermolecular carbopalladation with alkyne 2 generated intermediate 4. The palladium intermediate 4 then underwent a palladium migration process from the vinylic position to an aryl position to generate intermediate 7, which eventually afforded intermediate 5 via the palladium migration process. In this case, equilibrium between palladium (IV) hydride 7 and palladium (II) intermediate 8 was also possible. Intermediate 5 underwent either C-H bond insertion of the neighboring arene or electrophilic aromatic substitution to afford palladacycle intermediate 10. Finally, dibenzofuran 3 was obtained after reductive elimination. In fact, the electron-withdrawing effect of the oxygen atom played a prominent role in the ortho- other than in para-position selectivity by stabilizing the aryl palladium intermediate.

Direct annulation of heteroatoms with the proximate C-H bond represented a significant strategy to construct such versatile molecules in heterocyclic chemistry [25].

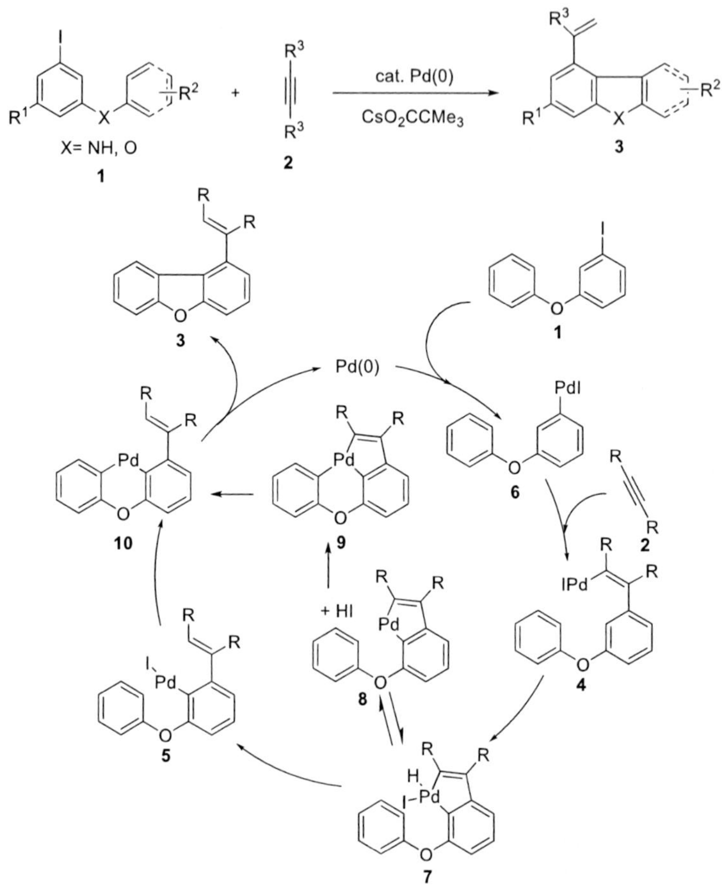

Scheme 1. Synthesis of substituted carbazoles via vinylic to aryl palladium migration.

In 2010, Yu and his coworkers disclosed a Pd(II)-catalyzed C-H activation/C-O cyclization reaction for the construction of dihydrobenzofurans 12 (Table 1) [26]. In their case, hydroxyl-directed C-H activation was promoted by Pd(OAc)2 as catalyst. Optimization conditions showed that 5 mol% Pd(OAc)2 combined 1.5 equiv. of Li2CO3 were beneficial for the transformation. PhI(OAc)2 acted as a two-electron oxidant to oxidize Pd(II) to Pd(IV) in the process. This reaction provided a direct method for the synthesis of various dihydrobenzofurans 12 in good to excellent yields employing different tertiary alcohols 11 as substrates.

Table 1. Palladium-catalyzed C-H activation/C-O cyclization

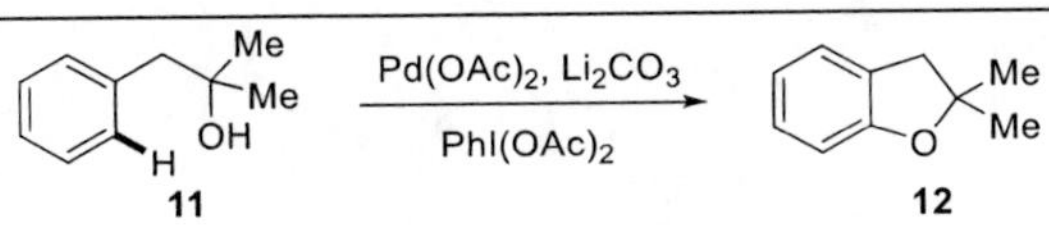

Product	Yield	Product	Yield
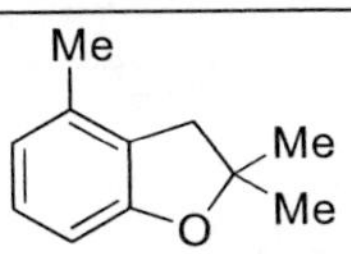	88%	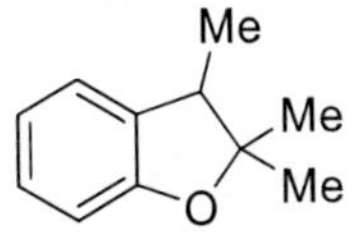	80%
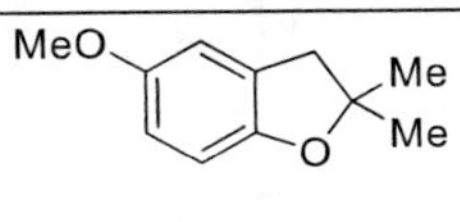	76%	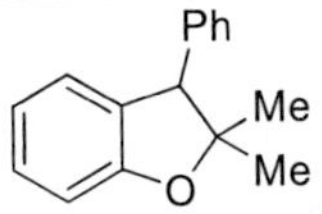	81%
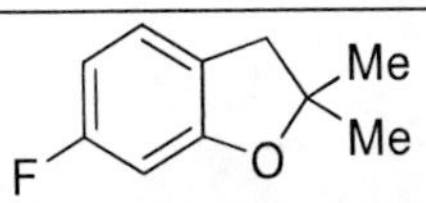	85%	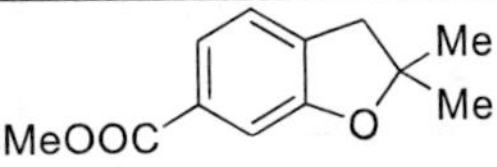	90%
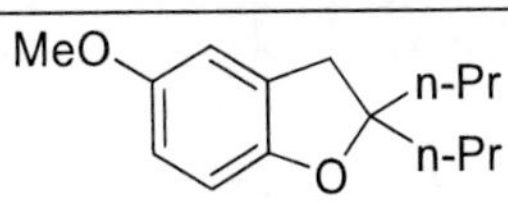	91%	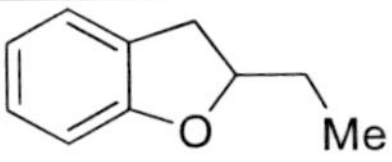	42%

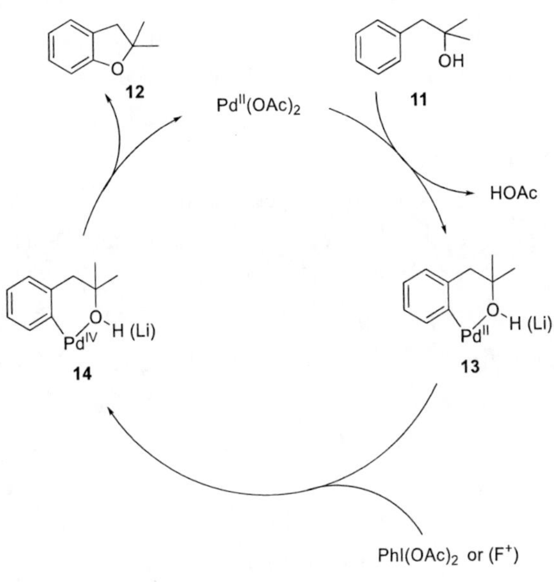

Scheme 2. Palladium-catalyzed C-H activation/C-O cyclization.

Table 2. Palladium-catalyzed dibenzofurans synthesis

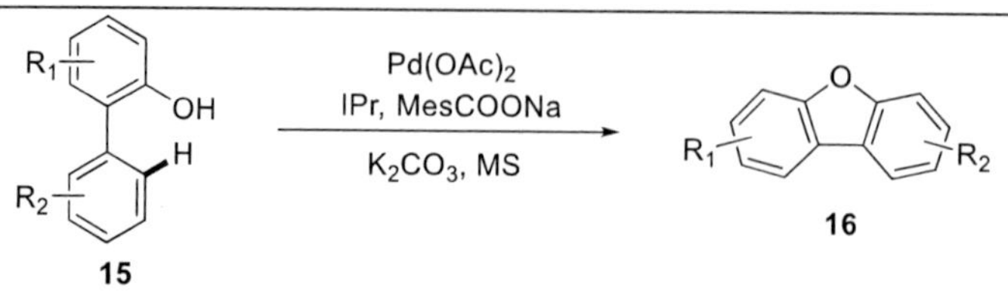

Product	Yield	Product	Yield
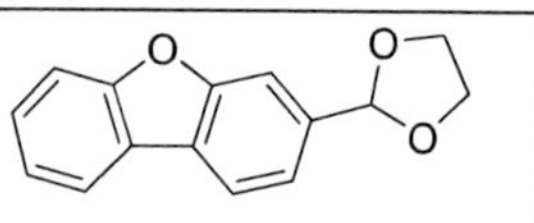	65%	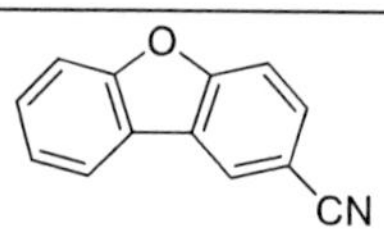	37%
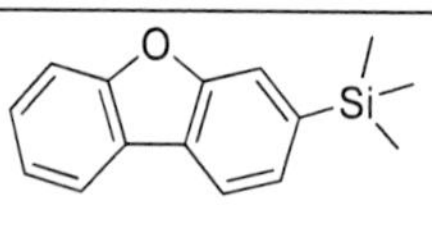	82%	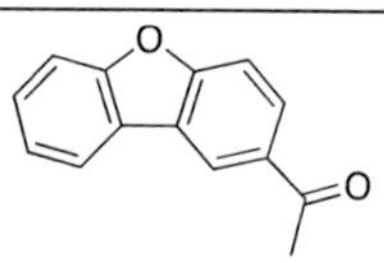	71%
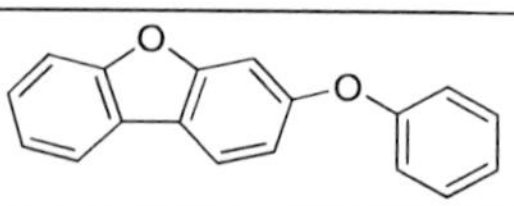	75%	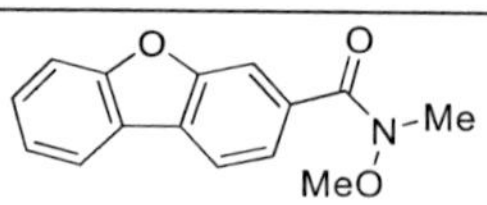	83%

The plausible mechanism indicated that an initially Pd(II)-catalyzed C-H activation of arenes was involved on the assist of the OH group of alcohols 11 followed by oxidation of [Pd(II)-OR] species 13 to a higher oxidation state (Scheme 2). Then C-O reductive elimination took place to furnish the final products, which also released the active Pd(II) species for the next cycle. Base would be beneficial to this transformation due to promoting the generation of the essential intermediate [Pd(II)-OR] species 13.

Lately, Liu and his coworkers have developed a practical palladium catalyzed phenol-directed intramolecular C-H activation/C-O coupling for the synthesis of dibenzofurans (Table 2) [27]. Although the Yu group has reported a similar protocol on the route to access dihydrobenzofurans by using the strong oxidant PhI(OAc)2, however, the method was not applicable due to the decomposition of phenol under such oxidative conditions. Accidently, they found that they got a 23% yield with target product 16 without inert gas protection, indicating air was an efficient oxidant in the transformation. The treatment of biphenyl phenol 15 with stoichiometric amount of Pd(OPiv)2 at 90°C for 15 min resulted a white crystalline solid. X-ray diffraction analysis revealed that Pd(II)-species was four-coordinated and carried an anionic pivalate ligand, which is consistent with the previous findings that anionic ligand could act as a proton shuttle during the C-H activation process. This method featured a broad substrate scope. Moreover, a range of functional groups including ketal, silyl, ether, cyano, ketone, amide and so on were well tolerated under the reaction condition. The basic mechanism was implemented and indicated that the turnover-limiting step was C-O reductive elimination instead of C-H activation.

Table 3. Palladium-catalyzed annulation between phenol and internal alkyne

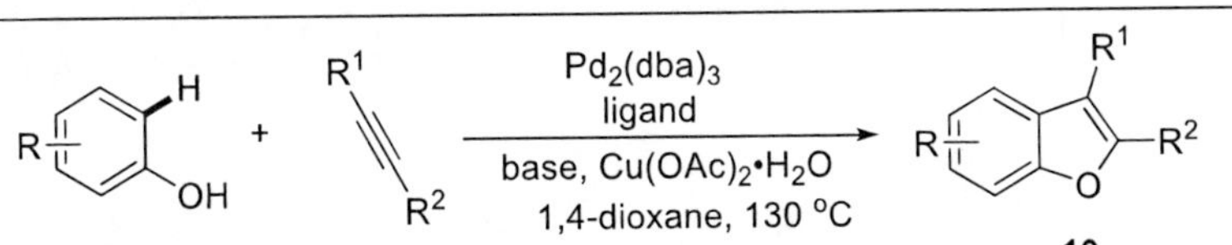

17 **18** **19**

Product	Yield	Product	Yield
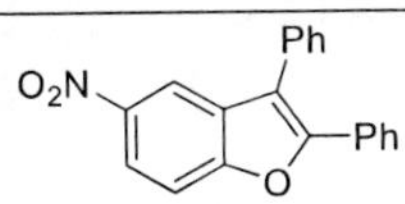	92%	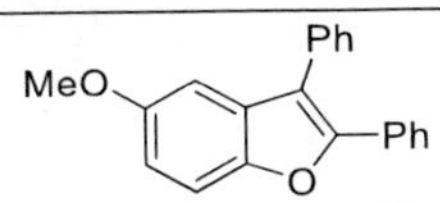	88%
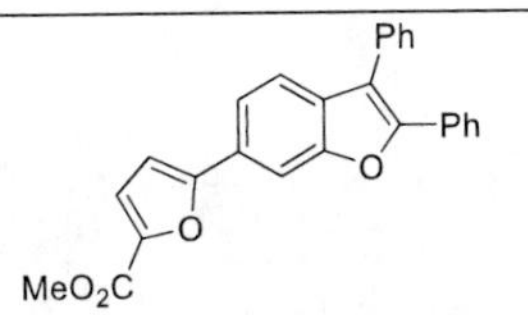	91%	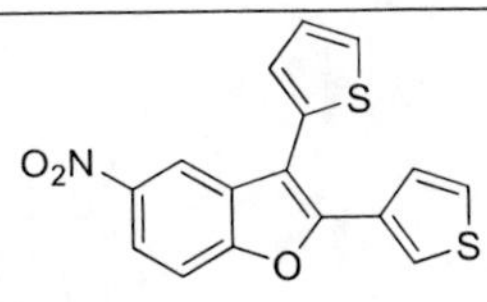	55%
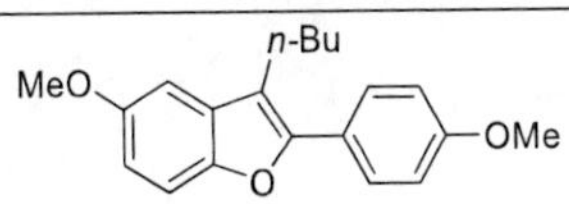	68%	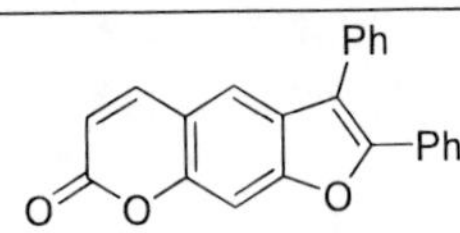	53%

In 2013, the Sahoo group reported a direct access to benzo[b]furans 19 through palladium-catalyzed oxidative annulation of phenols 17 and internal alkynes 18 (Table 3) [28]. The transition-metal-catalyzed annulations between aniline and alkynes providing indole and its derivatives have been extensive studied, however, the similar one step synthesis of benzofurans from readily available phenols and unactivated alkynes has remained elusive because of the following challenges: 1) the four-membered oxygen-containing metallacycle was unfavorable during the reaction process; 2) the formation of C-O bond from Pd(II) intermediate via reductive elimination was difficult; 3) phenols were sensitive to the strong oxidant. Sahoo group successfully furnished the highly substituted benzofurans on the condition of Pd2(dba)3 as the catalyst, bathophenanthroline as the ligand AgOAc as the base. They found that Pd(0) catalysts and the N-bearing bidented ligands are pivotal to this transformation. The method showed a good functional group tolerance. Moreover, alkyl- and aryl-substituted alkyne were also compatible for this transformation, giving good to moderate yields of the product. Pharmaceutically phenol substrates have a positive result under the optimal reaction conditions.

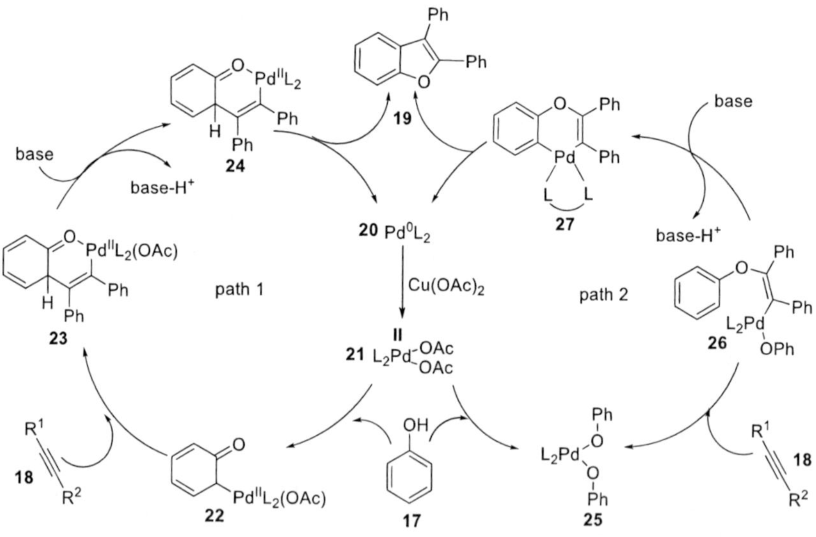

Scheme 3. Palladium-catalyzed annulation between phenol and internal alkyne.

A plausible mechanism was presented based on the Pd(0)/Pd(II) catalytic cycle (Scheme 3). The complex of the N-bearing bidentate ligand coordinating to Pd(0) catalyst 20 followed by oxidized to Pd(II) species 21 in the presence of oxidant Cu(OAc)2, two different ways take place: (1) C-palladation (path 1); (2) O-palladation (path 2). The C-Palladation formed a quinone-type moiety 22, then alkyne 18 was inserted into the C-Pd bond to generate intermediate 23 followed by a subsequent base-mediated aromatization. The final product 19 and Pd(0) species 20 were generated due to reductive elimination. In path 2, phenoxide Pd species 25 was generated on the phenol moiety through the electrophilic palladation, which reacts with alkyne 18 to achieve 26. Then the further cyclization occurred under the assist of the base. The desired benzofuran 19 was obtained via reductive elimination of cyclized intermediate 27.

As known, traditional methods for the synthesis of benzofurans relies on palladium-catalyzed tandem Sonagashira coupling/5-endo-dig cyclization starting from alkynes and *o*-halophenols [29-32]. However, simple protocols of applying phenols without halogen and simple alkynes as starting material to directly prepare such molecules remains rare [33, 34]. In 2011, Wang group reported a base-promoted addition reaction of phenols 28 with bromoalkynes 29 to generate (Z)-2-bromovinyl phenyl ethers 32 in high yields under metal-free reaction conditions (Table 4) [35]. Then, the obtained product could be transformed into the corresponding 2-substituted benzofurans 30 through using Pd(II)- or Pd(0)-catalyzed C-H bond functionalization which PdCl2 displayed a higher reactivity in the reaction. The base also played an essential role in annulation reaction and K2CO3 was the best one. Under the optional reaction conditions, 2-substituted benzofurans 30 were smoothly generated via the consecutive addition reaction/direct C-H bond functionalization. It was also worth mentioning that the transformation of phenols 28 with bromoalkynes into 2-submitted benzofurans 30 could be achieved in one pot with a simple procedure, providing an attractive alternative to Sonagashira coupling/ 5-endo-dig cyclization without use of ortho-halogen-substituted phenols.

Table 4. Palladium-catalyzed one-pot synthesis of benzofurans

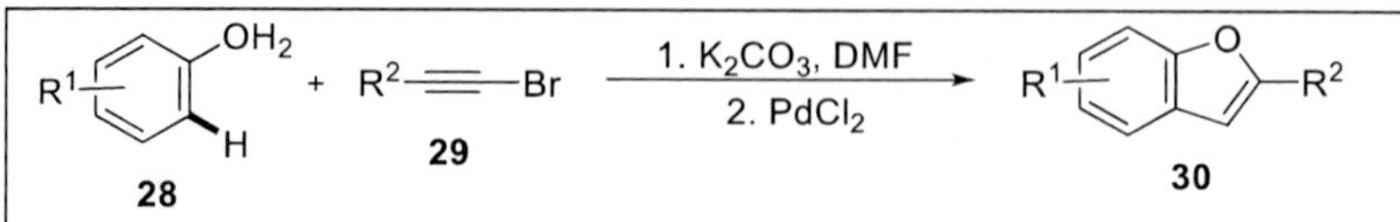

Product	Yield	Product	Yield
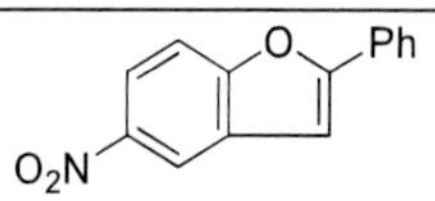	91%	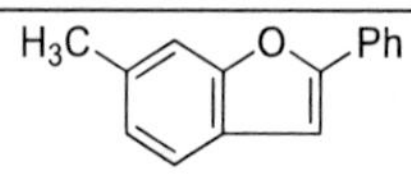	65%
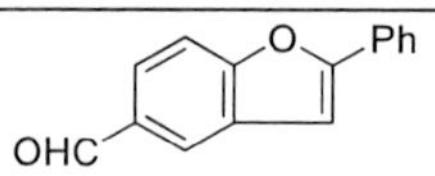	85%	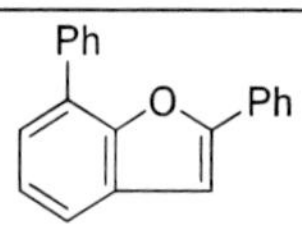	76%
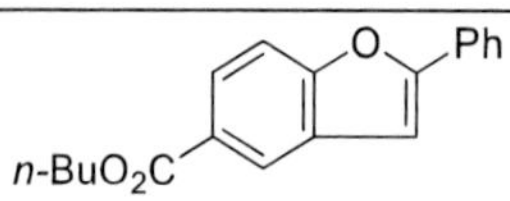	83%	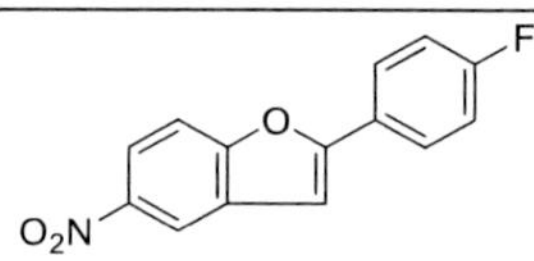	71%
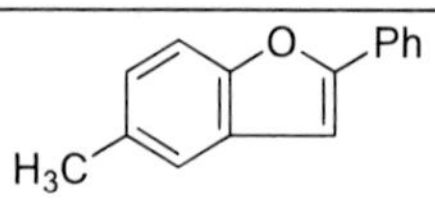	58%	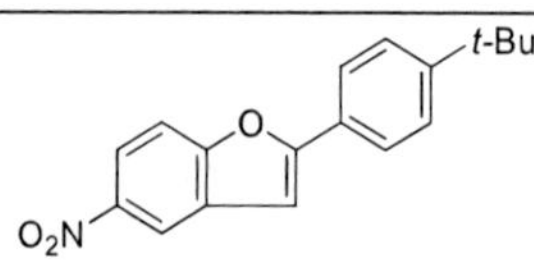	90%

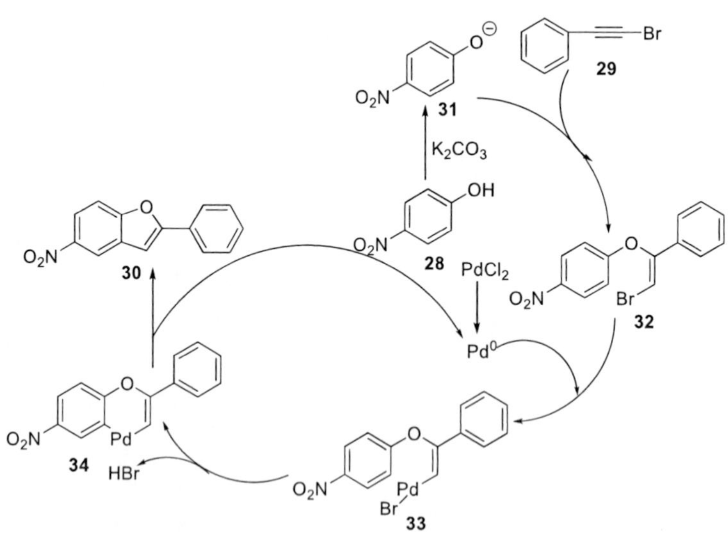

Scheme 4. Palladium-catalyzed one-pot synthesis of benzofurans.

The reaction occurs involving an intermolecular addition reaction and Pd-catalyzed intramolecular cyclization of bromovinyl phenyl ether 32 via C-H activation (Scheme 4). The base promoted intermolecular nucleophilic addition of phenol 28 to phenylethynyl bromide 29, generating (Z)-2-bromovinyl phenyl ether 32. The following oxidative addition took place to get intermediate 33 to react with a palladium catalyst. Subsequently, intermediate 33 underwent an intramolecular electrophilic palladation to furnish the essential intermediate 34 through C-H activation, which followed by the reductive elimination process to afford benzofuran 30 and regenerate palladium catalyst for the next catalytic cycle.

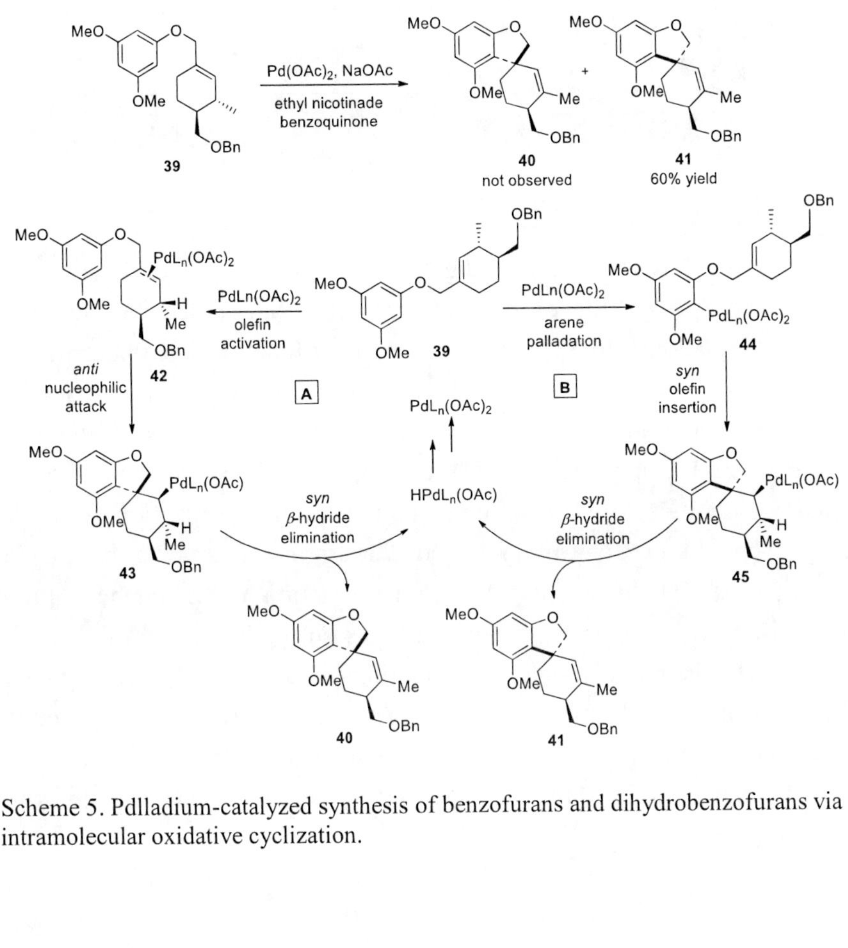

Scheme 5. Pdlladium-catalyzed synthesis of benzofurans and dihydrobenzofurans via intramolecular oxidative cyclization.

Table 5. Palladium-catalyzed synthesis of benzofurans via intramolecular oxidative cyclization

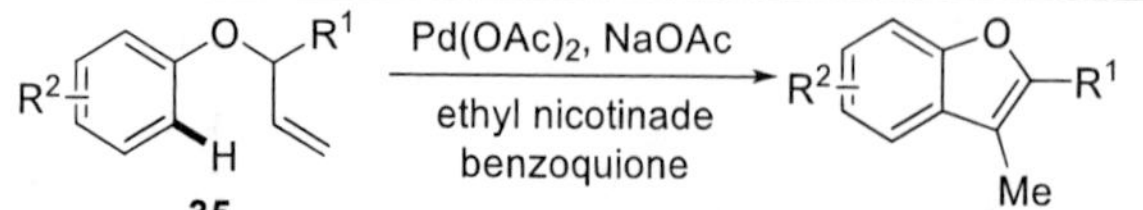

35 36

Product	Yield	Product	Yield
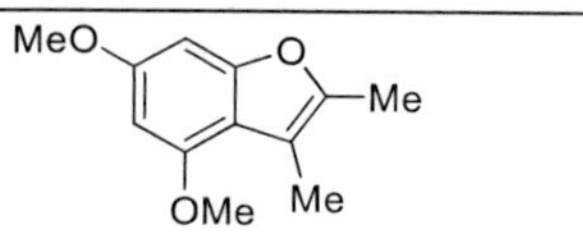	77%	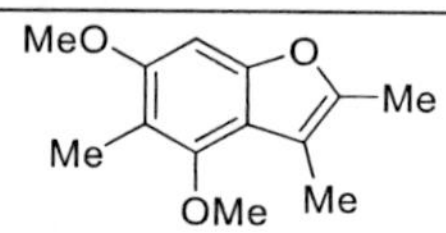	75%
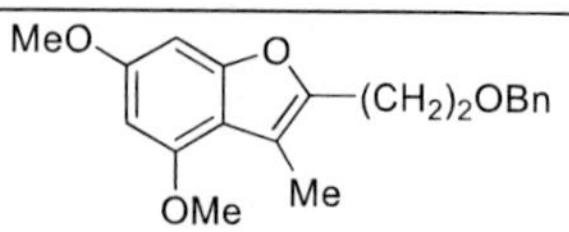	62%	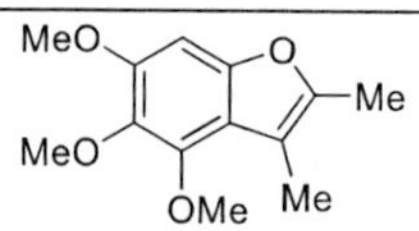	61%
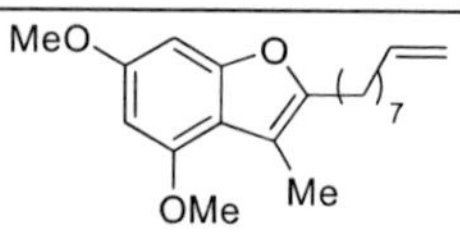	54%	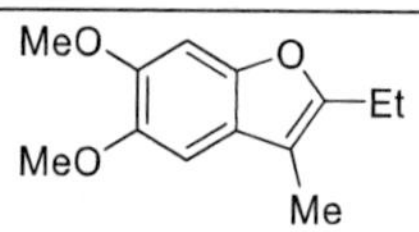	56%

On the basis of palladium-catalyzed oxidative cyclization of unsaturated indoles, the Stoltz group reported Intramolecular Fujiwara-Moritani arylations [36-38] for the synthesis of functionalized benzofurans 36 and dihydrobenzofurans 38 via direct intramolecular oxidative arene/olefin coupling reactions in 2004 (Table 5 and Table 6) [39]. They prepared allyl phenyl ether substrate 35, which was subjected to the reaction conditions for the aforementioned indole cyclization [40], the target benzofuran 36 was obtained during a series of experimental investigations. The generality of the transformation to produce benzofuran showed that the scope of methodology was currently limited to electron-rich arenes. The allylic moiety was functionalized at both the proximal and distal positions with various groups. The quaternary carbon-containing dihydrobenzofurans 38 could also be achieved on the use of tri- and tetrasubstituted olefins under the reaction conditions in the process to avoid olefin isomerization.

Table 6. Palladium-catalyzed synthesis of dihydrobenzofurans via intramolecular oxidative cyclization

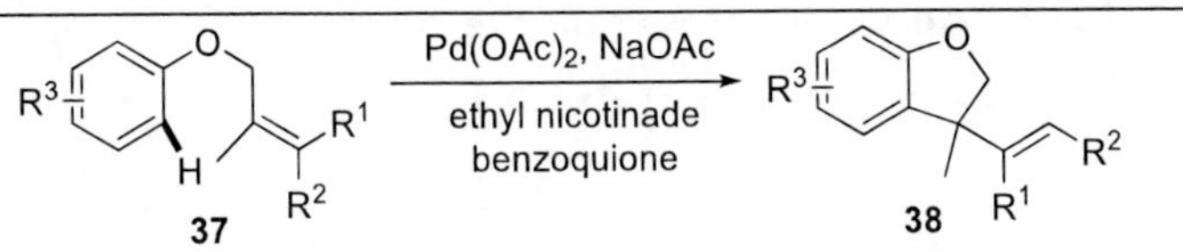

Product	Yield	Product	Yield
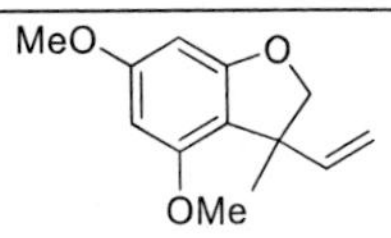	74%	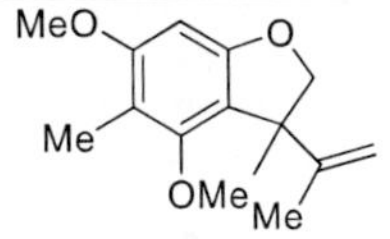	63%
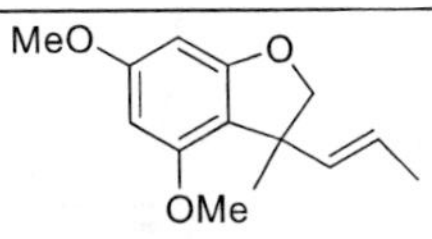	58%	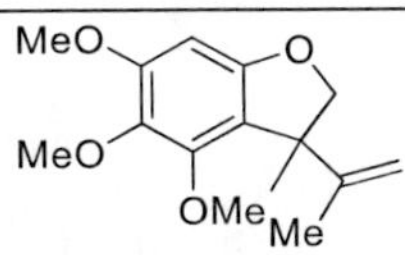	66%
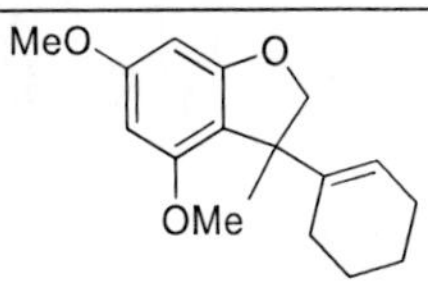	55%	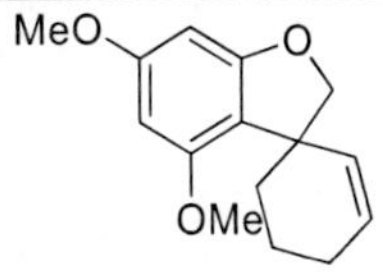	80%

In order to probe the corresponding mechanism, they prepared substrate 39 to distinguish between two mechanistically distinct pathways (Scheme 5). The dihydrobenzofuran 41 was produced after cyclization of 39, which suggested that pathway B was more possible and C-H bond functionalization preceded olefin insertion and β-hydride elimination.

In 2005, the Youn group developed a facile conversion of allyl aryl ethers 46 for the construction of the benzofurans 47 via Pd(II)-catalyzed oxidative cyclization (Table 7) [41]. In addition to Pd(II)-catalysts, Pd(PhCN)2Cl2 was promoted efficiently for this reaction, while 1,4-benzoquinone was the oxidant. The addition of base would promote the cyclization with Na2CO3 providing optimized yield. This method was compatible with a range of functional groups like methoxy and free hydroxyl. It should be mentioned that a higher catalyst loading and increased temperatures were required for relatively electron-deficient arenes, while taryl crotyl ether was used as the substrate, the desired benzofuran was yielded along with a small amount of chromene, which underwent 6-endo cyclization of the claisen derived 2-(R-methylallyl)phenol.

Table 7. Palladium-catalyzed oxidative synthesis of 2-substituted benzofurans

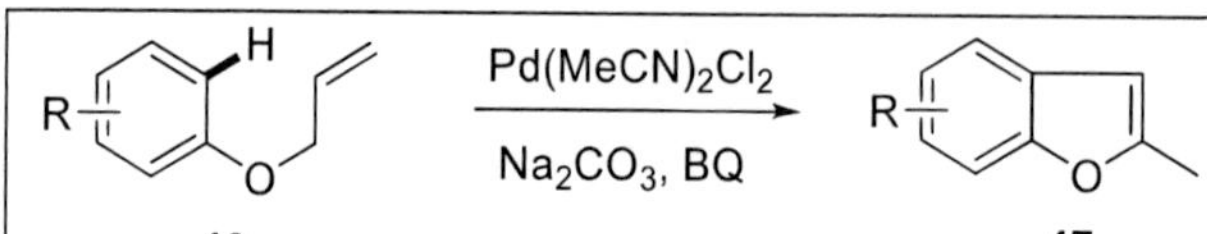

46 → 47

Product	Yield	Product	Yield
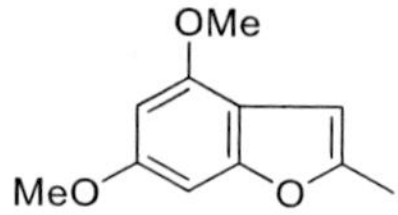	60%	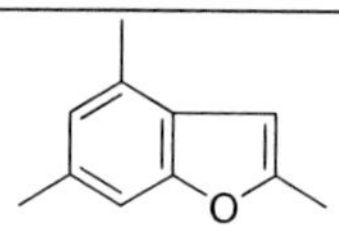	49%
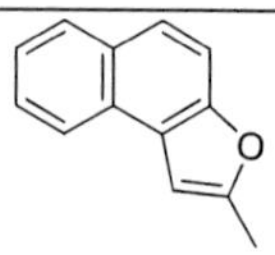	57%	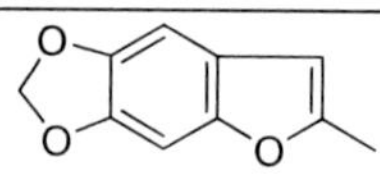	62%
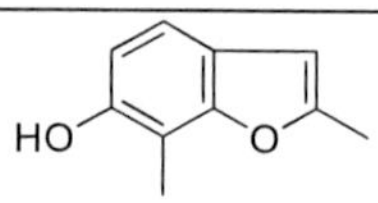	49%	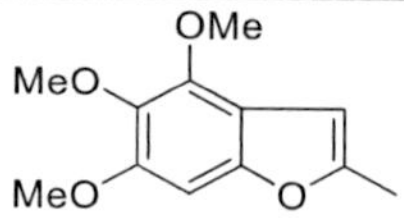	66%

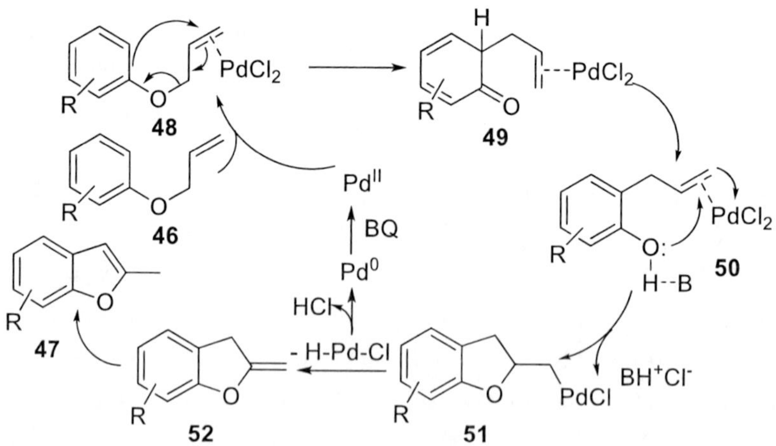

Scheme 6. Palladium-catalyzed oxidative synthesis of 2-substituted benzofurans.

For the mechanism, the Pd-complexed olefin 46 underwent claisen rearrangement to form the corresponding 2-allylphenol intermediate 48, followed by intramolecular annulation via intramolecular nucleophilic attack of the phenolic oxygen promoted by the palladium catalyst combination of a base (Scheme 6). Subsequently, β-H elimination produced the 2.3-dihydro-2- methylene benzofuran 52, which isomerized the final 2-methylbenzofuran 47. And the Pd(0)-species were reoxidized as the active Pd(II)-catalyst by BQ.

Table 8. Palladium-catalyzed addition of arylboronic acids to nitriles

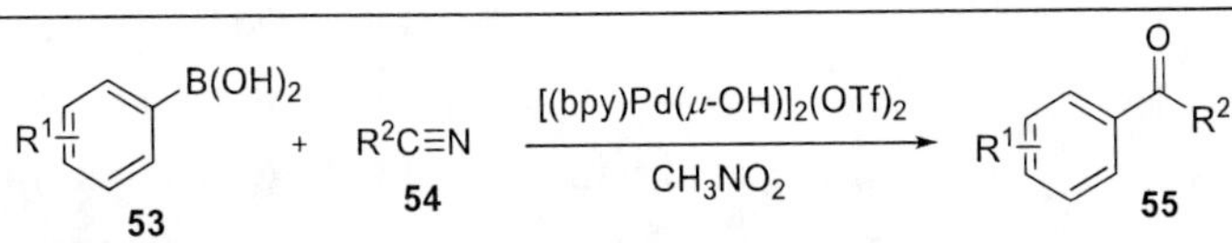

Product	Yield	Product	Yield
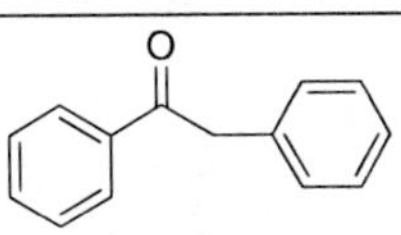	91%	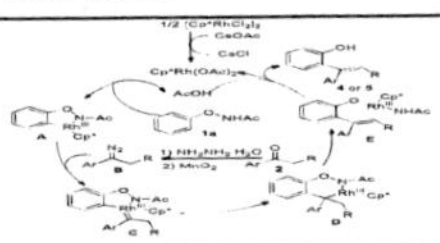	78%
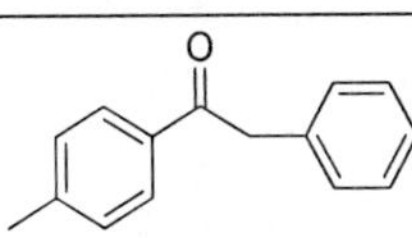	86%	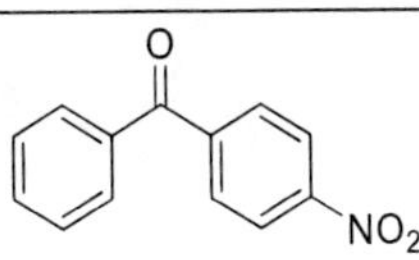	86%
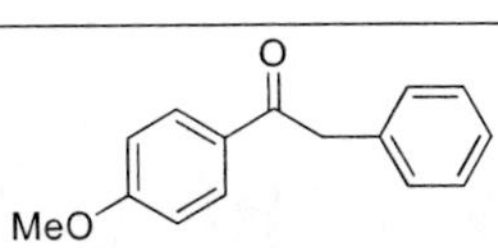	87%	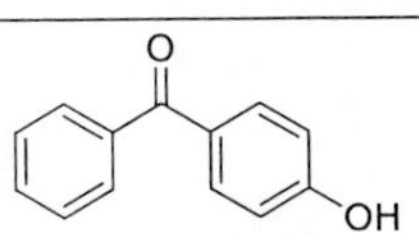	91%
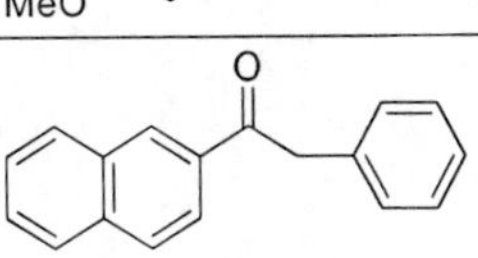	89%	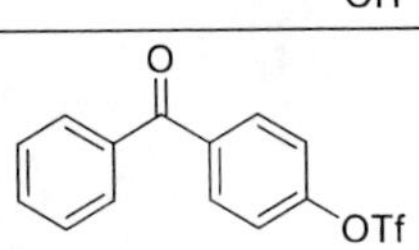	91%

In the organic synthesis area, the addition of carbon-metal species to carbon-heteroatom multiple bonds, such as nitrile group was an important reaction related to C-C bond construction [42, 43]. In 2006, the Lu group reported a one-step synthesis of benzofurans 57 from phenoxyacetonitriles 56 via cationic Pd(II)-catalyzed addition of arylboronic acids 53 to nitriles 54 (Table 8) [44]. In general, nitriles are stable to metal catalysts and

CH3CN or benzonitrile could be applied as the solvent in the transition-metal-catalyzed transformations. The Lu group has previously reported several additional transformations of palladium species to the nitrile groups. In their findings, 2.2-bipyridine (bpy) was crucial for the reactions, probably due to the increased stabilization and reactivity of the arylpalladium species [45, 46]. After the investigation of the optimized reaction conditions, the cationic palladium species [(bpy)Pd+(í-OH)]2(-OTf)2 was effective under air with the addition of PhB(OH)2 to phenylacetonitrile.

Then they synthesized a range of benzofurans 57 from the addition of arylboronic acids 53 to phenoxyacetonitriles 56 under this synthetic protocol. Various 3-substituted benzofurans 57 were obtained in medium to good yield (Table 9). In some cases, aryl ketones 58 were isolated, implying that the initial reaction proceeded with the addition of arylboronic acids 53 to nitriles 56. Moreover, another cationic palladium species [(bpy)Pd2+($H_2O)_2$](-OTf)2 was also effective in some transformations.

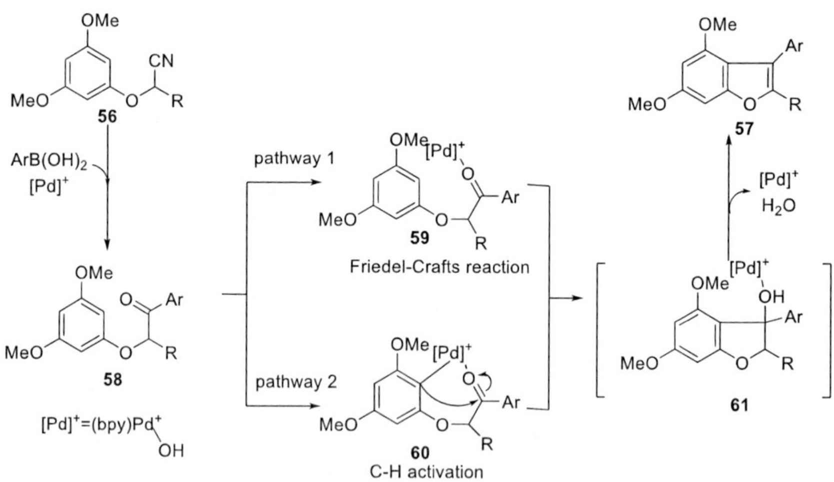

Scheme 7. Formation of benzofurans from 3.5-dimethoxy-phenoxyacetonitriles under the catalysis of palladium.

The mechanism of the synthesis of benzofurans 57 may proceed as follows (Scheme 7): first, PhB(OH)2 was added to the nitrile group of phenoxyacetonitrile 56, forming the corresponding aryl ketones 58 under the assist of the cationic palladium species, which played as a Lewis acid. Then intramolecular Friedel-Crafts reaction occurred to generate the intermediate 61. Finally, benzofuran derivatives 57 were obtained after dehydration. Another pathway involved the palladium(II)-catalyzed C-H activation of the aromatic ring and nucleophilic attack to the carbonyl group was also possible in this transformation.

Table 9. Synthesis of benzofurans from the addition of arylboronic acids to phenoxyacetonitriles

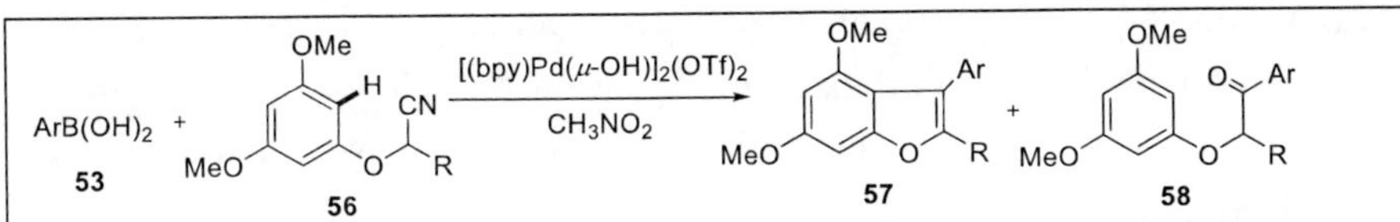

Product	Yield	Product	Yield
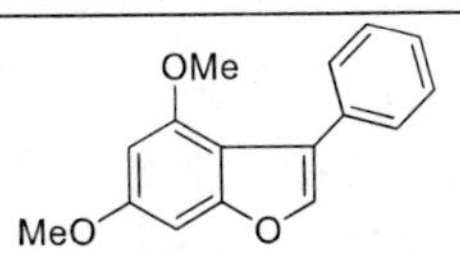	62%	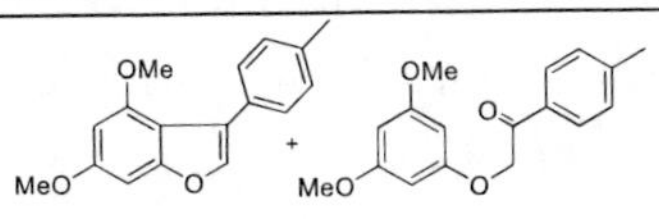	15%+56%
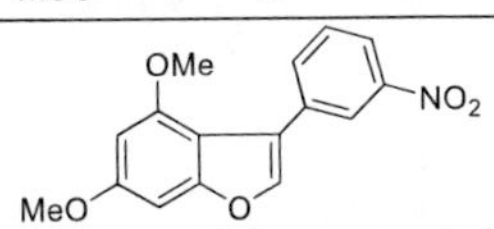	67%	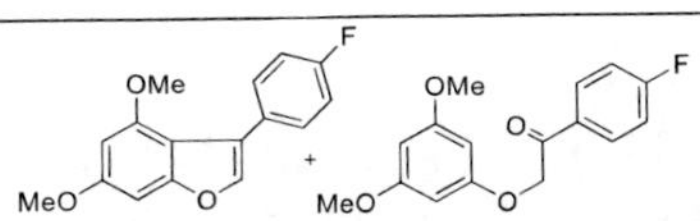	20%+80%
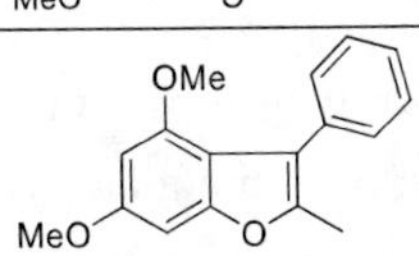	82%	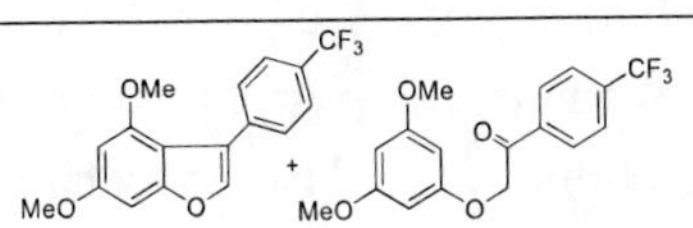	52%+35%
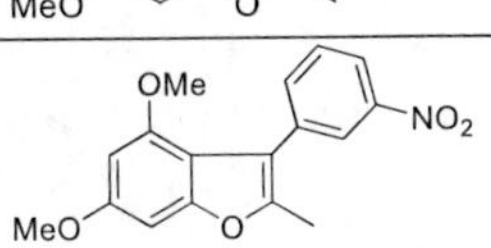	65%	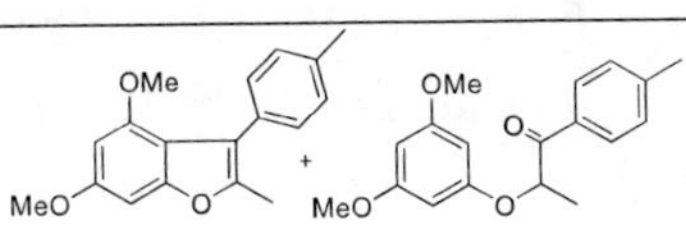	58%+42%

There are two classical approaches to the construction of polysubstituted benzofurans: 1) acylation of 3-arylbenzofurans; [47, 48] 2) the condensation of 2-hydroxybenzophenones with α-bromoarylketones [49-51]. However, the harsh reaction conditions, multistep synthesis of precursors or limited examples restricted their practical application. In 2012, a more efficient method to obtain such polysubstituted benzofurans was reported by the Bertounesque group [52]. They employed 2-aroylbenzofurans 62 as the substrate, which were readily synthesized through Rap-Stoermer condensation on the direct palladium-catalyzed arylation at the 3-position, generating 3-aryl-2-benzoylbenzofurans 64 (Table 10). After the optimized reaction conditions, they found that Pd(OAc)2 combination of P(t-Bu)2Me·HBF4 might efficiently promote the reaction in the presence of K_2CO_3, and catalytical PivOH at 150°C in mesitylene. Without PivOH, the yield of the target product significantly dropped, which confirmed its privotal role as a cocatalyst. The reaction scope of this method was investigated and a variety of functional groups were well tolerated on the aryl halide. The corresponding polysubstituted benzofurans 64 were obtained in good to excellent yields. Both electron-withdrawing and electron-donating aryl bromides 63 were found to be reactive. Other 2-substituted benzofurans 62 were also examined under the optimized conditions with good yields when benzofurans 62 were substituted by electron-withdrawing groups. However, no reaction took place with benzofurans 62 featuring electron-donating groups under the reaction conditions. To overcome the limitation, they adopted another method of KOAc as a base in DMA and the arylation could be accomplished with good yields.

A plausible mechanism has been proposed for direct arylation (Scheme 8). The difference of reactivity between 2-substituted benzofurans 62 featuring electrondonating groups and electron-withdrawing groups ruled out the possibility of a SEAr mechanism and a concerted metalation-deprotonation (CMD) pathway may be possible. Moreover, the addition of PivOH also demonstrated the CMD mechanism, probably due to the fact that the bromine of the complex Ar-Pd-X may be replaced by the pivalate anion during C-H bond cleavage process.

In 2007, the DeBoef group presented an efficient method for synthesizing heterocoupled biaryls via simultaneous C-H bond functionalization of two unfunctionalized arene substrates (Table 11) [53]. The aerobic oxidative coupling of benzofurans 74 with benzene 75 and derivatives exclusively took place at 2-position of bezofurans 74 by applying Pd(OAc)2 as the catalyst. The heteropolymolybdovanadic acid H4PMo11VO40 (HPMV) promoted the rate of reaction with an almost quantitive yield.

Table 10. Palladium-catalyzed direct arylation of 2-substituted benzofurans with aryl bromides

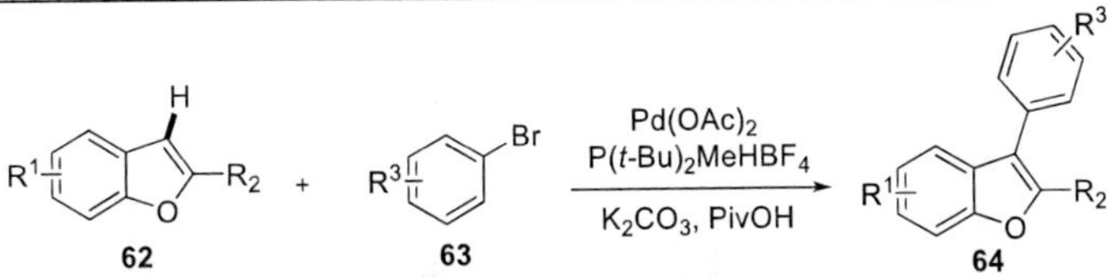

Product	Yield	Product	Yield
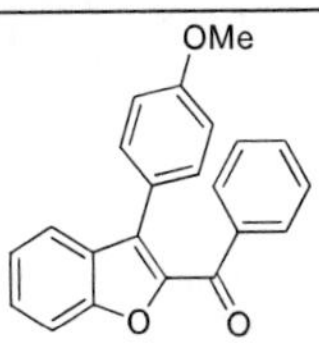	98%	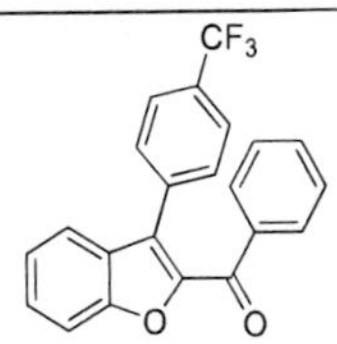	94%
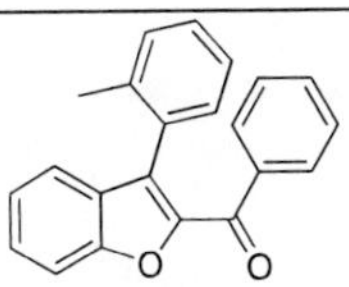	61%	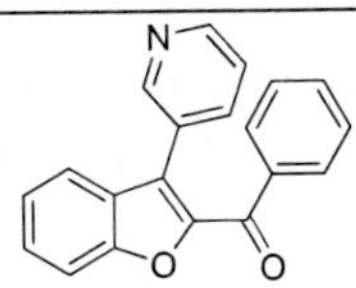	83%
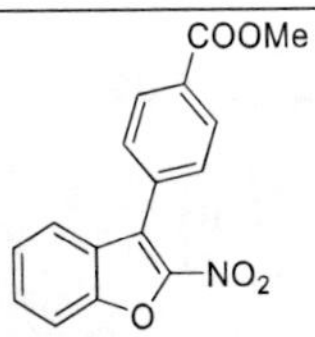	77%	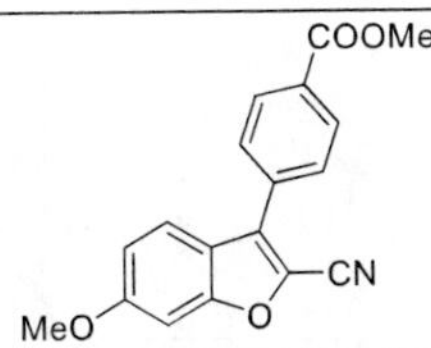	81%
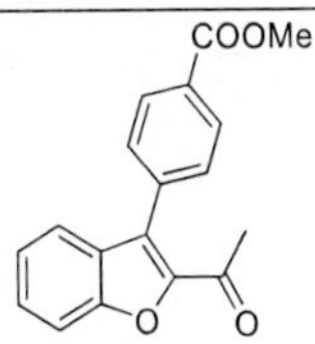	56%	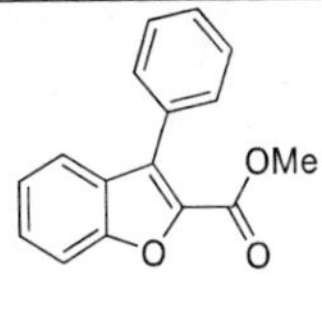	92%

In 2012, the Shi group developed a palladium-catalyzed direct lactonization of 2-arylacetic acids 77 to synthesize fully functionalized benzofuranone derivatives 78 through a reaction sequence which involves

C-H activation and C-O formation process (Table 12) [54]. In their findings, PhI(OAc)2 was a generally effective oxidant for the Pd(II)-Pd(IV) catalytic pathway. The combination of NaOAc, CsOAc and AgOAc was also beneficial to the transformation.

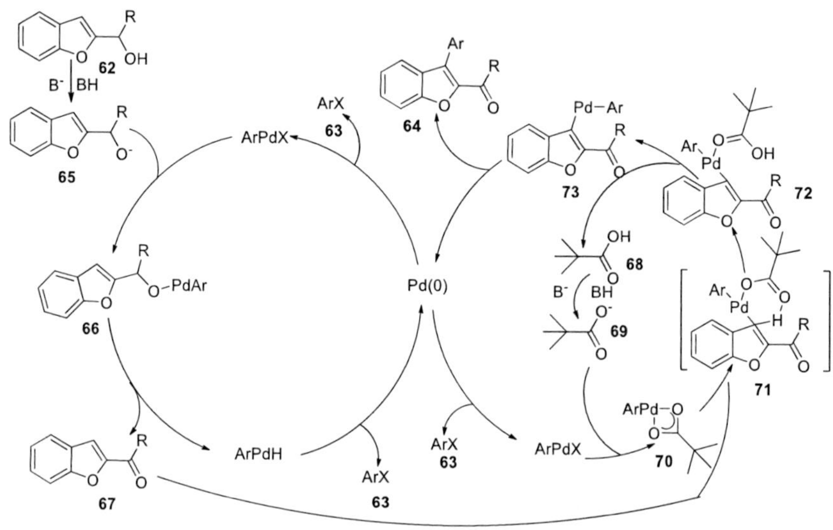

Scheme 8. Palladium-catalyzed direct arylation of 2-substituted benzofurans with aryl bromides.

Table 11. Palladium-catalyzed synthesis of heterocoupled biaryls

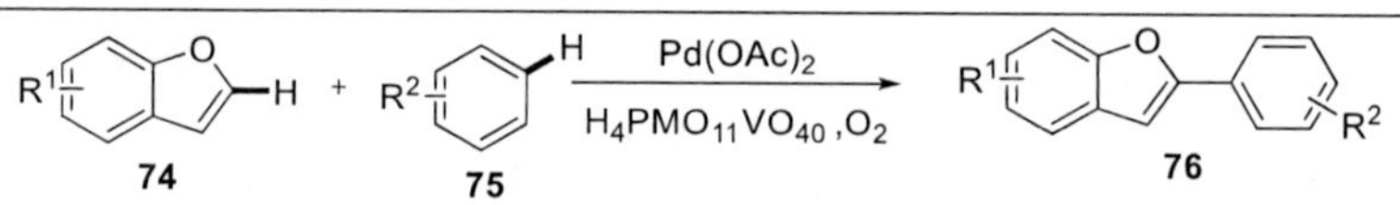

Product	Yield	Product	Yield
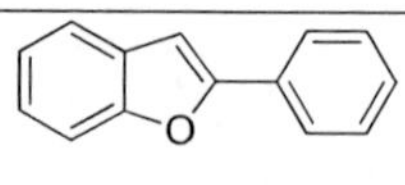	84%	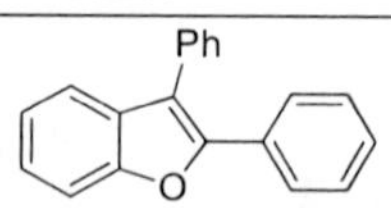	32%
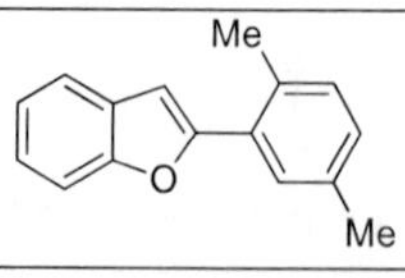	49%	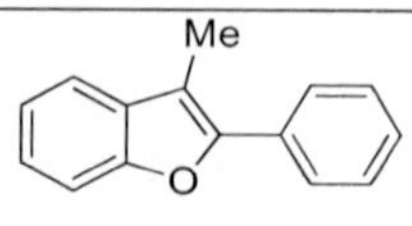	0%

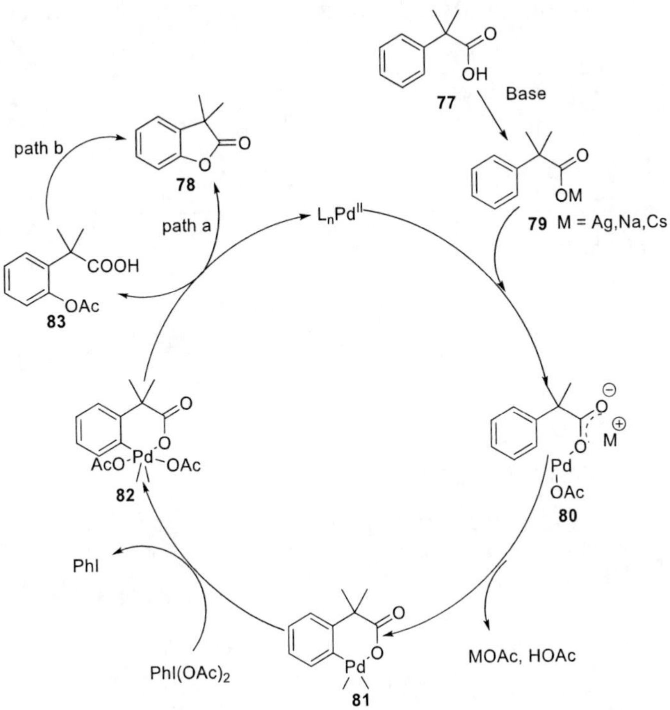

Scheme 9. Palladium-catalyzed direct lactonization of 2-arylacetic acids.

The proposed reaction mechanism was depicted as follows: the Pd species 80 was generated from the carboxy salts 79 and palladium catalyst which was formed through the deprotonation of carboxylic acids (Scheme 9). The concerted metalation deprotonation (CMD) process took place due to the dissociation of C-H bond, and the six-membered palladacycle mediate 81 was achieved. After that, the key Pd(IV) intermediate 82 was generated under the oxidation of the Pd(II) species by PhI(OAc)2. Finally, reductive elimination took place to produce the product 78.

In general, it is well documented that the synthesis of benzofuran synthesis depends on using alkyne as a coupling partner, however, the poor regioselectivity and limited substrate might be the main limitation. The synthesis method involving olefin was a challenging task due to its high chemo- and regioselectivity.

Table 12. Palladium-catalyzed direct lactonization of 2-arylacetic acids

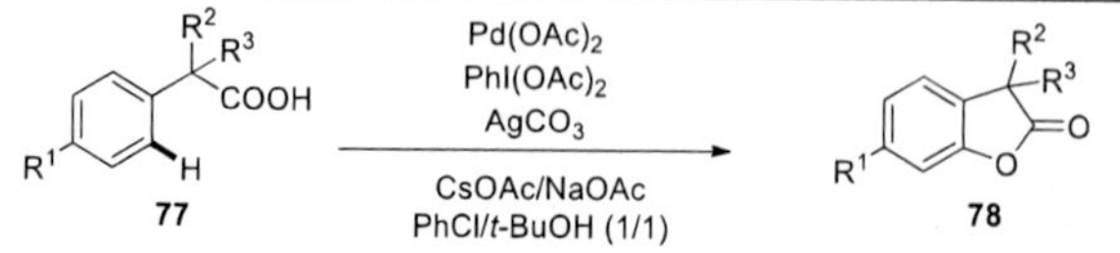

Product	Yield	Product	Yield
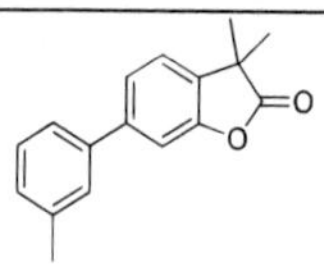	75%	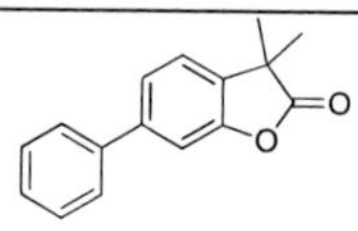	76%
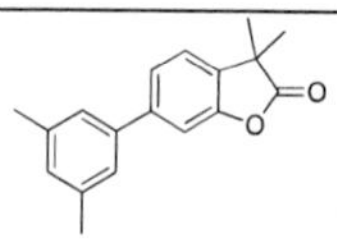	77%	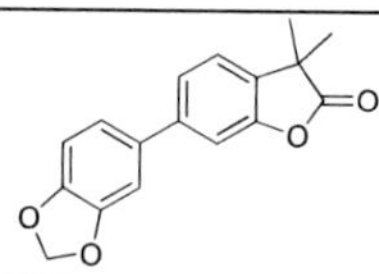	62%
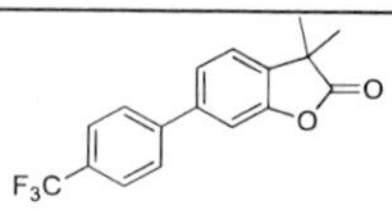	89%	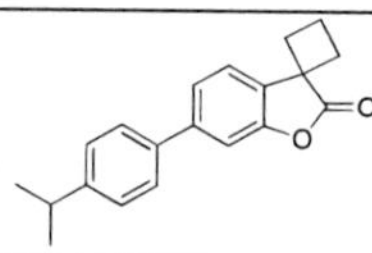	17%
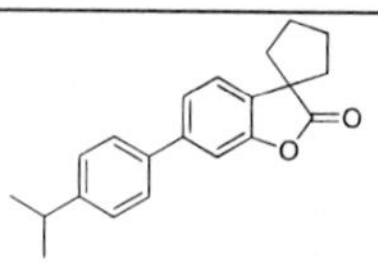	70%	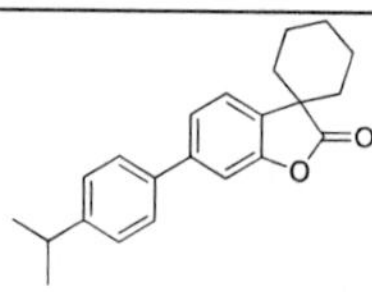	69%
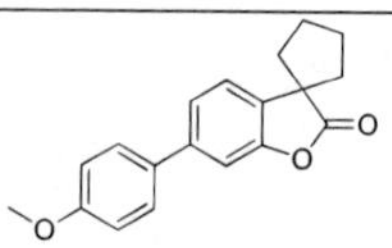	64%	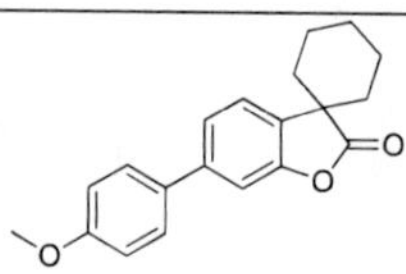	66%

In 2013, the Maiti group reported the first palladium-catalyzed benzofuran 86 synthesis by introducing styrenes as the coupling partner through the C-H functionalization of phenols 84 (Table 13) [55]. In their study, 2-substituted benzofurans 86 were realized by reacting phenol 84 with styrenes 85. The optimized reaction conditions were as shown: Pd(OAc)2 as the catalyst, 1.10-phenanthroline as the ligand, 1.0 equivalent of $Cu(OAc)_2$ was used as the oxidant and NaOAc as the base in DCE at 110°C for 24 hours. Without a base, the yield of the benzofuran 86 decreased significantly in the phenol 84 have a better performance compared to the neutral group. After testing the scope with styrene 85, they also evaluated dif*f*erent terminal and internal aliphatic olefins. For example, the reaction of 2-naphthol with 10-bromodec-1-ene afforded a single product, while in the case of 1-octene, the 2-substituted benzofuran installing an exocyclic double bond at C2 was finally achieved along with the observation of the product (A: B = 1: 1.3). The cyclic olefin was also compatible and resulted in the formation of the benzofuran product. Moreover, 2,3-disubstituted benzofuran was produced with using 2,3-methylenedioxy phenol, where the methyl group was present in the 3-position. The chain length of the olefin substrates did not inhibit the efficiency of the reaction. This efficient method allowed the synthesis of unsymmetrical 2,3-disubstitutedbenzofurans, representing a significant feature.

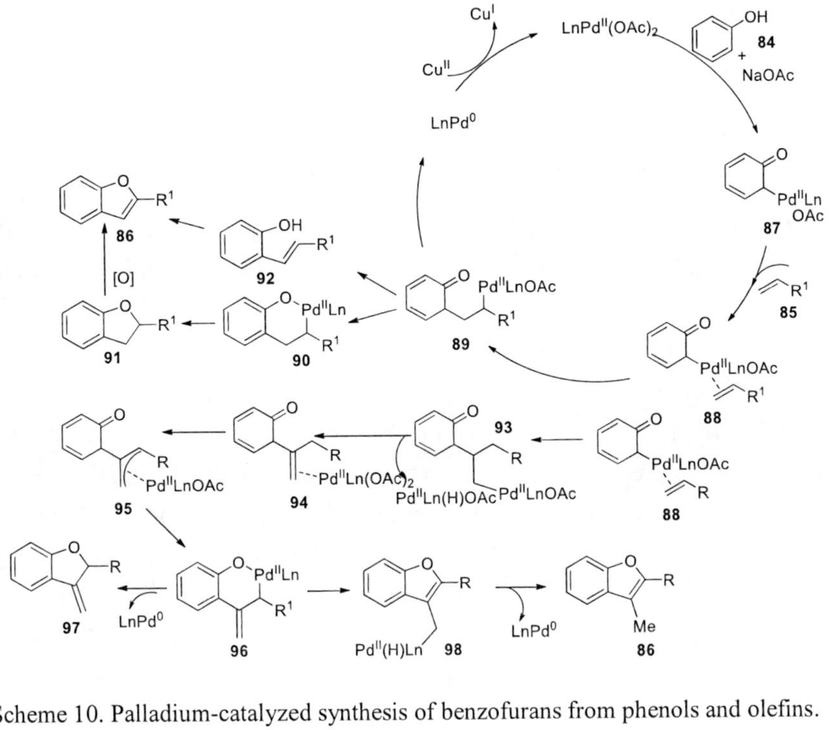

Scheme 10. Palladium-catalyzed synthesis of benzofurans from phenols and olefins.

Table 13. Palladium-catalyzed synthesis of benzofurans from phenols and olefins

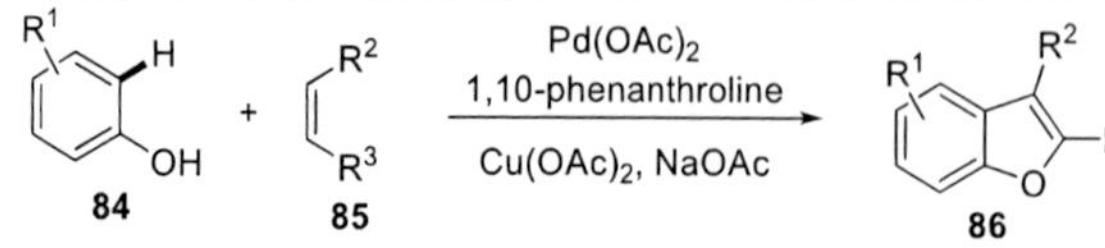

Product	Yield	Product	Yield
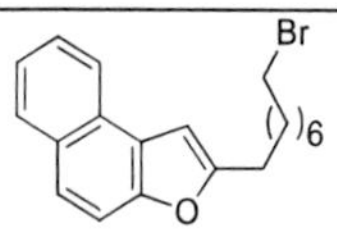	76%	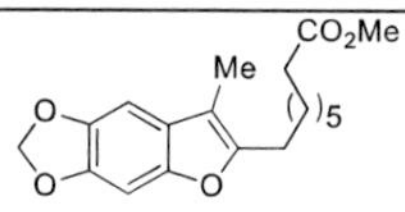	55%
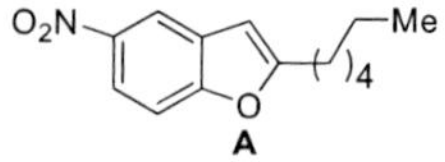 A 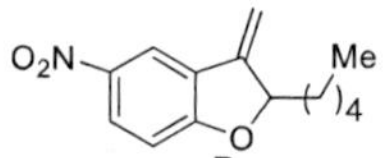 B	72% (A:B=1:1.3)	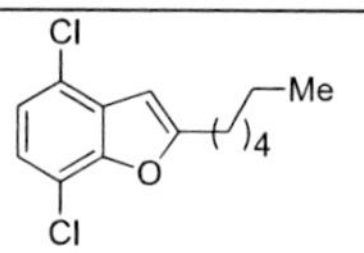	37%
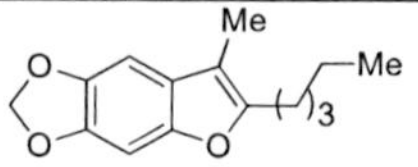	54%	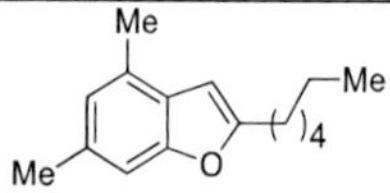	46%

A plausible catalytic cycle mechanism was provided (Scheme 10). First, the ortho palladation of phenol 84 provided the quinone-type moiety 87, which produced the species 88 after interaction with the olefin 85. Subsequently, intermediate 89 could be formed and underwent reductive elimination to lead to 2-hydroxystilbene 92. The essential intermediate 92 might be finally converted into the benzofuran 86 under the reaction conditions and the released Pd(0) residue was regenerated Pd(II)-catalyst through the oxidation of Cu(OAc)2. Alternatively, the final product 86 which can be produced via F. plausible pathways are also depicted in Scheme 10. The pattern of aliphatic olefin 85 was slightly different. Insertion of aliphatic olefin 85 lead to intermediate 93. Then the β-hydride elimination occurred, producing the intermediate 94. The stable complex 95 was generated after the allylic shift of 94, which converted a six-membered cyclic palladium complex 96. As a result, aliphatic olefins 85 could produce disubstituted benzofuran 86 in some cases.

Table 14. Palladium-catalyzed synthesis of 3-substituted benzofurans from phenols and cinnamic acids

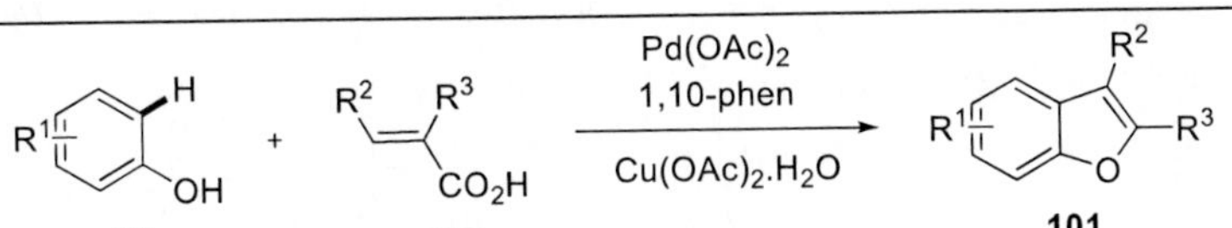

99 **100** **101**

Product	Yield	Product	Yield
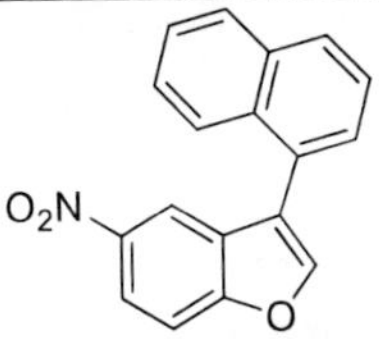	82%	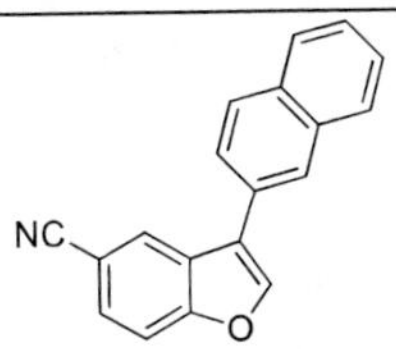	50%
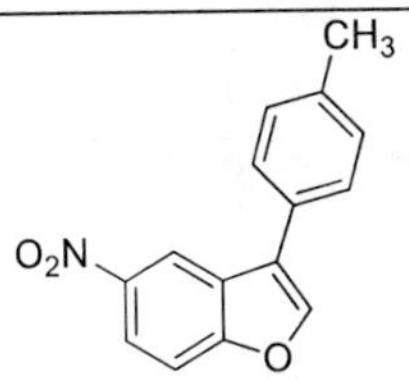	75%	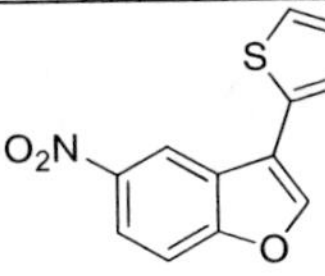	45%
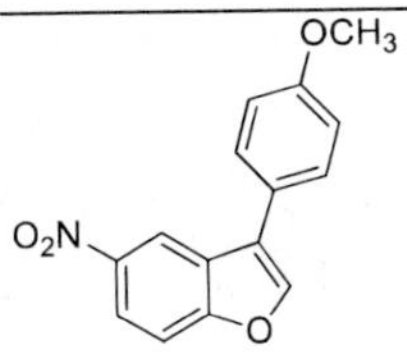	70%	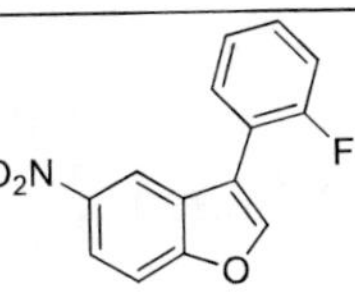	42%
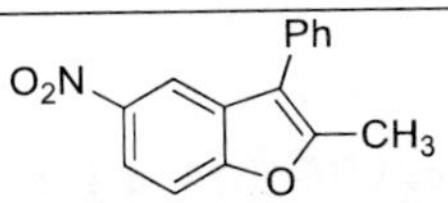	41%	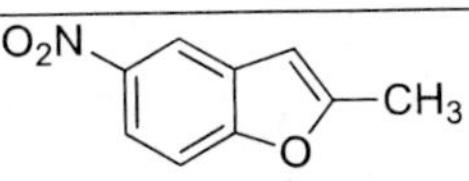	44%

3-Substituted benzofurans represent one of the most prevalent structural motifs in natural products, which endow a range of biological activities [56-58]. In 2015, Maiti and coworkers reported an effective synthetic method of 3-substituted benzofurans 101 in combination with our previous report on 2-substituted benzofuran 86 synthesis (Table 14) [59]. In this regard, they achieved the corresponding products by the treatment of the phenols with cinnamic acids, which the carboxylic acid moiety could play as a traceless directing group [60, 61]. After extensive experimental investigation, 3-arylbenzofurans 101 in DCE were obtained by using phenols 99 and cinnamic acids 100 on the condition of Pd(OAc)2 with 1.10-phenanthrolinium under an oxygen atmosphere. Notably, the products were isolated as the single regioisomeric formation, which the products of 2-arylbenzofurans were not observed, probably due to the fact that carboxylic acid moiety reversed the electronic nature of two carbon atoms across the double bond. The protocol has a good functional tolerance and various substituted phenols 99 and cinnamic acid derivatives 100 were compatible in the transformation. Morever, internal alkenyl carboxylic acids were also tested and the corresponding substituted benzofurans 101 were obtained in moderate yields by employing α,β-unsaturated carboxylic acids as the substrates.

A plausible mechanism of 3-arylated benzofuran 101 synthesis was proposed based on the experimental details (Scheme 11). Mechanistic findings showed that the reaction started from the formation of a quinone type moiety 102 during this process. It is irreversible since no D/H exchange was observed with d5-PhOH. Insertion of cinnamic acids 100 into the C-Pd bond lead to the intermediate 103, which was the rate-determining step in the reaction and the aromatic group would be at the β-position with respect to the Pd-center. The removal of CO2 from intermediate 103 would result in α,α-disubstituted olefin 104, which was isolated and characterized as the essential intermediate. The intermediate 104 would transform into 105 in the presence of Pd species, which underwent β-hydride elimination to provide give the final product 101.

In 2015, the same group reported a similar strategy to synthesize 3-methylene-2.3 dihydrobenzofuran 108 and 2.3-dialkyl benzofuran 109 by replacing the cinnamic acid with α,β-unsaturated alkenyl carboxylic acids 107 (Table 15) [62]. More interestingly, the reaction showed the substituent on the substrate has a profound effect on the product: strongly electron withdrawing substituents such as nitro group and cyano group produced 2,3-dihydrobenzofuran 108; while the keto group and naphthyl furnished 2,3-dialkyl benzofuran 109.

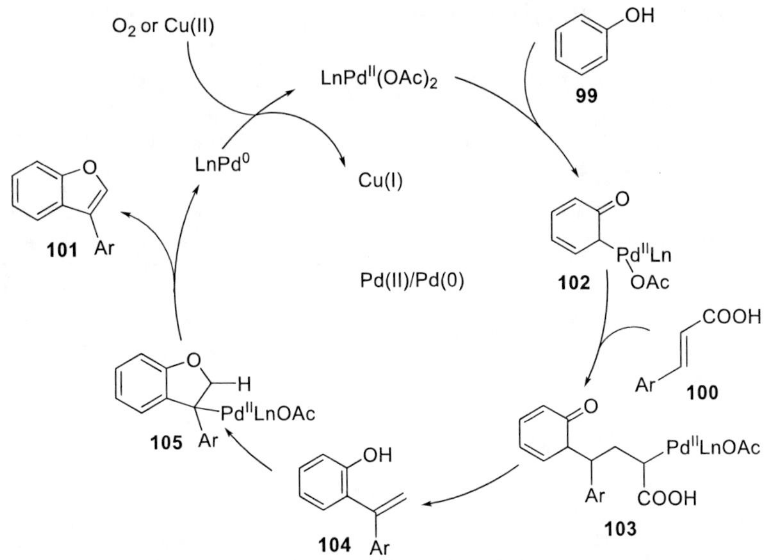

Scheme 11. Palladium-catalyzed synthesis of 3-substituted benzofurans from phenols and cinnamic acids.

Table 15. Palladium-catalyzed synthesis of 2,3-disubstituted benzofurans

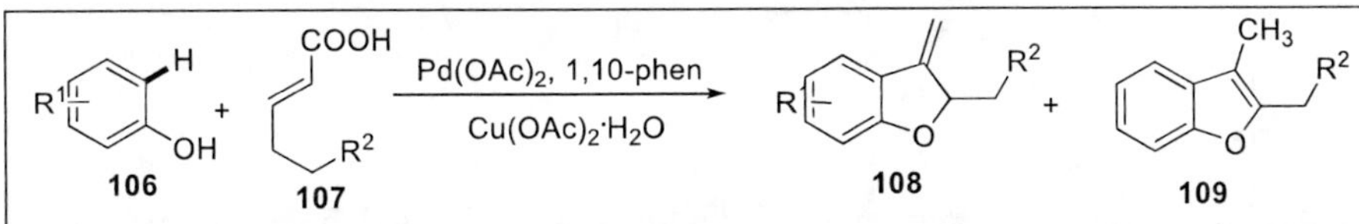

Product	Yield	Product	Yield
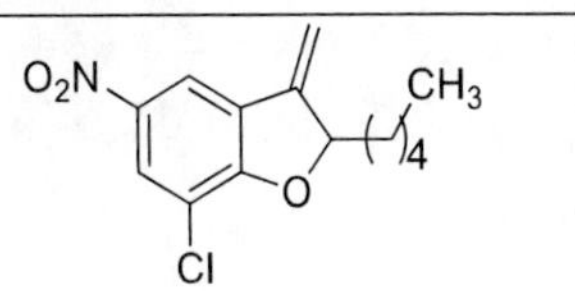	66%	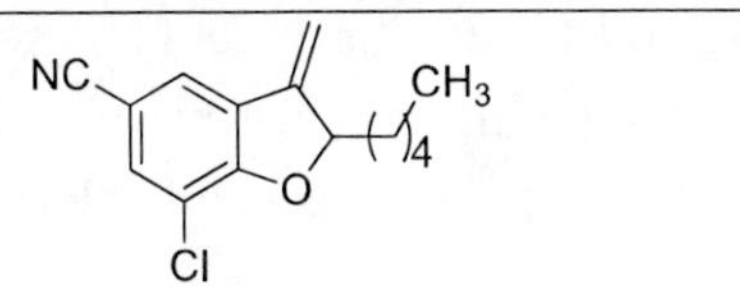	69%
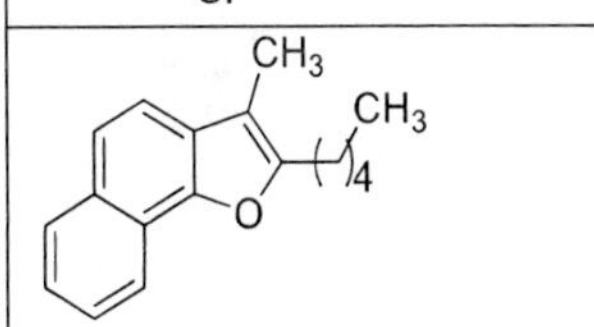	69%	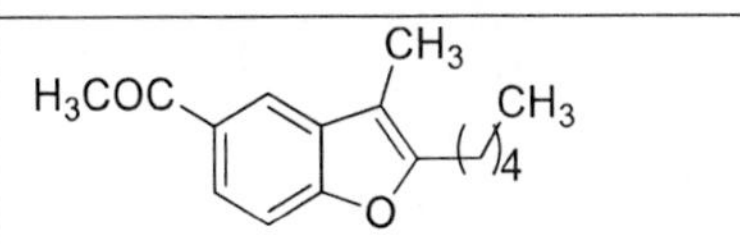	59%

Table 16. Synthesis of deuterated 2.3-dihydro benzofurans

Product	Yield	Product	Yield	Product	Yield
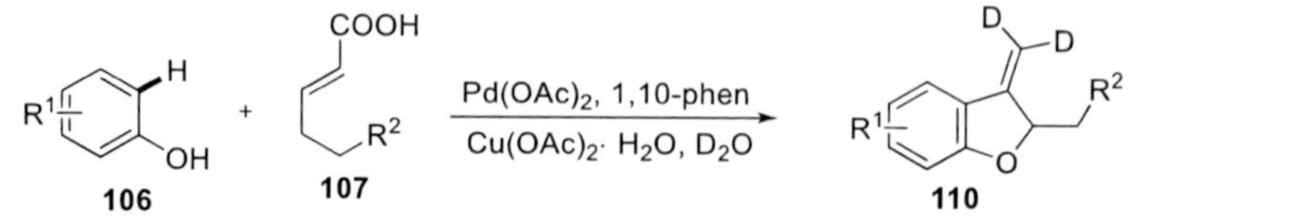					
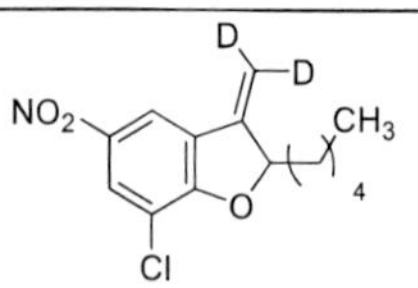	65%	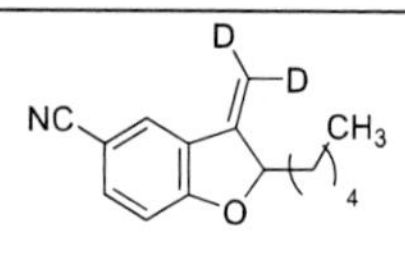	47%	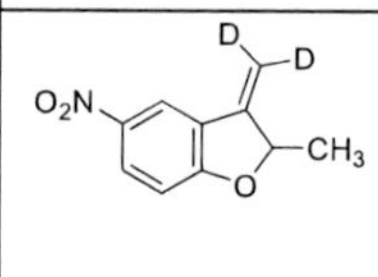	43%
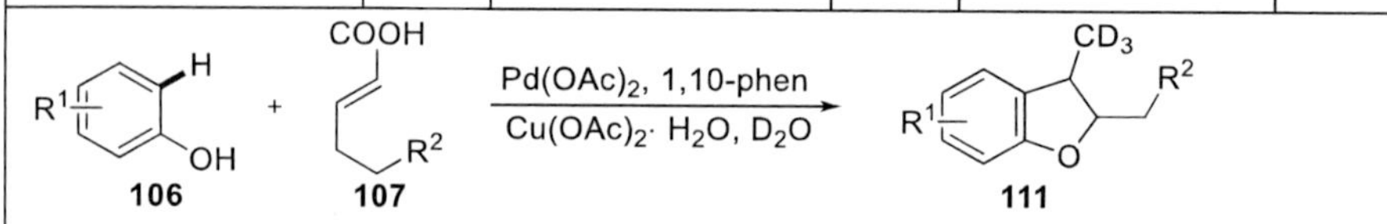					
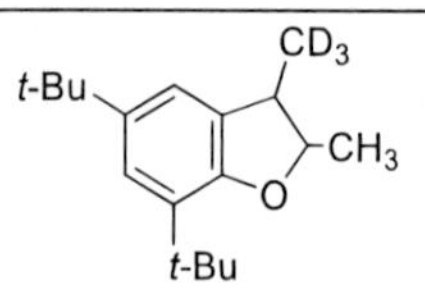	53%	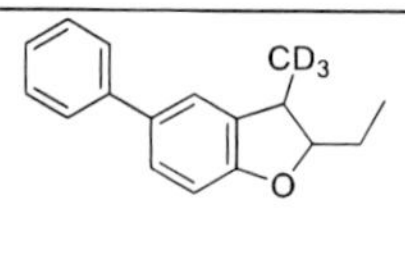	42%	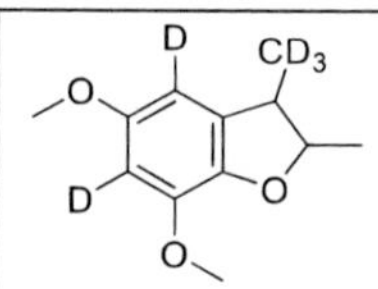	50%
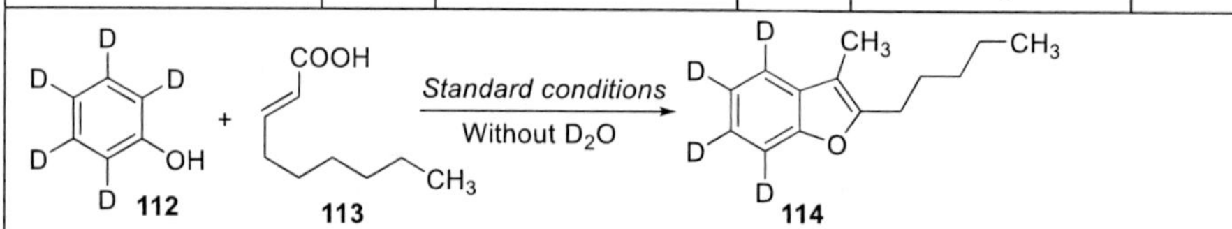					

The maximum percentage of deuterium-substituted benzofurans were accessed in one step by the addition of a stoichiometric amount of D2O (about 0.5 mL) into the reaction system (Table 16). Moreover, non-deuterium scrambling product was observed on the use of PhOH-d5 as a coupling partner in the absence of D2O.

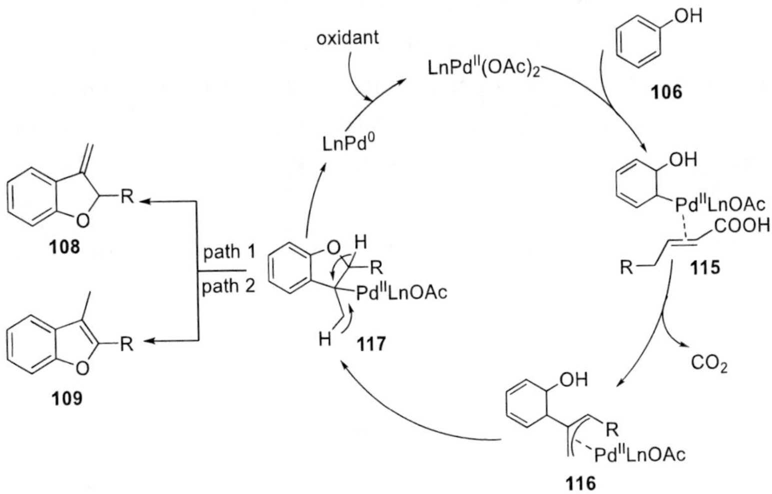

Scheme 12. Palladium-catalyzed synthesis of 2.3-disubstituted benzofurans.

A plausible mechanism of related to 3-methylene-2,3-dihydrobenzofuran 108 and 2.3-disubstituted benzofuran 109 synthesis was proposed based on the experimental observations (Scheme 12). A phenanthroline-palladium(II) complex 115 was formed due to the C-H bond activation at the ortho-position of the phenol 106 followed by the insertion of α,β-unsaturated carboxylic acids 107, which gave the Pd-allyl intermediate 116. Then, intermediate 116 took place the cyclization to form an intermediate 117, which is essential for the formation of 108 and 109. Thus, β-hydride elimination lead to the accesses to the final products along with Pd(0) species.

Rh-Catalyzed Benzofuran Synthesis on the Use of Oxidizing-Directing Group via C–H Functionalizations

The transition-metal-catalyzed oxidative annulation has been developed as a powerful tool for the construction of complex molecular [63-65]. In general, achieving the C-H oxidative annulation in combination with directing group (DG) and stoichiometric or excess amounts of external oxidants [66-68]. Recently, C-H functionalization develops an innovative strategy with oxidizing-directing group (ODG), which acts as both a directing group DG and an internal oxidant [69-72]. In this section, Rh-catalyzed C-H bond functionalizations to construct benzofurans on the use of oxidizing-directing group will be summarized and several cases involving Ir- or Ru-catalyzed reactions were also showcased in the end.

Table 17. Rhodium-catalyzed reaction of *N*-phenoxyacetamides and alkynes under reaction condition A

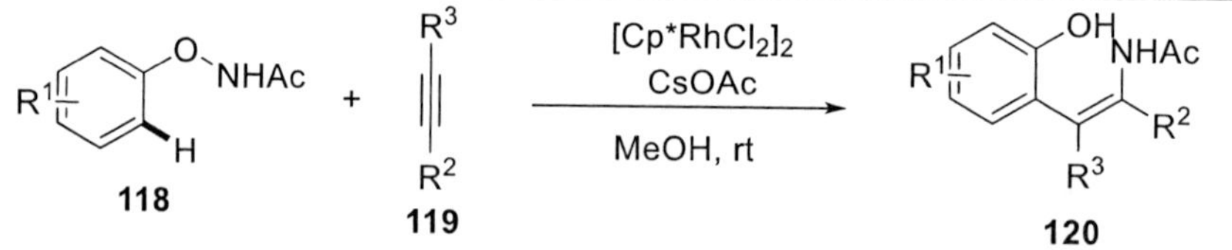

Product	Yield	Product	Yield
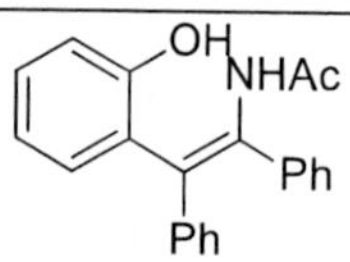	82%	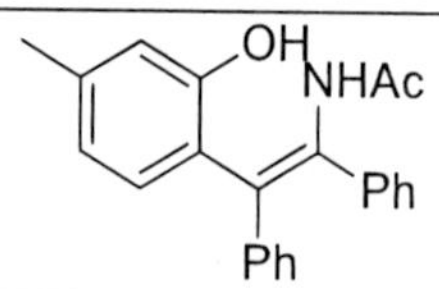	80%
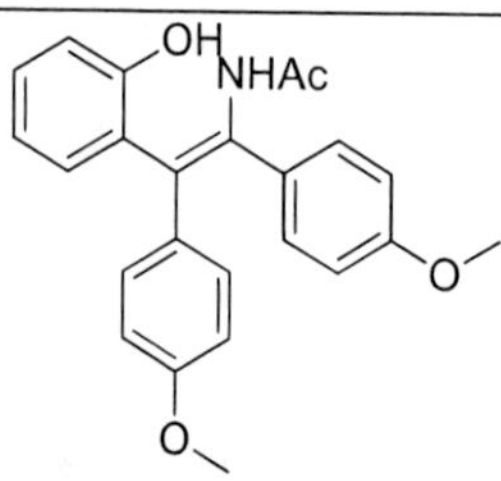	77%	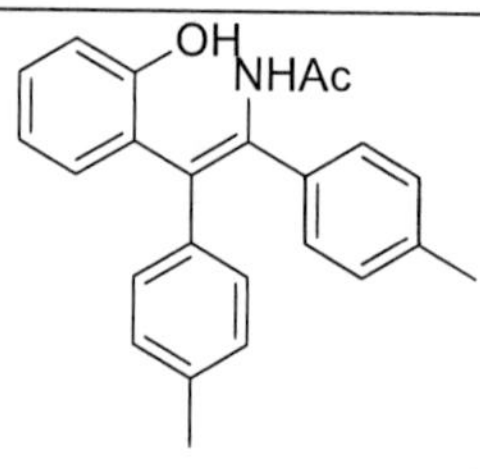	79%
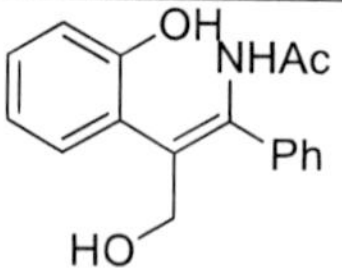	70%	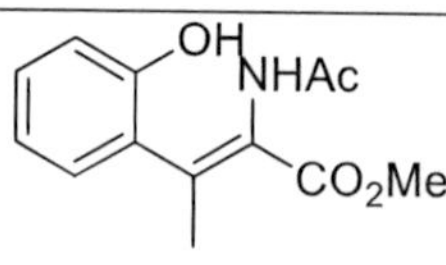	69%

In 2013, Lu and co-workers disclosed the development of a mild rhodium(III)-catalyzed redox-neutral C-H functionalization of N-phenoxyacetamides 118 with alkynes 119 for the synthesis of ortho-hydroxyphenyl-substituted enamides 120, and an unexpected synthesis of valuable benzofuran derivatives 121 through C-C/C-O bond formation was accomplished by simply switching the reaction conditions (Table 17 and Table 18) [73]. Symmetrical diaryl acetylenes bearing various substituents and unsymmetrical disubstituted acetylenes were compatible for this reaction, affording the corresponding product in moderate to good yields. The synthesis for benzofuran 121 under different condition B, employing CH2C12 as solvent and AcOH as an additive. In the case of 3-phenyl-2-propyn-1-ol, a mixture of regioisomers were observed. It was different from a single product 120 was obtained with the same substrate under reaction condition A, using methanol as solvent and CsOAc as an additive.

Table 18. Rhodium-catalyzed reaction of *N*-phenoxyacetamides and alkynes under reaction condition B

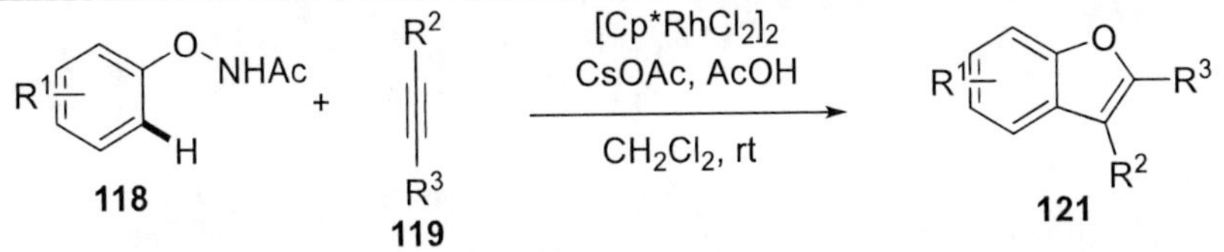

118 119 121

Product	Yield	Product	Yield
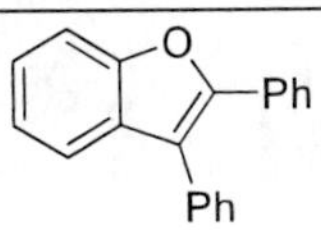	88%	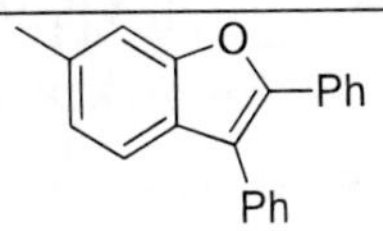	72%
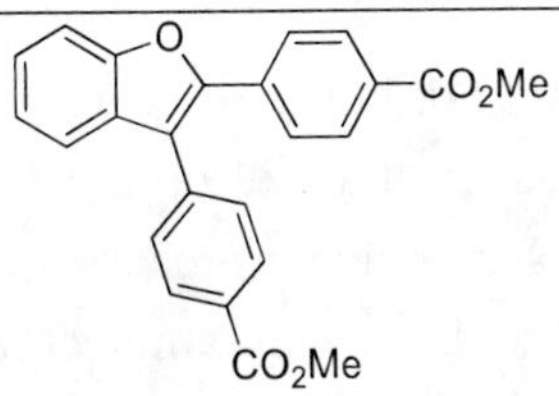	64%	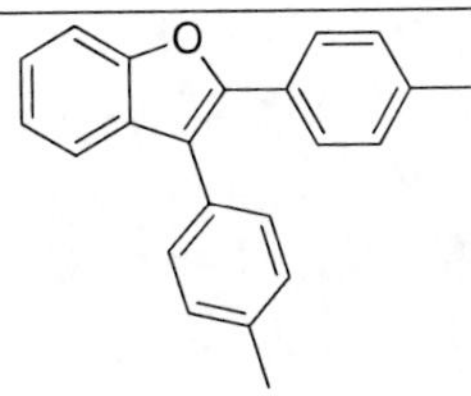	71%
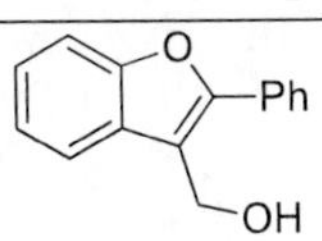 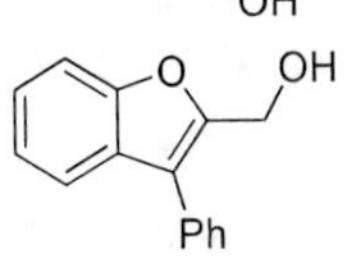	79% (2:1)	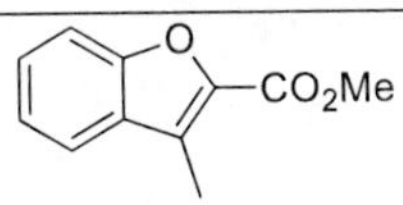	90%

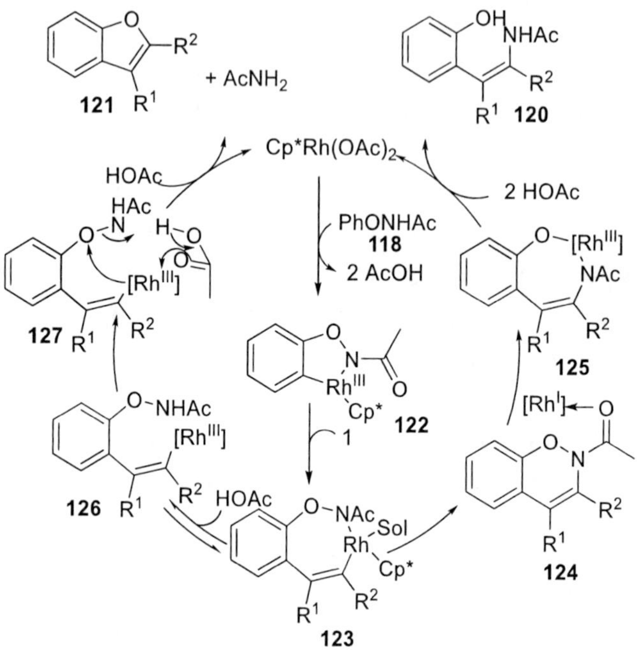

Scheme 13. Rhodium-catalyzed reaction of N-phenoxyacetamides and alkynes.

The authors found that the proposed mechanism, the highly selectivity was controlled mainly by the solvent (Scheme 13). Weakly coordinating solvents, such as methanol, might stabilize 123 by forming an 18-electron species, which could undergo reductive elimination/oxidative addition to deliver the enamide product 120. In noncoordinating or bulky solvents, the N-Rh bond in 123 might be protonated more easily, thus affording the intermediate 126. Since N-phenoxyacetamide 118 is a known precursor of the phenoxemium ion in the presence of acid, an intramolecular substitution might occur to form the C-O bond and break the N-O bond with the aid of acetic acid.

Subsequently, in their previous work, they designed a novel oxidizing directing group for mild Rh-catalyzed C-H functionalizations of N-phenoxyacetamide with alkynes to furnish ortho-alkenyl phenols or benzofuran products selectively (Table 19) [73]. The author and co-workers reported N-phenoxypivalamide 128 with internal alkynes 129 for the synthesis of benzofurans 130 without any external oxidant by using a less expensive ruthenium as catalyst to achieve an analogous but more economic process [74]. Various alkynes 129 bearing different functional groups, such as ester, ketone and amide, were well tolerated in this reaction to give good yields of the expected products 130 with a high regioselectivity. However, N-phenoxypivalamide 128 bearing with electric-withdrawing group resulted in a lower yield. A similar mechanism to that for the C-H oxidative annulation reaction, as described above.

Table 19. Synthesis of benzofurans via ruthenium-catalyzed redox-neutral C-H functionalization

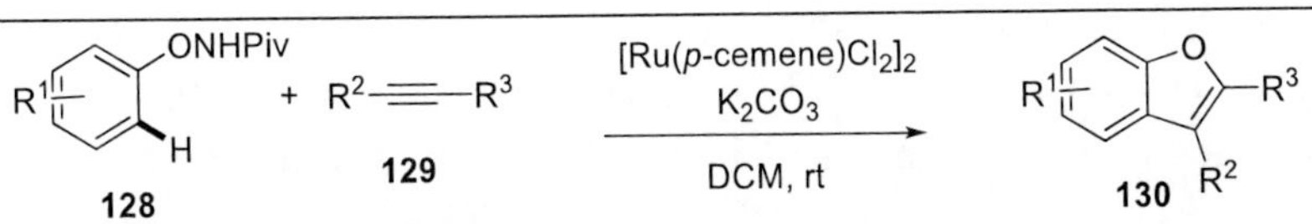

Product	Yield	Product	Yield
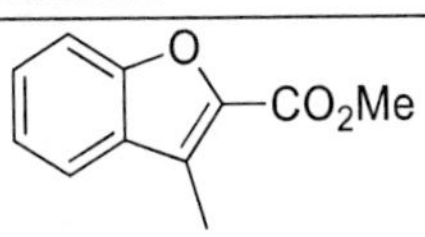	99%	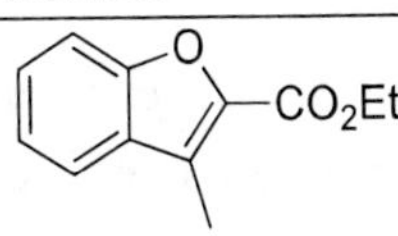	96%
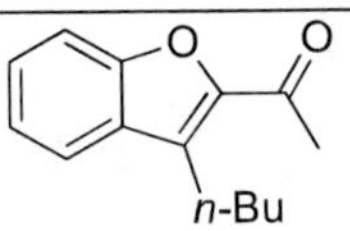	93%	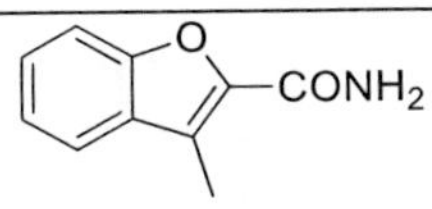	60%
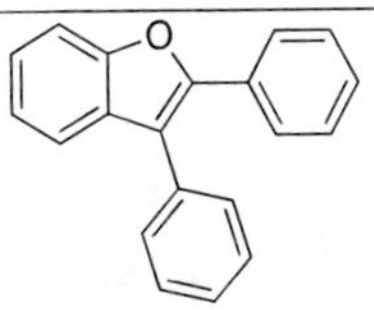	57%	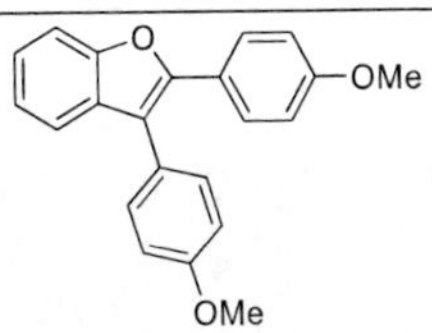	54%
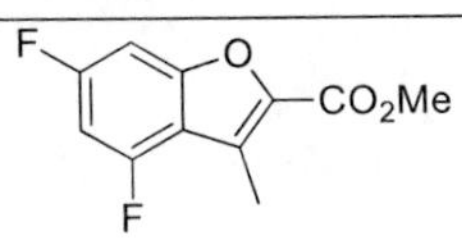	13%	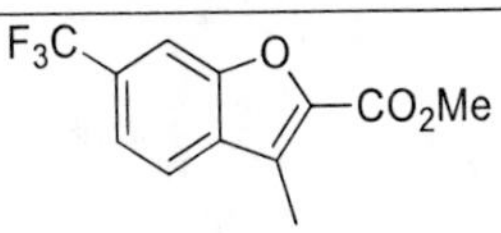	23%

Zhao and Huang's group demonstrated that rhodium catalyzed C-H vinylation reaction between an N-aryloxyacetamide 131 and an internal alkyne 132 to explore the cascade strategy (Table 20) [75]. The oxyacetamide group is such a multifunctional warhead that enables a series of C-H functionalization cascades and allows direct access to structurally diverse polycyclic heterocyles in one pot [76, 77]. The double cascade, designed between N-phenoxyacetamide 131 and diphenylacetylene 132 obtained the enamide product 134. Treatment of 134 with Ag2CO3 in the presence of Rh obtained the desired dihydrobenzofuro[2,3-d]oxazole 133, however, combined the dioxygenative cyclization with the initial C-H vinylation condition in a one pot cascade. Gratifyingly, 5 mol% [RhCp*Cl2]2 and 1.2 equiv. of Ag2CO3 in MeOH at 25°C smoothly converted N-phenoxyacetamide 131 to the desired dihydrobenzofuro[2,3-d]oxazole 133 directly in 97% isolated yield skeleton bearing two adjacent quaternary stereogenic centres as a single diastereomer and this is the first example of the heterocycle containing two chiral centres benzofuran derivatives prepared in one pot via a C-H activation process. A series of substituted aryloxyacetamides 131 bearing electronic-donating substituents were isolated in high yield. Unsymmetrical internal alkynes 132 were also generated single regioisomeric products effectively.

Table 20. Rhodium-catalyzed one-pot synthesis of dihydrobenzofuro[2,3-*d*]oxazoles

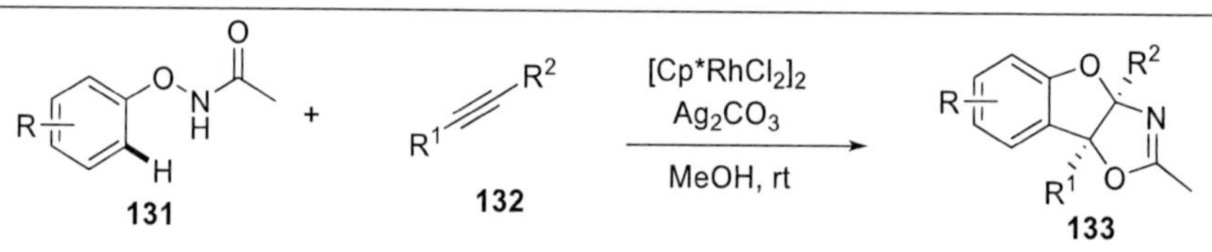

Product	Yield	Product	Yield
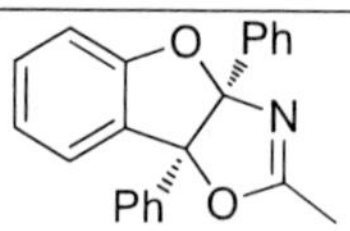	99%	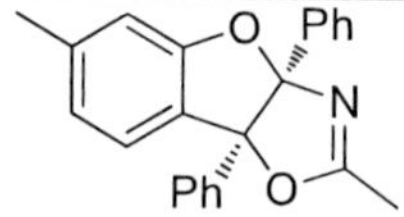	96%
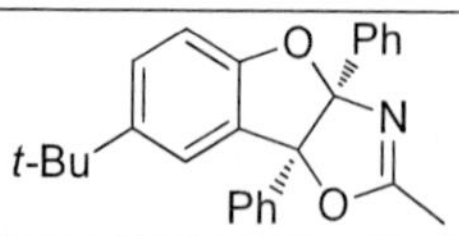	91%	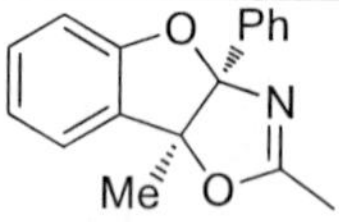	70%
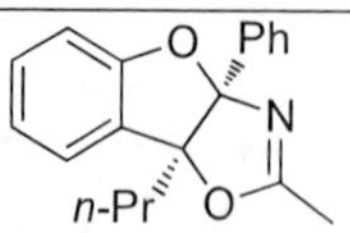	75%	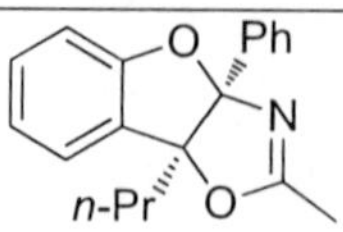	40%

Next, the triple cascade was achieved in one pot, using [Cp*RhCl2]2, AgTFA and CsOAc at 120°C (Table 21). Simply increasing the amount of AgCO3 to 6.0 equiv afforded the corresponding dimethyl ketal product in 80% yield, a quadruple cascade (Scheme 14). This reaction cascades, up to four steps, features unprecedented dioxygenation, oxidation and rearrangement mechanisms that are fully controlled by the choice of silver oxidant, its equivalence and reaction temperature.

Table 21. Rhodium-catalyzed synthesis of polysubstituted isoquinoline

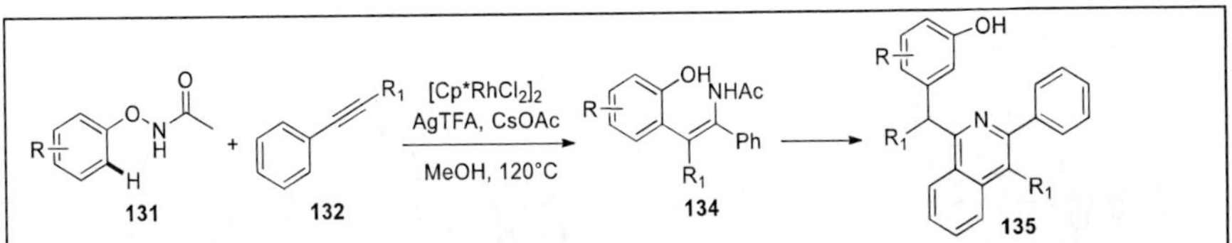

Product	**Yield**	**Product**	**Yield**
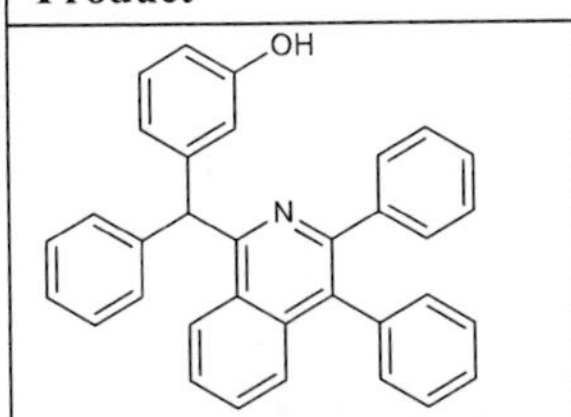	99%	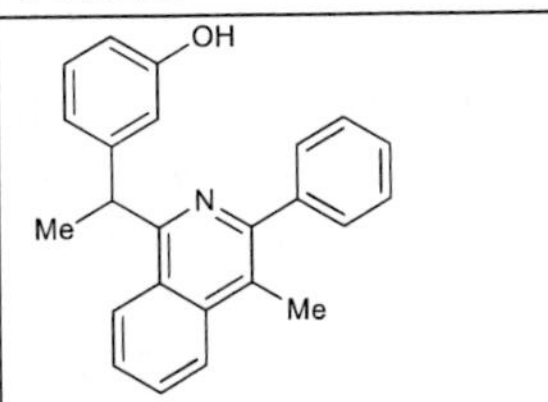	96%

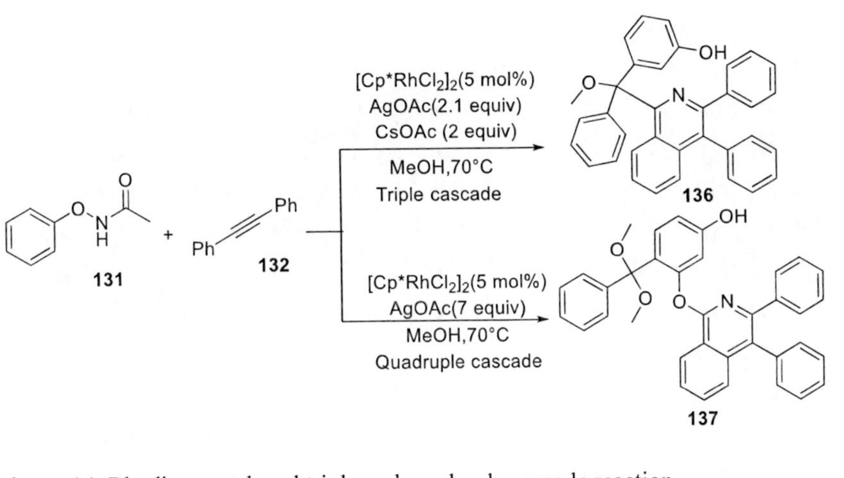

Scheme 14. Rhodium-catalyzed triple and quadruple cascade reaction.

Table 22. Rhodium-catalyzed alcohol-involved C-H functionalization

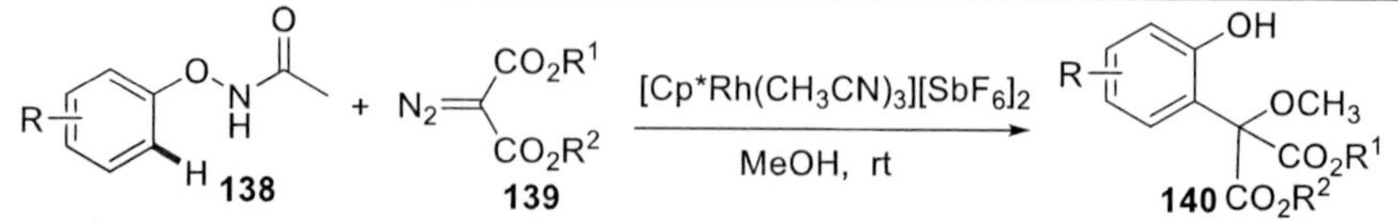

Product	Yield	Product	Yield
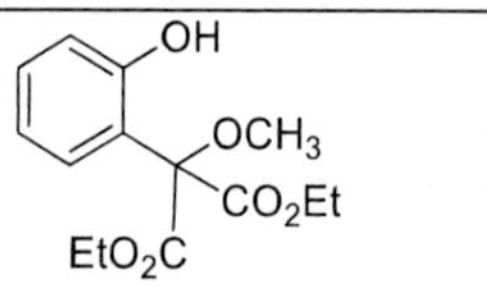	81%	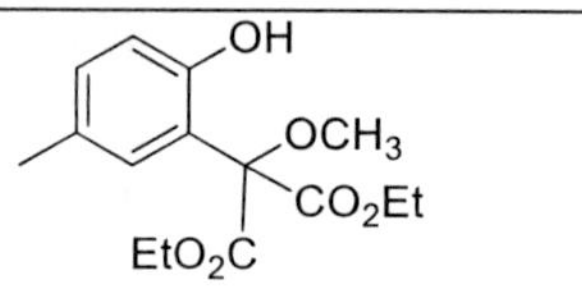	78%
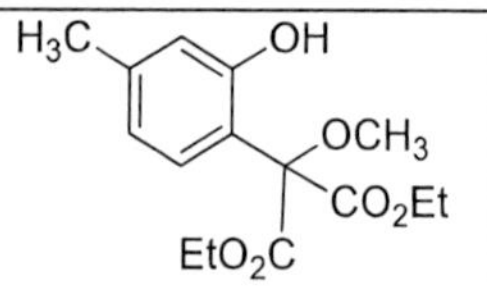	73%	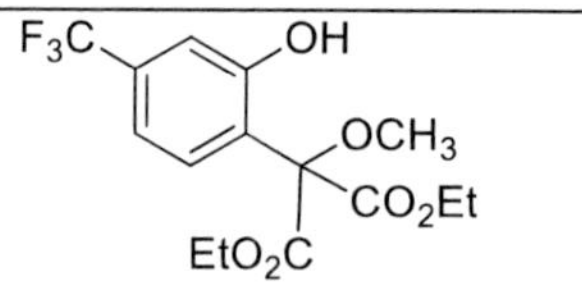	70%
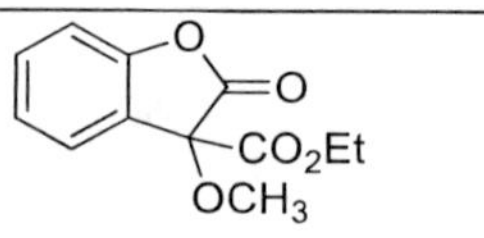	55%	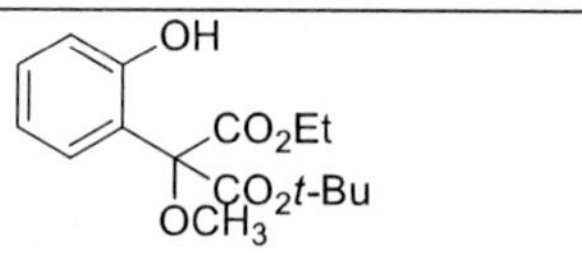	78%
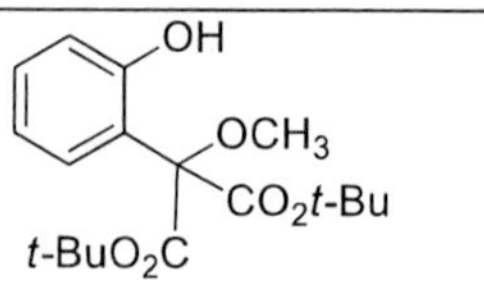	72%	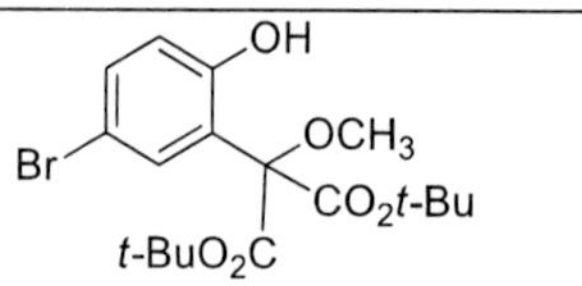	70%

Li and co-workers reported Rh(III)-catalyzed and alcohol-involved carbenoid C-H insertion into N-phenoxyacetamides 138 using α-diazomalonates 139 for the direct synthesis of 2-(2-hydroxyphenyl)-2-alkoxymalonates 140 with an α-quaternary carbon center and a free-OH moiety (Table 22) [78]. In this reaction the O-NHAc group was used as the ODG and MeOH was used not only as the solvent but also as the reagent to mediate the alcoholysis of the intermediate via a similar 1,4-addition pathway. Interestingly, substrates bearing methyl and trifluoromethyl groups at the meta-position, respectively, provided the corresponding products in moderate yields with exclusive regioselectivity. In addition, tert-butyl diazomalonates were also investigated in the Rh(III) system. They coupled efficiently with N-phenoxyacetamides 138 to offer the corresponding ortho-alkylated products in synthetically useful yields, where the tert-butyl moiety was retained perfectly.

Since MeOH was used not only as the solvent but also as the reagent, several alkyl alcohols were subsequently evaluated with the current catalytic system, the reactions were occurred successfully giving the corresponding products (Table 23). Transformation of the product esterlysis/decarboxylation, giving the valuable 2-hydroxy-α-methoxy-benzenacetate 141 in the presence of LiOH (Scheme 15). In addition, the important 142 was obtained from product 140 through a standard intramolecular-transesterification. Further transformation of 142 via an esterlysis/decarboxylation process could yield the 3-substituted benzofuran-2(3H)-one 143, it is a very valuable skeleton in natural products and biologically active compounds. This reaction provided a straightforward way for installing both an α-quaternary carbon center and a free-OH moiety into the phenyl ring, thus giving access to useful 2-(2-hydroxyphenyl)-2-alkoxymalonates 140 with good substrate/functional group tolerance.

Table 23. Rhodium-catalyzed C-H functionalization with different alcohols

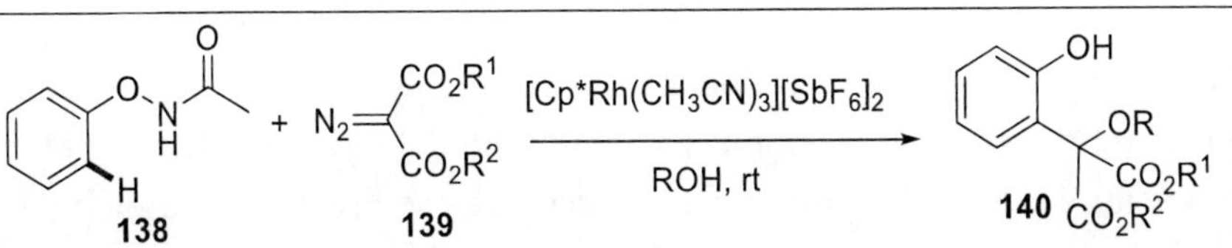

Product	Yield	Product	Yield
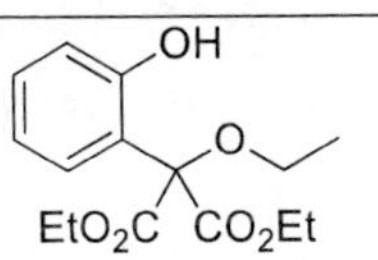	64%	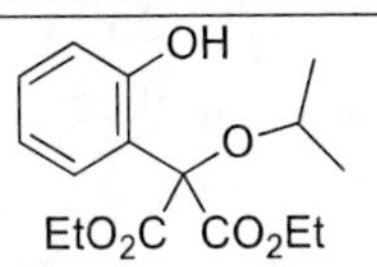	51%
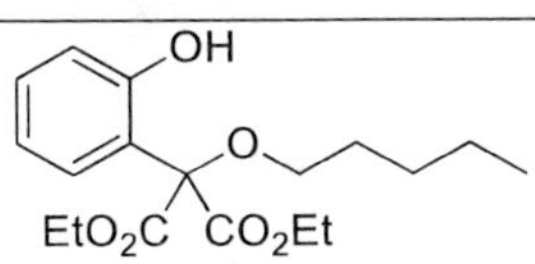	57%	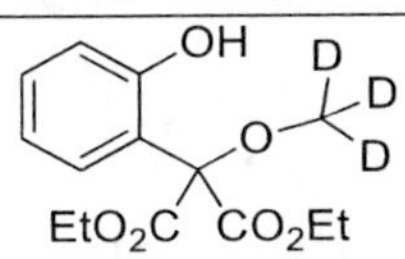	76%

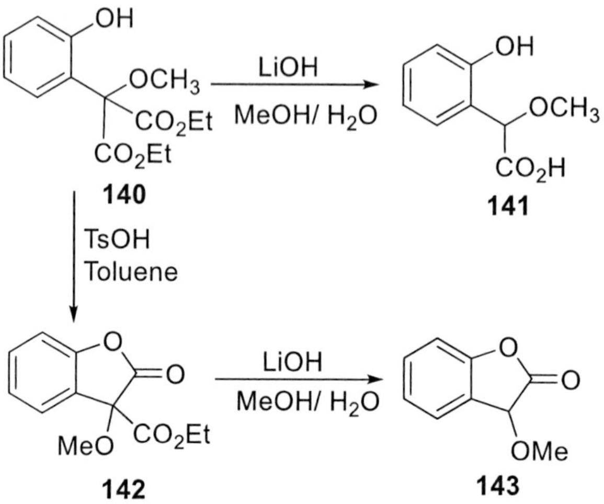

Scheme 15. Rhodium-catalyzed alcohol-involved synthesis of 3-substituted benzofuran-2(3H)-one.

Li and co-workers reported an efficient rhodium(III)-catalyzed coupling reaction of N-phenoxyacetamides 144 with propargyl carbonates 145 to yield 3-alkylidene dihydrobenzofuran derivatives 146 via C-H functionalization/cascade cyclization (Table 24) [79]. This protocol features both redox-neutral C-H functionalization and cascade reaction, in which three new bonds are formed in one step. A range of substituted N-phenoxyacetamides 144 and tert-butyl carbonate 145 were employed, and the corresponding 2,3-dihydrobenzofuran derivatives 146 were obtained effectively under the optimized conditions. Moreover, the electron-deficient N-phenoxyacetamides 144 seemed to be less effective giving the desired products in moderate yields even at higher temperature. Next, the effect of the substituents on tert-butyl propargyl carbonates 145 was examined under the standard conditions, the reaction was also compatible in various propargyl carbonates 145.

A plausible mechanism shown that Cp*Rh(OAc)2 acted as the active catalyst, followed by the N-H/C-H cleavage to yield a five-membered rhodacycle intermediate 147 (Scheme 16). Regioselective insertion of alkyne 145 generated intermediate 148, which underwent reductive elimination to give intermediate 149 and Rh(I) species. Subsequently, O-N bond cleavage and protonolysis regenerated the active Rh(III) species together with the enamide 150. With -OBoc as the leaving group, an intramolecular substitution proceeded to form the C-O bond and the exocyclic double bond simultaneously.

Table 24. Rhodium-catalyzed synthesis of 3-alkylidene dihydrobenzofuran via C-H functionalization/cascade cyclization

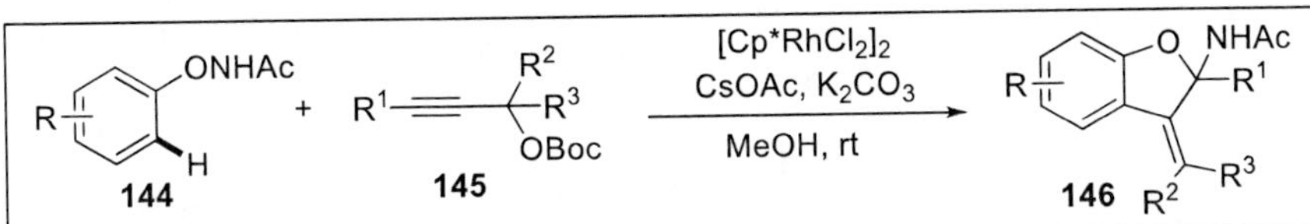

Product	Yield	Product	Yield
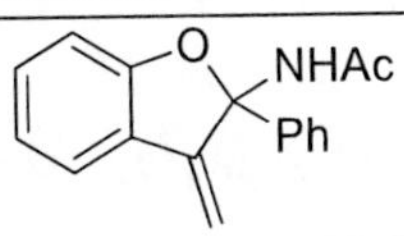	88%	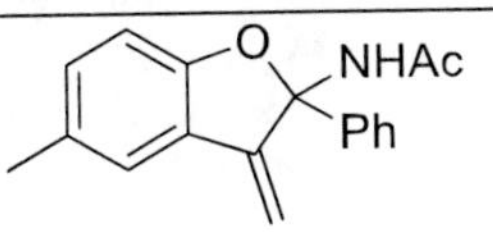	81%
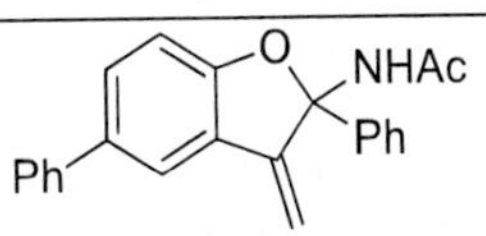	84%	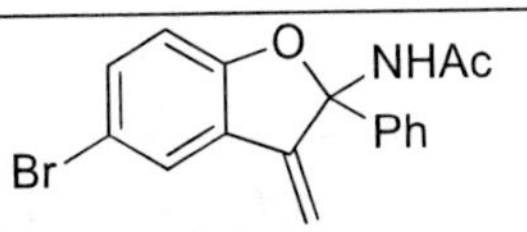	74%
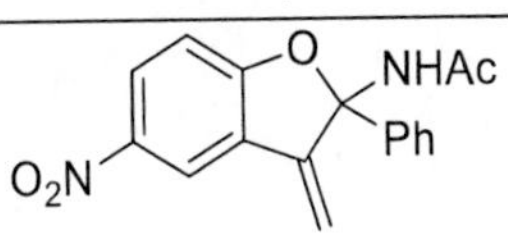	37%	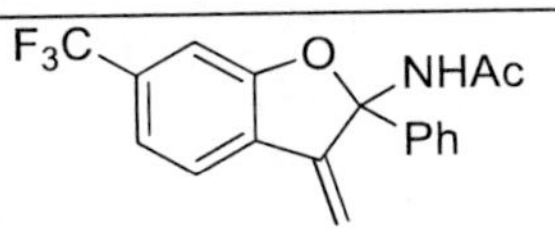	39%
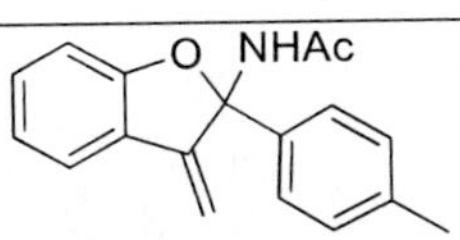	81%	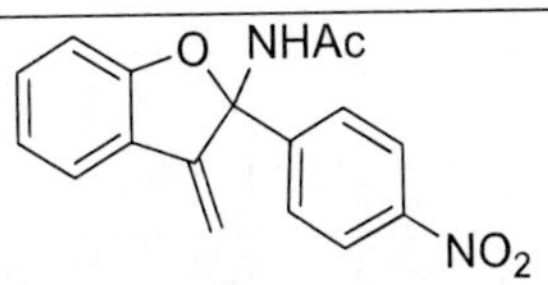	94%
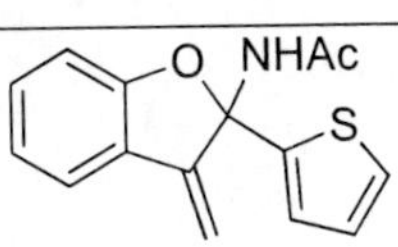	85%	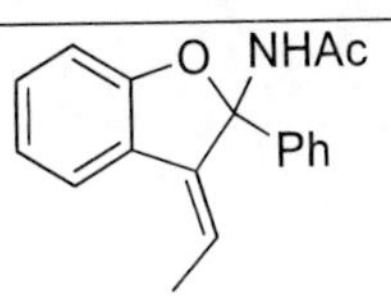	81%

Xu's group reported Cp*Rh(III)-catalyzed carbooxygenation of 1,1-disubstituted alkenes with N-aryloxyacetamides 152 afford 2,3-dihydrobenzofuran derivatives 154 with a quaternary carbon center in good to excellent yields under mild redox-neutral conditions (Table 25) [80].

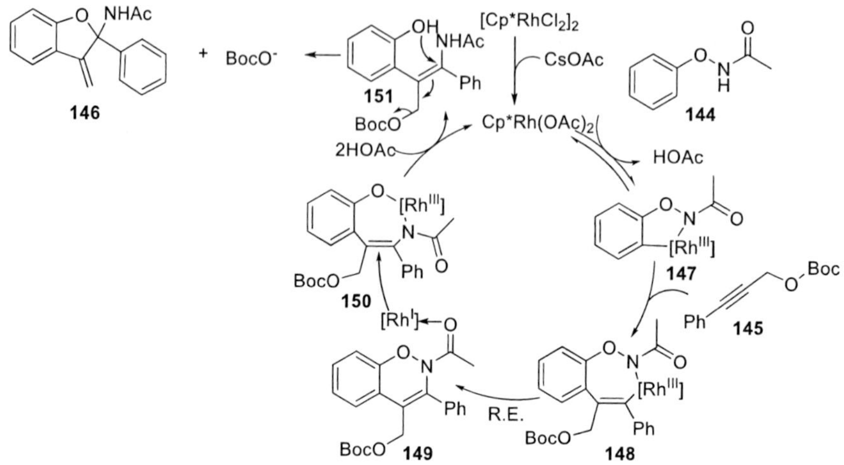

Scheme 16. Rhodium-catalyzed synthesis of 3-alkylidene dihydrobenzofuran via C-H functionalization/cascade cyclization.

Table 25. Rhodium-catalyzed synthesis of 2,3-dihydrobenzofuran via carbooxygenation

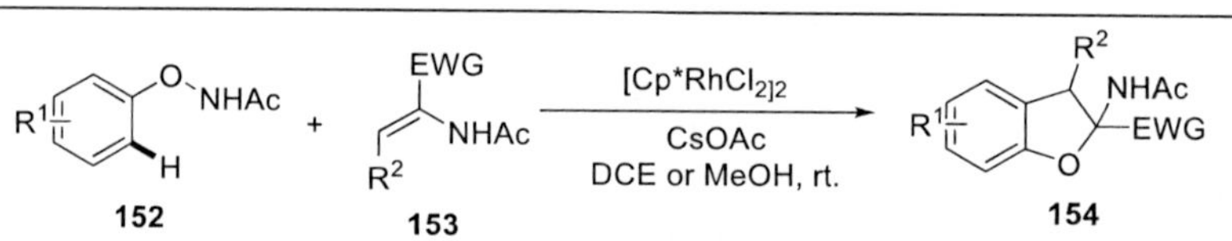

Product	Yield	Product	Yield
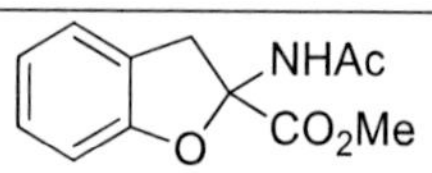	91%	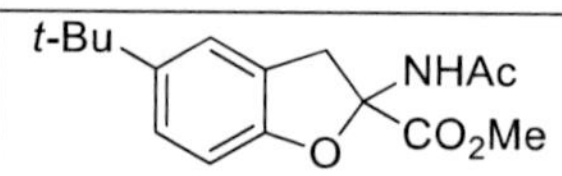	91%
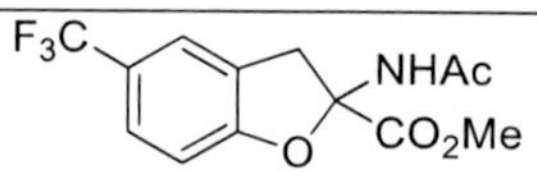	80%	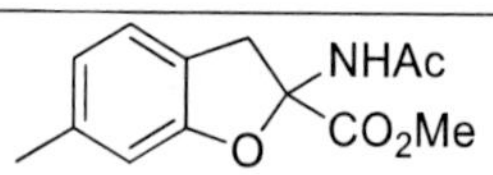	92%
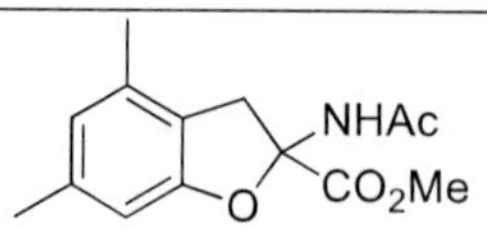	80%	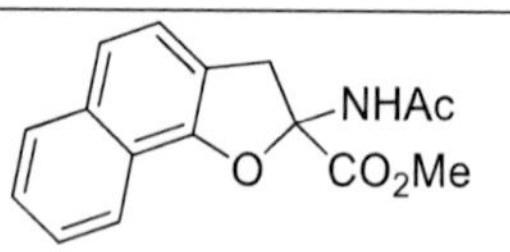	82%
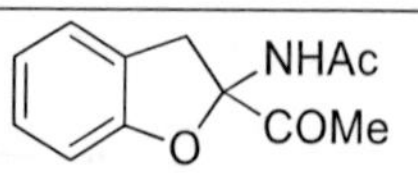	60%	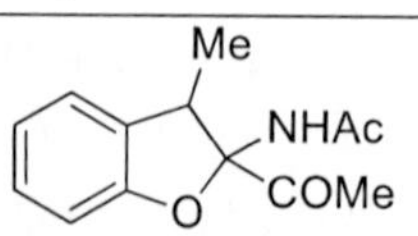	0%

Various N-aryloxyacetamides 152 were tested in the reaction with 2-acetamidoacrylate 153 and the corresponding products 154 could be isolated in good to excellent yields. And acetyl-substituted N-vinylacetamide was compatible for this reaction. However, trisubstituted alkene was incompatible under the optimized conditions, maybe due to the bulkiness that hinders binding to the Rh center.

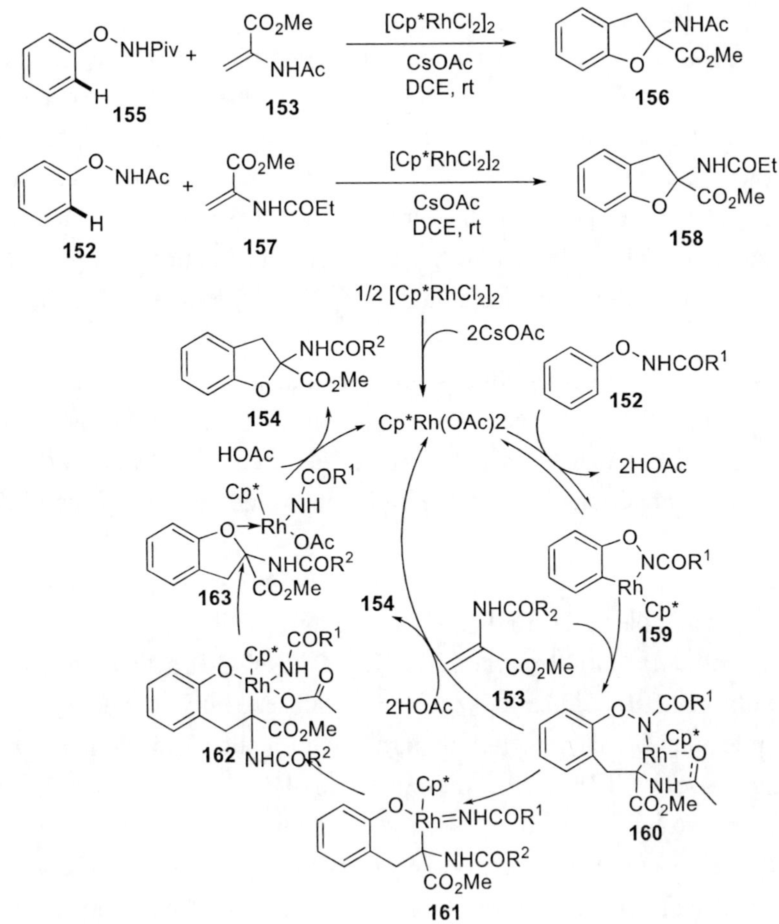

Scheme 17. Rhodium-catalyzed synthesis of 2,3-dihydrobenzofuran via carbooxygenation.

To further demonstrate the scope, N-phenoxypivalamide 155 was tested in the reaction with alkene 153, generating product 156 in 80% yield without the incorporation of the pivalamido group (PivNH-) (Scheme 17). On the other hand, the reaction of methyl 2-propionamidoacrylate 157 with N-phenoxyacetamide 152 produced product 158 with a propionamido group (EtCONH-) in 79% yield. The experiment results indicated that the amide group in annulative products 154 comes from the alkenes 153 rather than the N-aryloxyamides 152, which is essential for the process. Furthermore, mechanistic insights obtained from control experiments suggest a mechanism involving a Rh(III)-Rh(V)-Rh(III) catalytic cycle.

Zhou and co-workers have developed an unprecedented Cp*Rh(III)-catalyzed unsymmetrical C-H alkylation and amidation reaction of N-phenoxyacetamides 164 with diazo compounds 165 in one pot under mild and redox-neutral conditions (Table 26) [81]. The reaction was performed with various N-phenoxyacetamides 164 include acetyl, isobutyryl, cyclopropylcarbonyl, and phenylacetyl substituents on nitrogen, and electron-rich and electron-withdrawing substituents at the para positions all afforded good yields. When diazo-acetoacetate 165 was used, no desired product was obtained, suggesting that the electronic effect of diazo substrates is important for this C-H alkylation and subsequent amidation process.

Experimental and computational studies suggest that N-phenoxyacetamides 164 most likely occur an initial ortho C-H alkylation with diazo compounds 165 via a Rh(III)-catalyzed C-H activation, and the resulting Rh(III) intermediate 168 subsequently undergoes an intramolecular oxidative addition into the O-N bond to form a Rh(V) nitrenoid species 171 that is protonated by HOAc and further directed toward electrophilic addition to the second ortho position of the phenyl ring (Scheme 18). This reaction represents the first example of one-step, unsymmetrical difunctionalization of two ortho C-H bonds. In addition, transformation of 166 with LiOH in the presence of MeOH/H2O also resulted in decarbonylation and esterification to afford 7-amino-2(3H)-benzofuranone 173 in good yield.

Table 26. Rhodium-catalyzed unsymmetrical C-H alkylation and amidation reaction

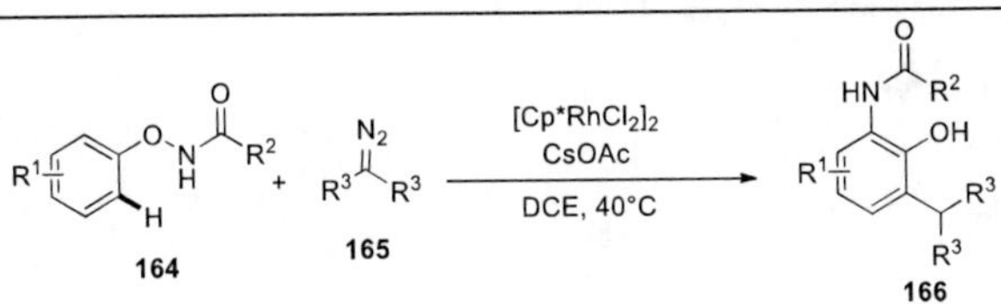

Product	Yield	Product	Yield
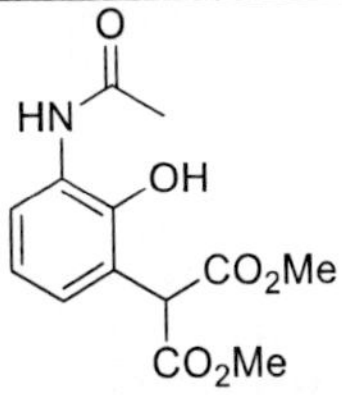	87%	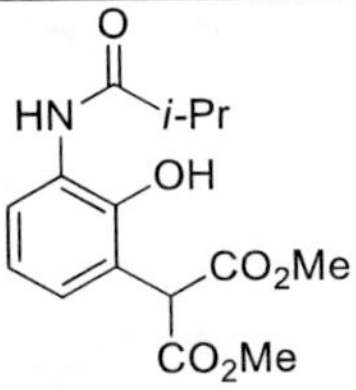	76%
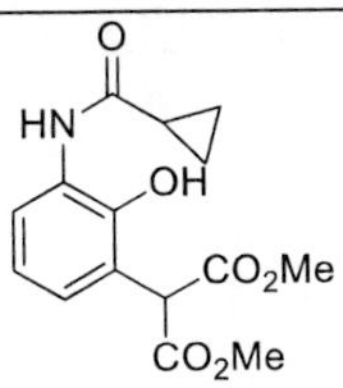	81%	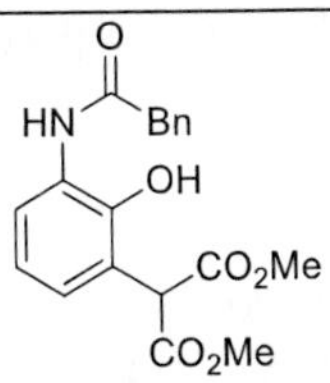	82%
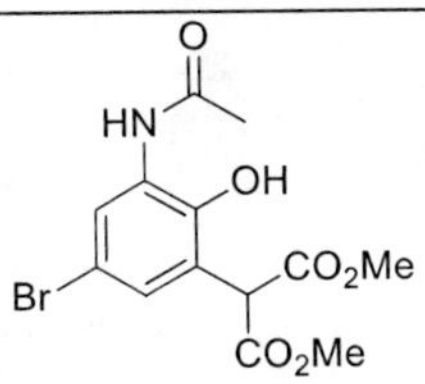	82%	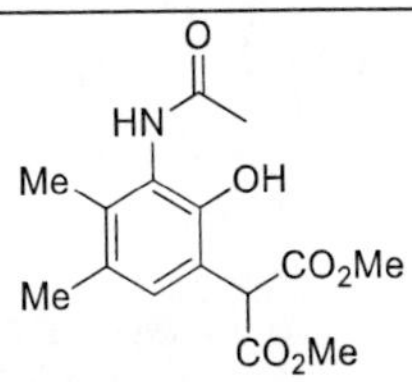	78%
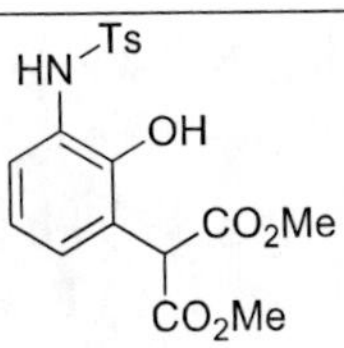	0%	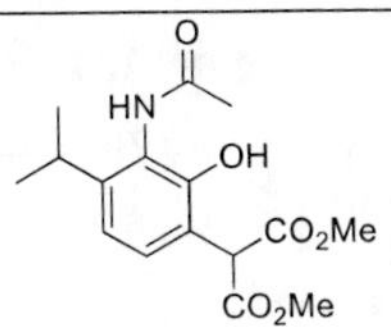	78%
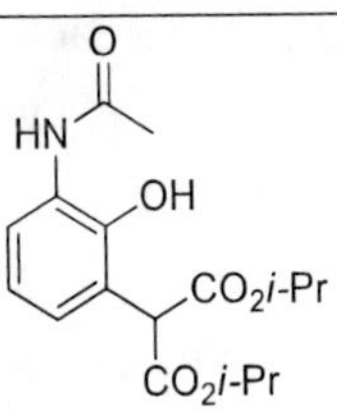	87%	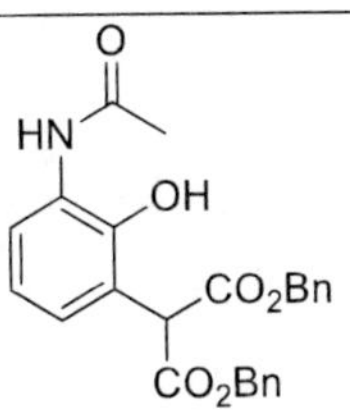	81%

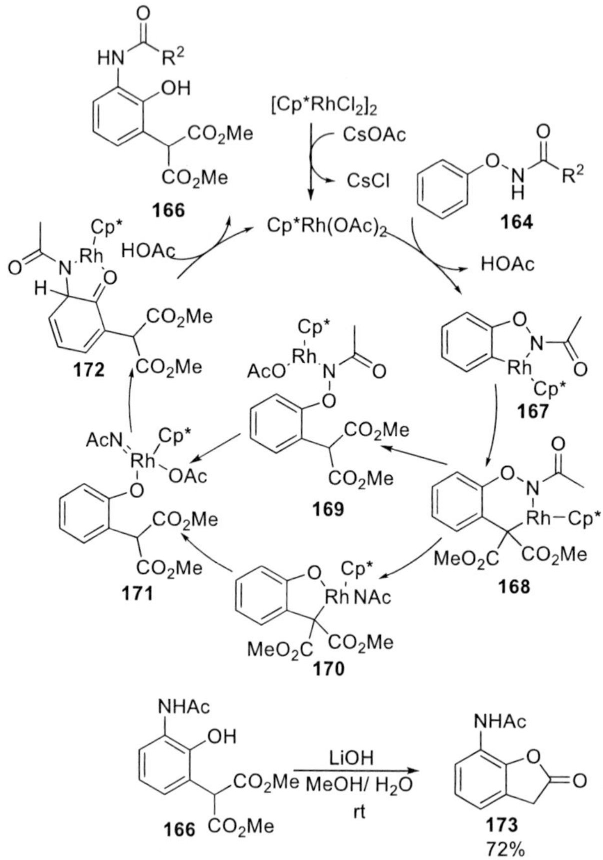

Scheme 18. Rhodium-catalyzed unsymmetrical C-H alkylation and amidation reaction.

Gao's group has demonstrated that the first Rh(III)-catalyzed, solvent-controlled, redox-neutral, regioselective C-H functionalization of N-phenoxyacetamides 174 with a new type of chelating propargyl alcohols 175 for synthesis of benzofurans 177 and chalcones derivatives 176 (Table 27 and Table 28) [82]. With highly polar solvent being the reaction medium, such as methanol and CH3CN, chalcone derivative 176 was obtained as the main product via ortho-alkenylation. While the lower polarity solvent DCE or DCM leading to synthesize benzofuran derivative 177 via an alternative [3+2] annulation process. Experimental and computational studies reveal that the hydroxyl group in controlling the regioselectivity and two solvent-controlled switchable reaction pathways controlled delivered the corresponding products with excellent chemoselectivity.

Table 27. Rhodium-catalyzed synthesis of chalcones via regioselective C-H functionalization

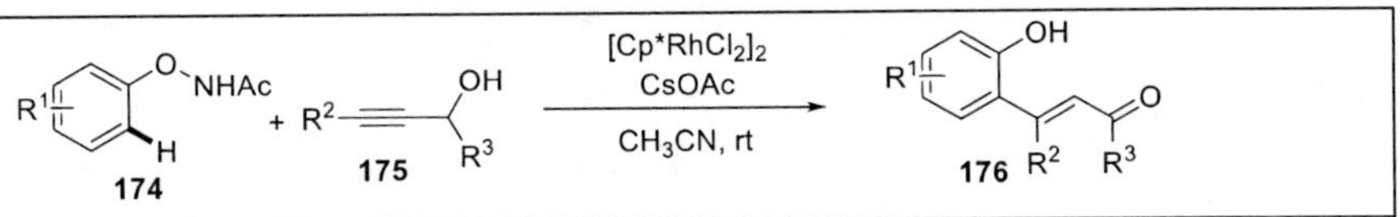

Product	Yield	Product	Yield
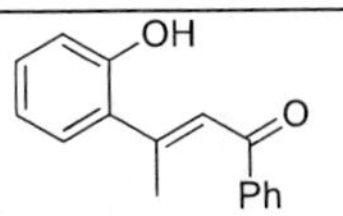	81%	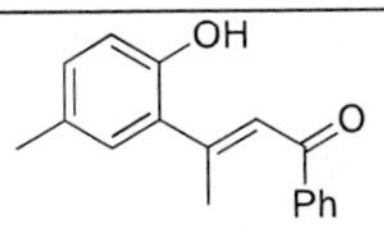	80%
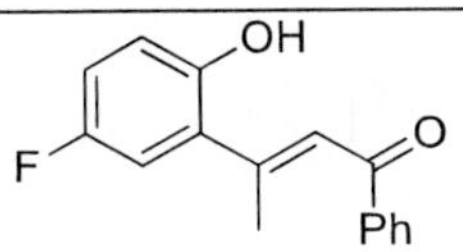	62%	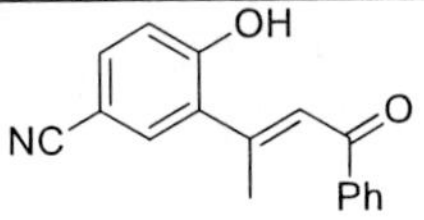	45%
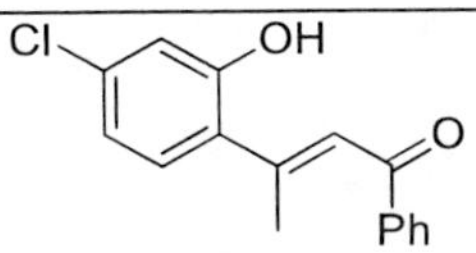	60%	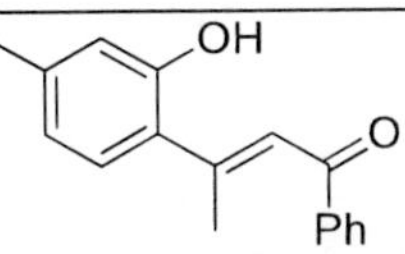	77%
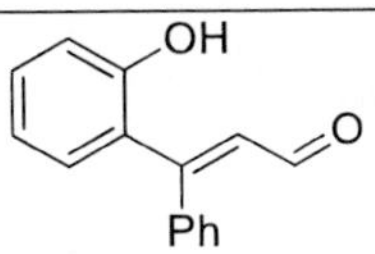	32%	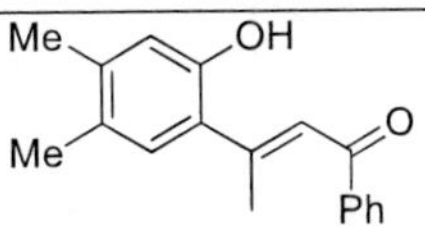	56%
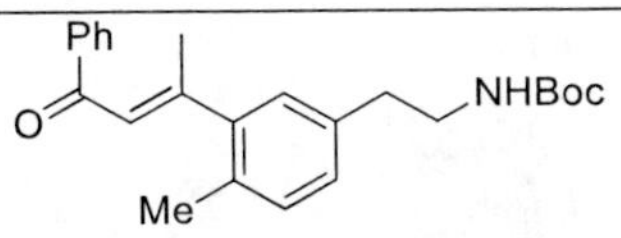	55%	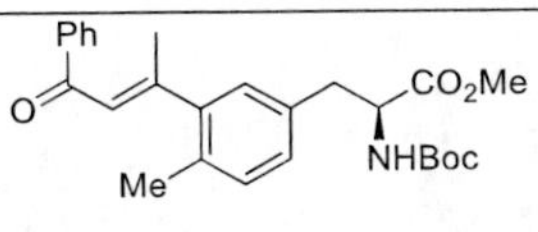	51%

A plausible mechanism involving the synergistic O-NHAc part and hydroxyl group directed regioselective C-H activation process is proposed (Scheme 19). The Cp*Rh(OAc)2 actiing as an active catalyst is generated through anion exchange, followed by facile irreversible ortho rhodation via a CMD mechanism to deliver the five-membered rhodacycle intermediate 178. Subsequent regioselective migratory insertion of the alkyne moiety 175 into the Rh-C bond of 178 forms the seven-membered intermediate 179, which is stabilized by the Rh-O coordination and the hydrogen bonding interaction. For the low-polarity solvents, further oxidation of 185 under air provides the final benzofuran product 177 followed by path b. For the high-polarity solvents, 179 undergoes a turnover-limiting oxidative addition process to give the Rh(V) species 180. Further β-H elimination of 180 and then hydrogen transfer yield an intermediate 181, followed by the protonolysis of 181 with the aid of HOAc to deliver the chalcone derivative 176 and regenerate the active Rh(III) catalyst by path a. This work is giving remarkable and meaningful progress in the area of TM-catalyzed and O-NHAc-assisted C-H activation and provide a rational basis for future development of synergistic dual DGs-directed C-H functionalization reactions.

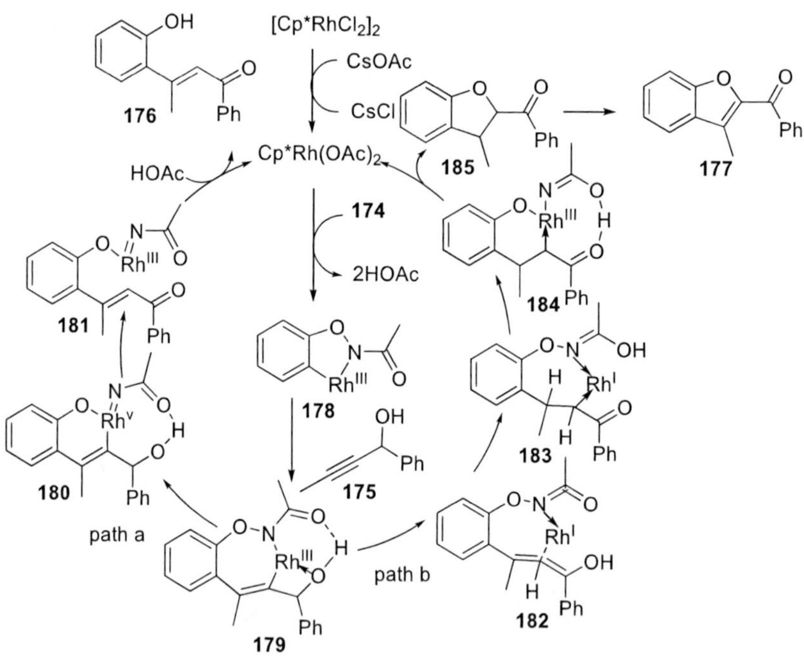

Scheme 19. Rhodium-catalyzed synthesis of benzofurans and chalcones via regioselective C-H functionalization.

Table 28. Rhodium-catalyzed synthesis of benzofurans via regioselective C-H functionalization

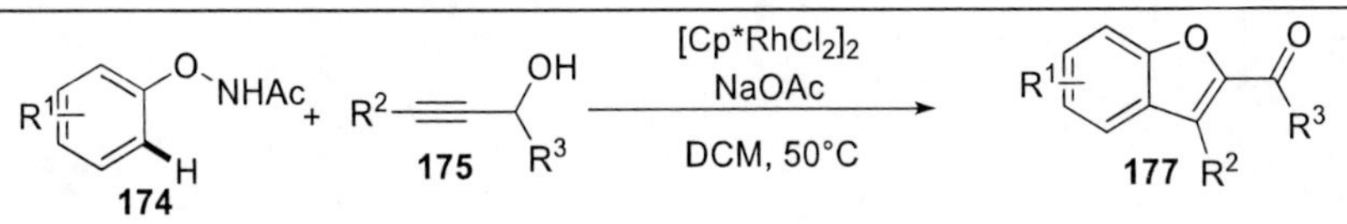

Product	Yield	Product	Yield
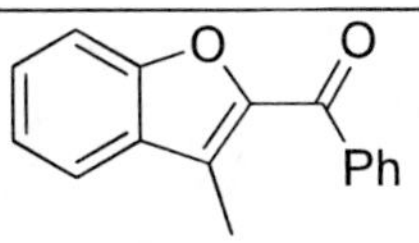	84%	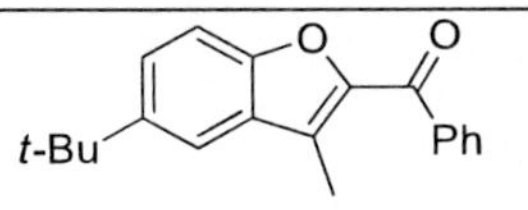	85%
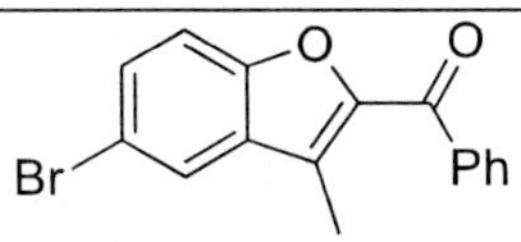	66%	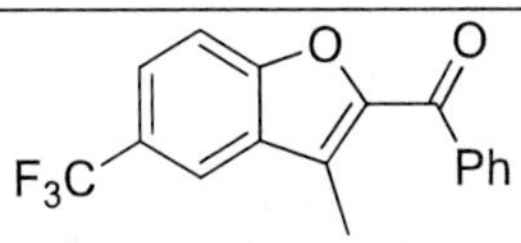	39%
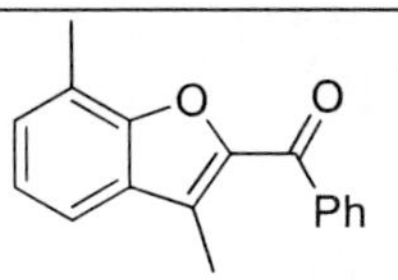	85%	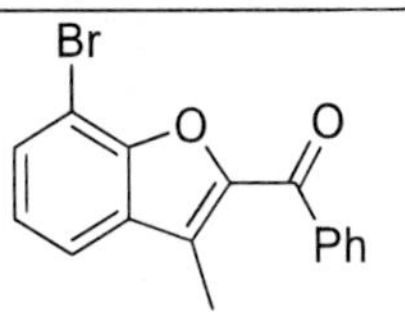	55%
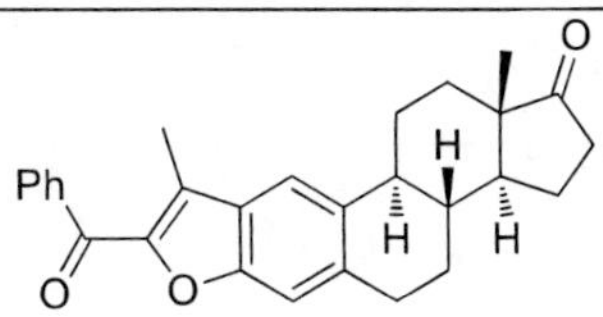	55%	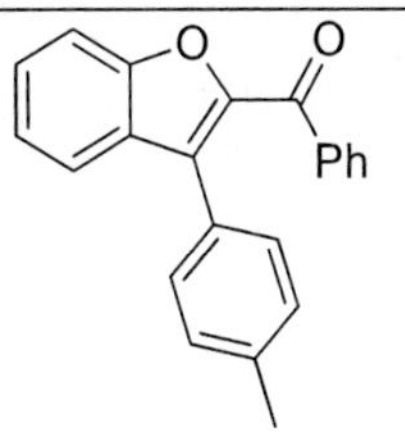	37%
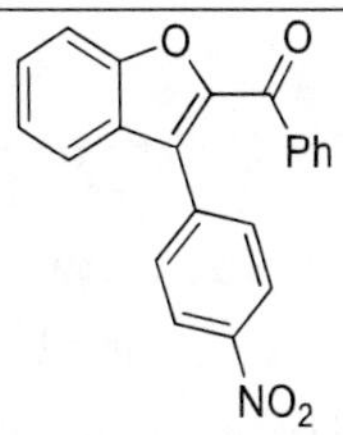	56%	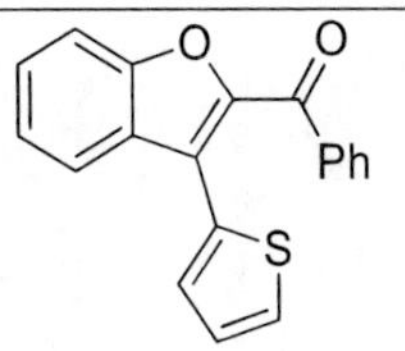	69%
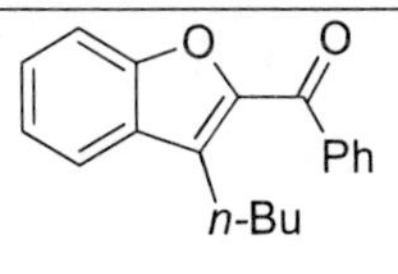	74%	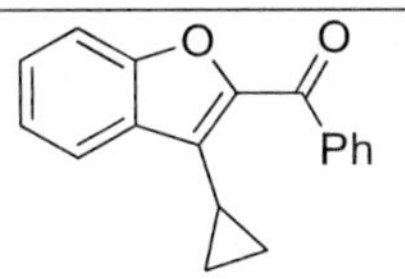	65%

Table 29. Rhodium-catalyzed redox-neutral cascade [3+2] annulation

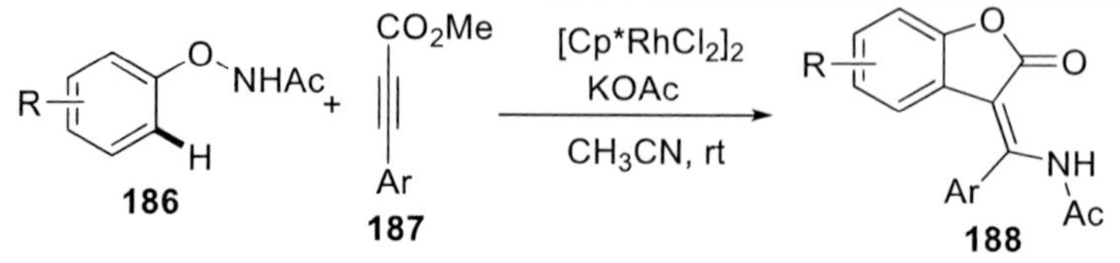

Product	Yield	Product	Yield
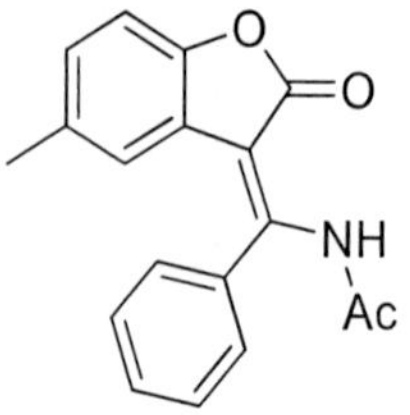	62%	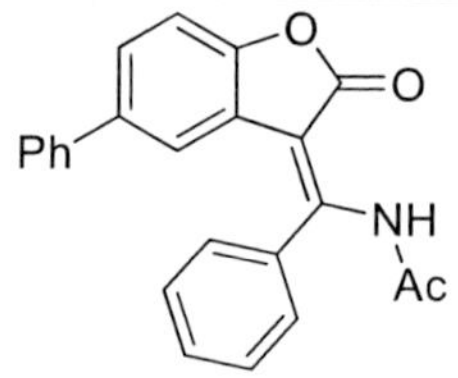	52%
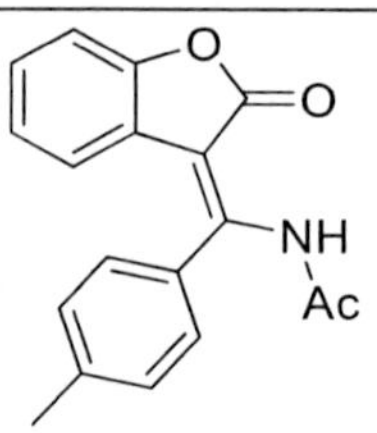	63%	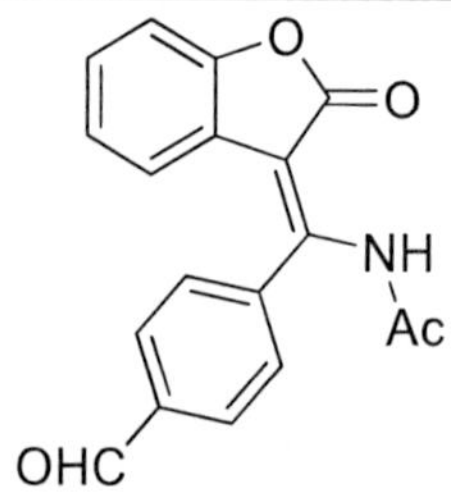	67%
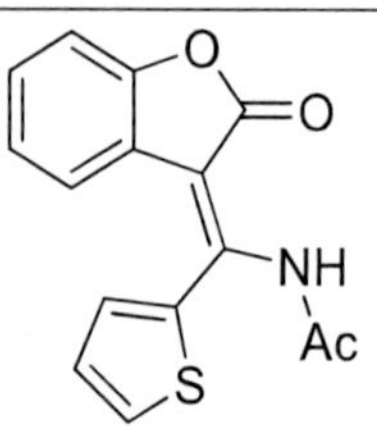	78%	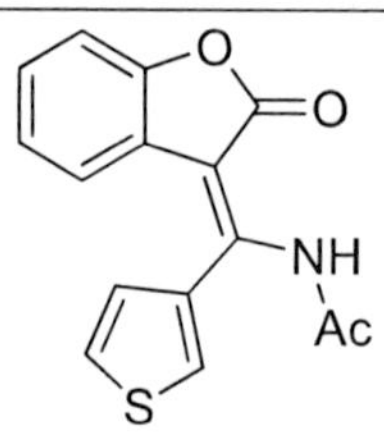	69%

Zhang and co-workers reported rhodium-(III)-catalyzed redox-neutral cascade [3+2] annulation of N-phenoxyacetamides 186 and propiolates 187 to provide a straightforward approach to benzofuran-2-(3H)-one skeletons 188 bearing exocyclic enamino motifs, by means of O-NHAc serving as an internal oxidative directing group (Table 29) [83]. A wide range of N-phenoxyacetamides 186 were compatible for the Cp*Rh(III)-catalyzed redox-neutral C-H functionalization/isomerization/lactonization cascade reactions, however, substrates containing electron-withdrawing groups showed poor reactivity. Exploring various propiolates 187 under standard conditions, electron-donating and eletron-withdrawing functionalities on the para- or meta-positions of the aromatic rings were well-tolerated. The reaction provides an access to benzofuran-2(3H)-ones 188 containing exocyclic enamino motifs featuring exclusive Z-configuration selectivity, moderate to good yields, broad substrate scope, excellent functional group compatibility, and mild reaction conditions. Experimental and computational mechanistic investigations revealed that the cascade reaction is possibly involved in a C-H functionalization/ isomerization/lactonization sequence.

Table 30. Rhodium-catalyzed C-H alkylation with allylic alcohols

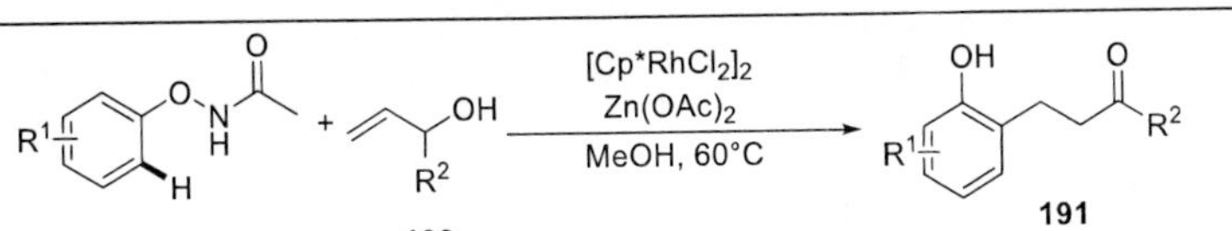

189 190 191

Product	Yield	Product	Yield
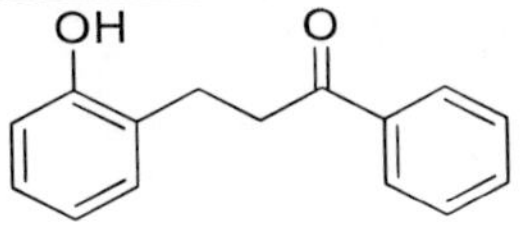	90%	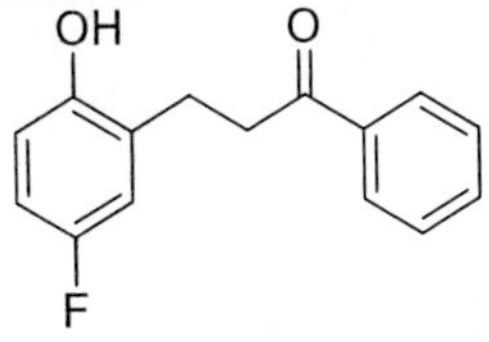	81%
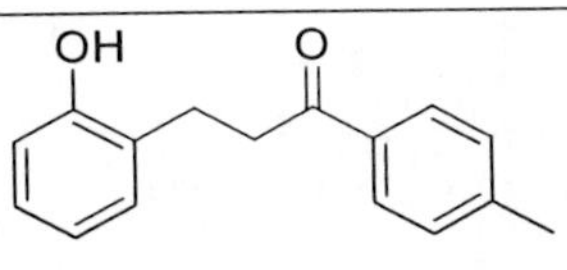	85%	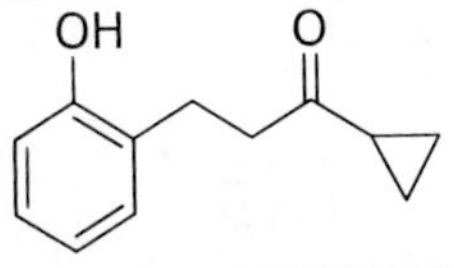	72%
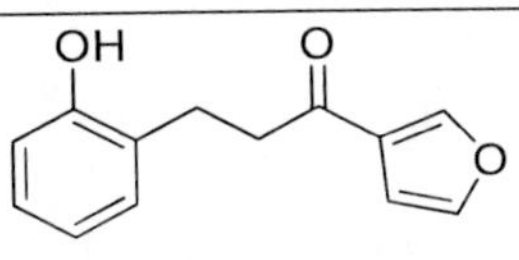	79%	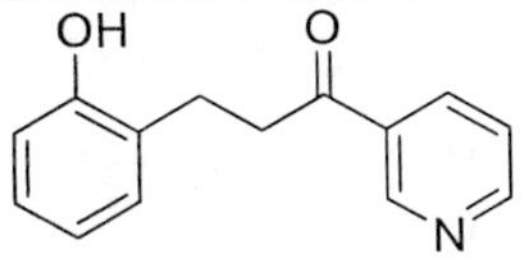	61%

Zhou et al. reported Rh(III)-catalyzed C-H alkylation with allylic alcohols 190 by using a traceless oxidizing directing group, leading to the construction of a series of β-aryl ketones 191 under mild and redox-neutral conditions (Table 30) [84]. The reaction proceeded with a wide substrate scope and good compatibility with functional groups. Considering the effect of substituents on N-phenoxyacetamide 189 or different allylic alcohols 190 all obtained good to excellent yield. Notably, heterocyclic substrates such as furan or pyridine, proved to be reactive, providing the products with good yield.

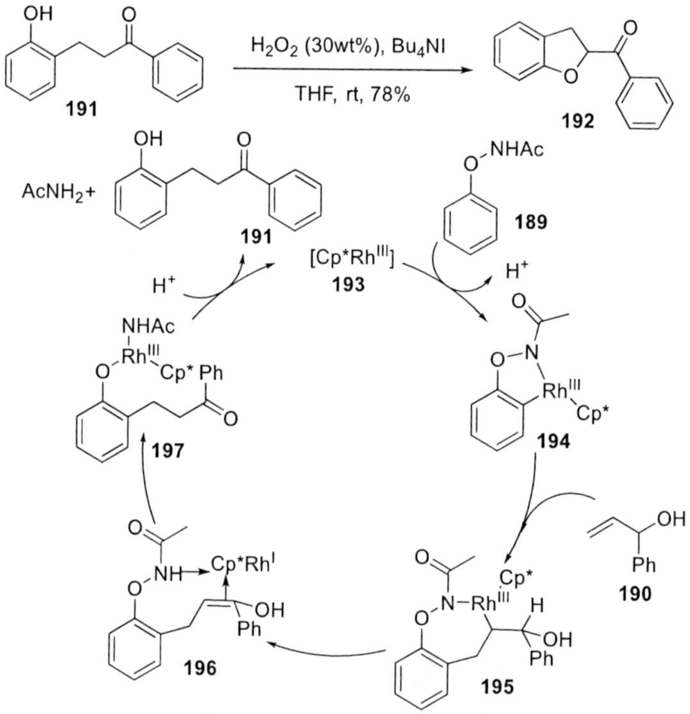

Scheme 20. Rhodium-catalyzed C-H alkylation with allylic alcohols and oxidative cyclization synthesis of 2-acyl-2,3-dihydrobenzofurans.

In addition, oxidative cyclization of phenol 191 in the presence of H2O2 and Bu4NI as catalyst gave a biologically active 2-acyl-2,3-dihydrobenzofuran derivative 192 in good yield (Scheme 20). A proposed reaction mechanism as shown, Cp*Rh(III) species involving an initial C-H activation to form rhodacycle 194. The olefin 190 insertion of 194 produces a seven-membered rhodacycle intermediate 195. Subsequently, β-H elimination and reductive elimination occur, affording intermediate

196 with the Rh(I) complex. Finally, the Rh(I) complex can be oxidized by the intramolecular N-O bond to afford intermediate 197 which can be protonated to give product 191 and regenerate the Cp*Rh(III) active catalyst.

Xu reported an efficient Cp*Rh(III)-catalyzed redox-neutral C-H activation/annulation of N-aryloxyacetamides 198 with alkynyloxiranes 199, which afford 2,3-dihydrobenzofurans 200 bearing a useful exocyclic (E)-allylic alcohol, and a tetrasubstituted carbon center with good yields with a broad substrate scope (Table 31) [85].

Table 31. Rhodium-catalyzed redox-neutral C-H activation/annulation synthesis of 2,3-dihydrobenzofurans

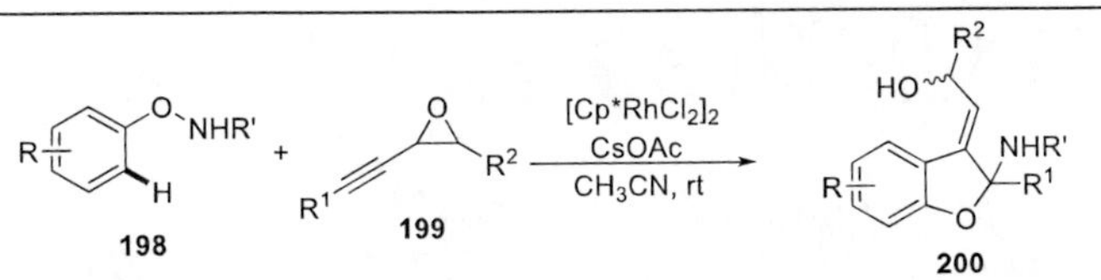

Product	Yield	Product	Yield
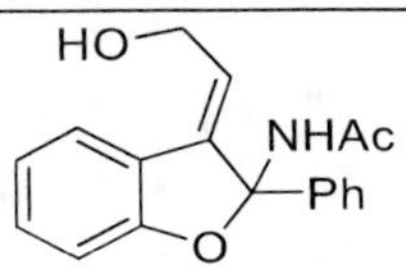	87%	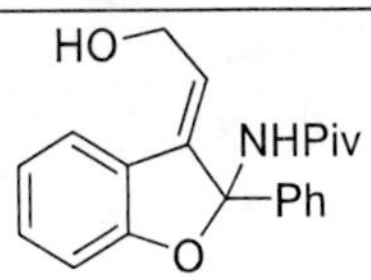	61%
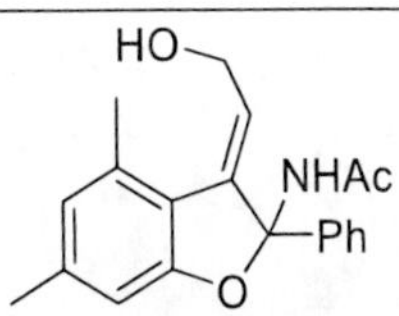	67%	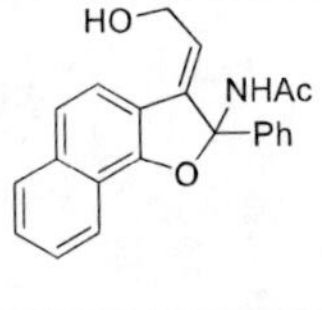	57%
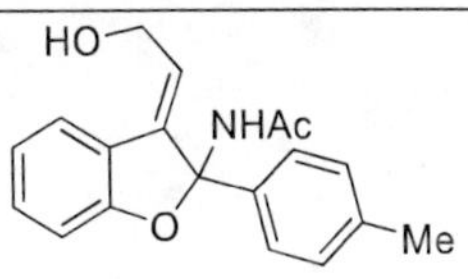	74%	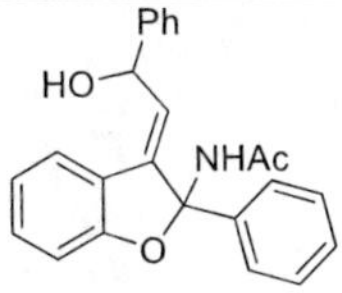	54%

Transformation of 200 with triethyl orthoformate and propionic acid could easily obtain a tricyclic heterocycle 201 (Scheme 21). In addition, the product 200 via a two-step sequence of oxidization and NHC-catalyzed rearrangement furnished a coumarin product 202. The authors proposed that a catalystic cycle mechanism, the product 200 was delivered via an intramolecular SN2′ ring opening of oxirane from enamide compound 207.

Zhu reported Rh(III)-catalyed one-pot reaction of N-phenoxyacetamide 208, ketones 209, and hydrazines for efficient synthesis of diaryl- and triaryl-substituted olefins 210 (Table 32) [86]. Under the optimized conditions, the one-pot reaction of N-phenoxyacetamide 208 and 2-phenylacetophenone 209 with a catalytic amount of [Cp*RhCl2]2 and AcOH (accelerate the formation of the hydrazine intermediate) in THF at room temperature, using MnO2 as an oxidant.

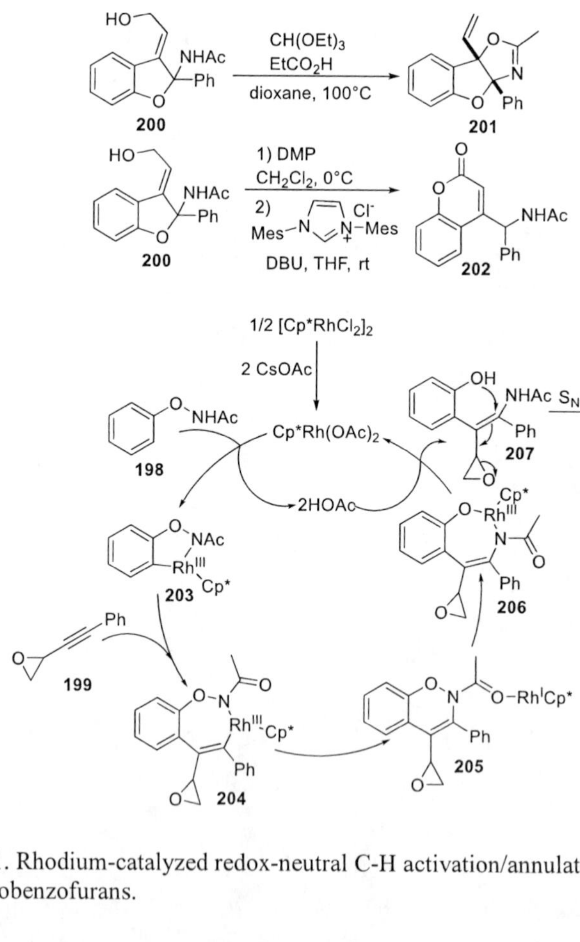

Scheme 21. Rhodium-catalyzed redox-neutral C-H activation/annulation synthesis of 2,3-dihydrobenzofurans.

Table 32. Rhodium-catalyzed one-pot reaction synthesis of diaryl- and triaryl-substituted olefins

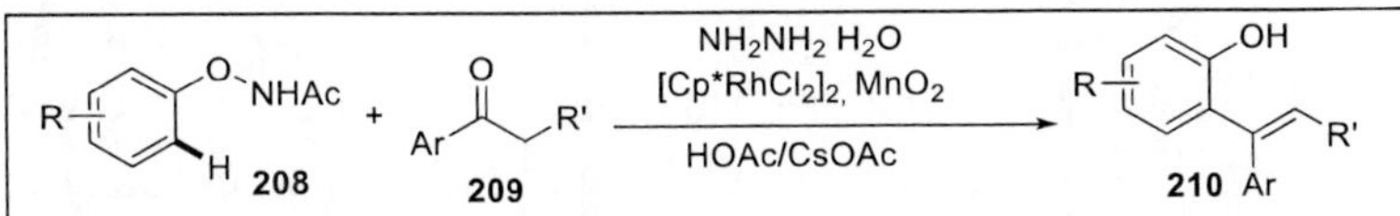

Product	Yield	Product	Yield
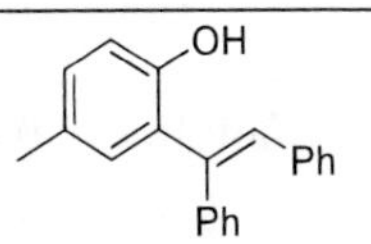	70%	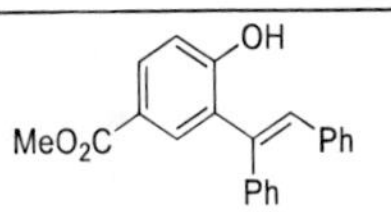	50%
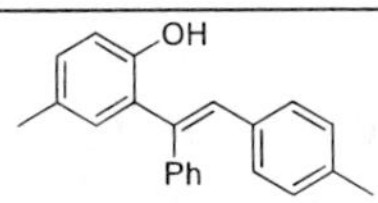	74%	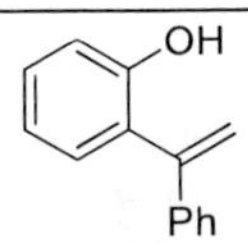	57%
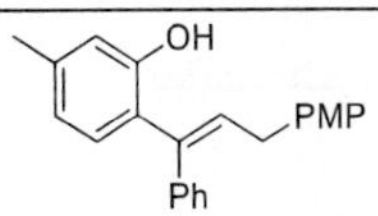	51%	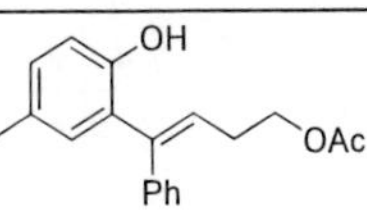	65%

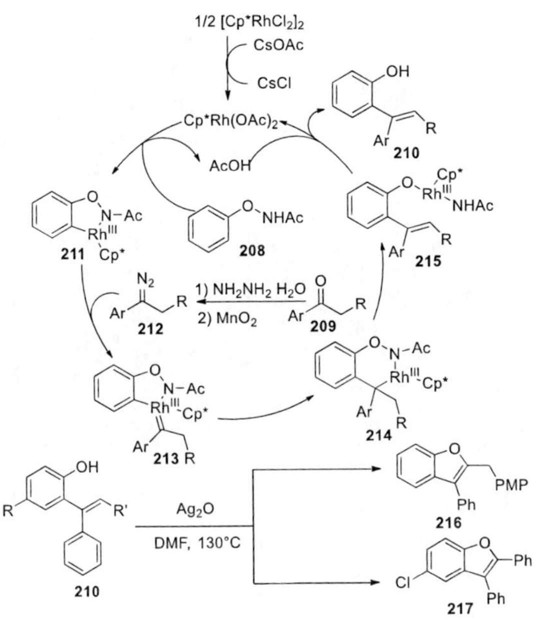

Scheme 22. Rhodium-catalyzed synthesis of substituted olefins and oxidative cyclization synthesis of 2,3-disubstituted benzofurans.

In this reaction, various ketones 209 were transformed into donor-donor diazo compounds, subsequently inserting into N-phenoxyacetamide 208, following by β-H elimination to generate the polysubstituted olefins 210. In addition, the diaryl- and triaryl-substituted olefins 210 could be controllable obtained by the ketones 209 bearing different functional groups. This transformation features a wide of substrate scopes and simpler starting materials. To further illustrate the synthetic utility of this reaction, diaryl-substituted olefin or triaryl-substituted ethylene 210 could convert to 2,3-disubstituted benzofuran derivatives 216 and 217 at high temperatures using Ag_2O as an oxidant (Scheme 22).

Table 33. Rhodium-catalyzed C-H activation/annulation of *N*-phenoxyacetamides with 1-alkynylphosphine sulfides and oxides

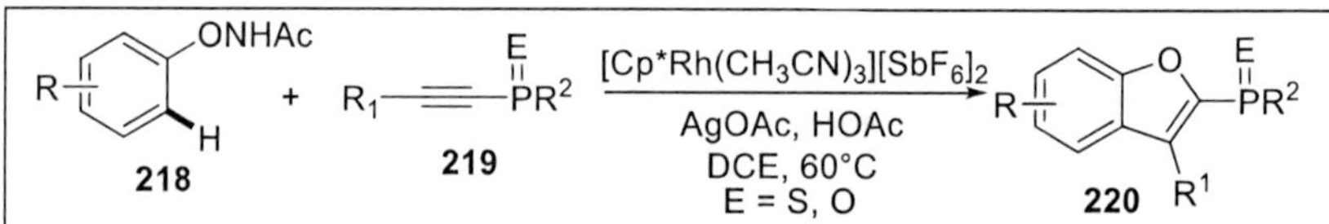

Product	Yield	Product	Yield
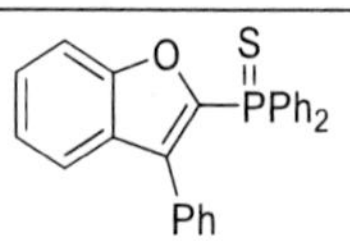	93%	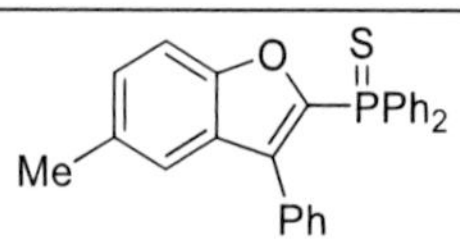	74%
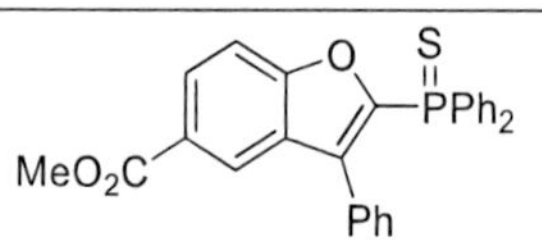	60%	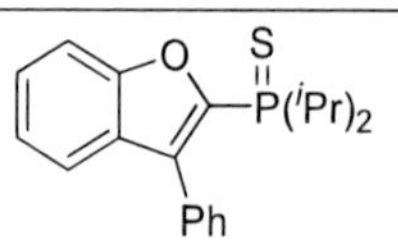	67%
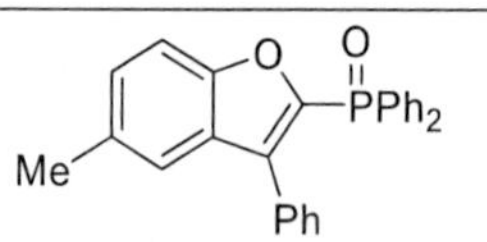	98%	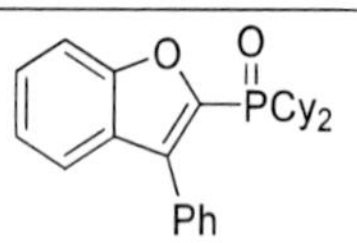	98%

Li reported Rh(III)-catalyzed C-H activation/annulation of N-phenoxyacetamides 218 with 1-alkynylphosphine sulfides and oxides 219 (Table 33) [87]. The conventional synthetic methods for heteroaryl phosphines are limited to the reactions of chlorophosphines with metalated heteroaromatic compounds. Transition metal catalyzed oxidative coupling of C-H bonds with 1-alkynylphosphine sulfides and oxides has been proven to be a straightforward and efficient approach to construct heteroarylphosphines compounds. Both electron-rich and electron-poor N-phenoxyacetamides 218 were nearly equally effective in this reaction under the optimal conditions. Notably, Alkyne with a bulky dicyclohexylphosphine sulfide and diisopropylthiophosphino moiety 219 were also a facile reactant for this catalytic process. Not only were phosphine sulfides applicable to the reaction, but phosphine oxides worked as well. All the products 220 were formed as a single regioisomer, therefore highlighting the excellent regioselectivity of the present catalytic system.

Reduction of the annulated products 2-benzofuranylphosphine sulfides and oxides 220 under the radical desulfidation reaction conditions afford the corresponding trivalent phosphines 221 quantitatively (Scheme 23). A combination of the air-stable trivalent-phosphine 3-benzylbenzofuran-2-yldicyclohexylphosphine 221 with Pd(OAc)2 provide active catalysts for Suzuki-Miyaura coupling of aryl chlorides in the presence of a stoichiometric amount of CsF in dioxane (Table 34). On the other hand, this new catalytic system was further examined in Buchwald-Hartwig amination reactions, both electron-rich and ortho-substituted aryl chlorides were successful, affording 226 in 83-90% yields (Table 35).

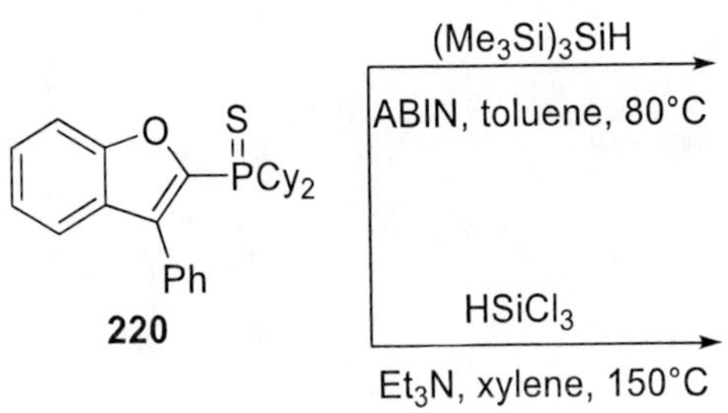

Scheme 23. Synthesis of air-stable trivalent-phosphine 3-benzylbenzofuran-2-yldicyclohexylphosphine.

Table 34. Palladium/221-catalyzed Suzuki-Miyaura coupling of aryl chlorides

$$\underset{\mathbf{222}}{\text{Aryl-Cl}} + \underset{\mathbf{223}}{\text{R-B(OH)}_2} \xrightarrow[\text{CsF, dioxane, 110°C}]{\text{Pd(OAc)}_2,\ \mathbf{221}} \underset{\mathbf{224}}{\text{Aryl-R}}$$

Product	Yield	Product	Yield
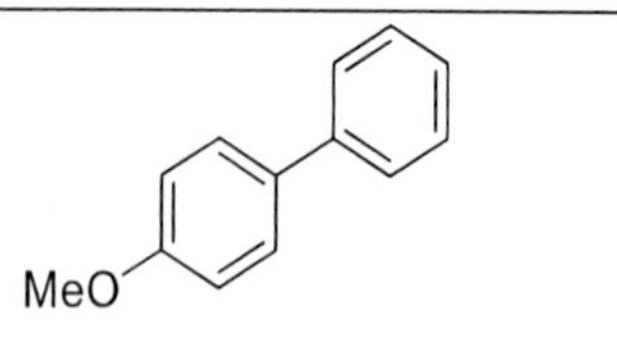	98%	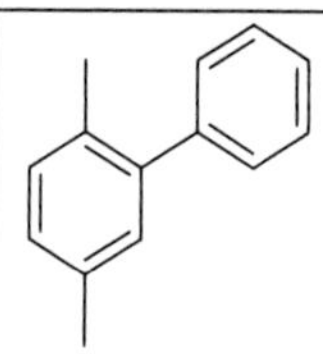	93%
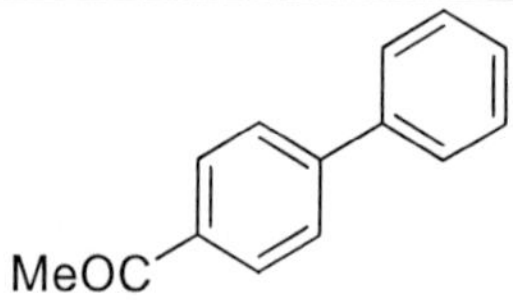	96%	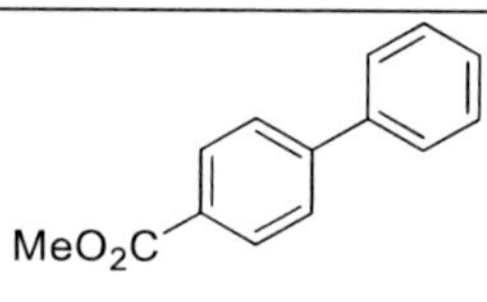	99%

Table 35. Palladium/221-catalyzed Buckwald-Hartwig amination of aryl chlorides

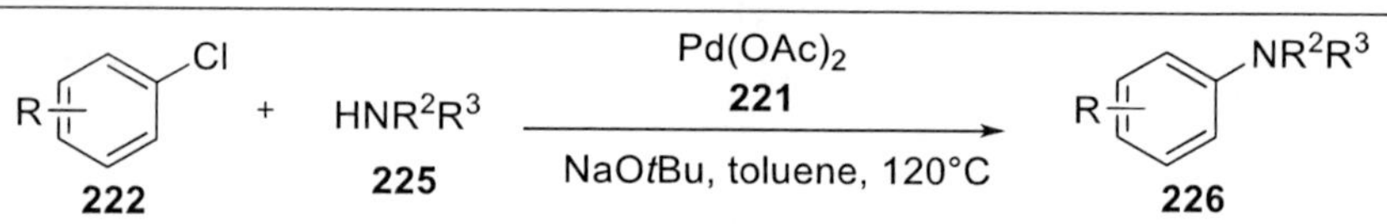

Product	Yield	Product	Yield
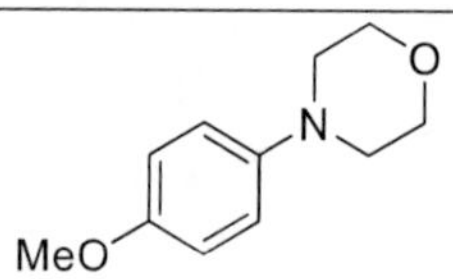	84%	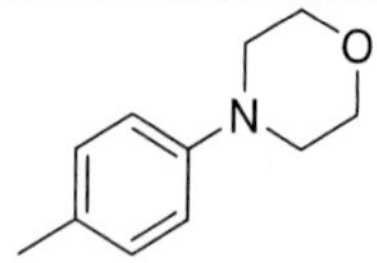	83%
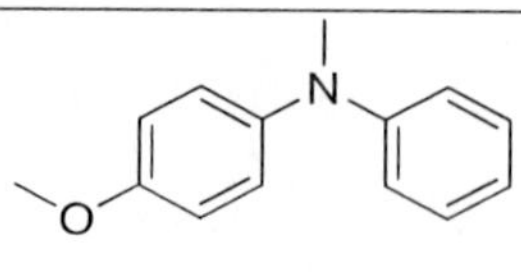	90%	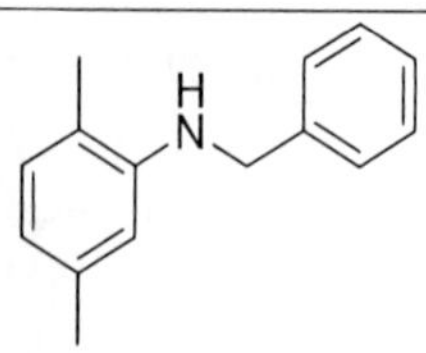	83%

Table 36. Rhodium-catalyzed regioselective annulation of alkynylcyclopropanes with *N*-aryloxyamides

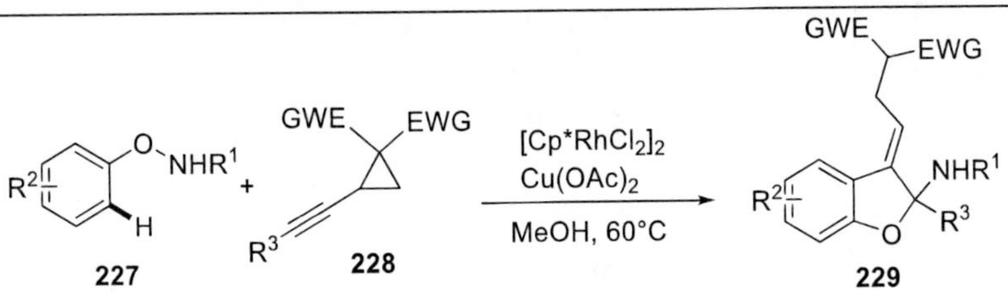

Product	Yield	Product	Yield
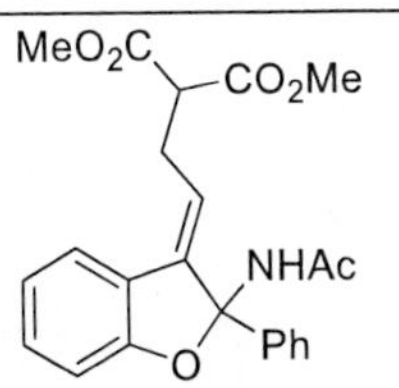	76%	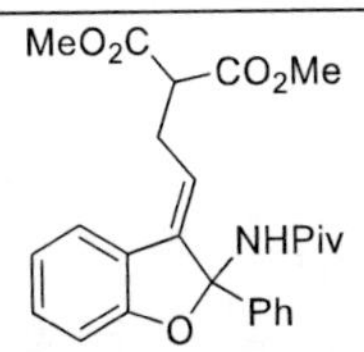	48%
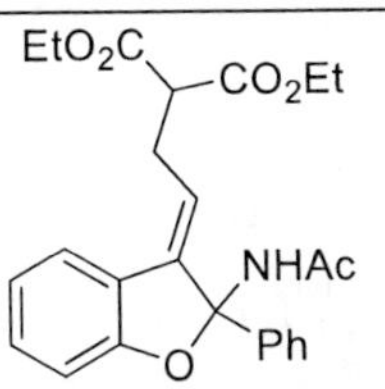	66%	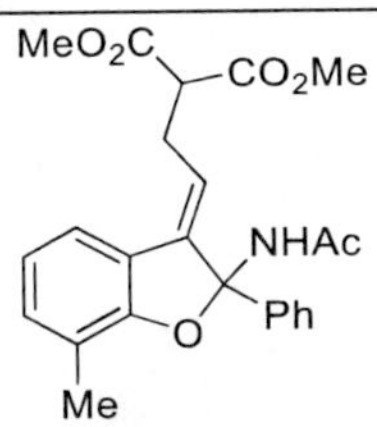	36%
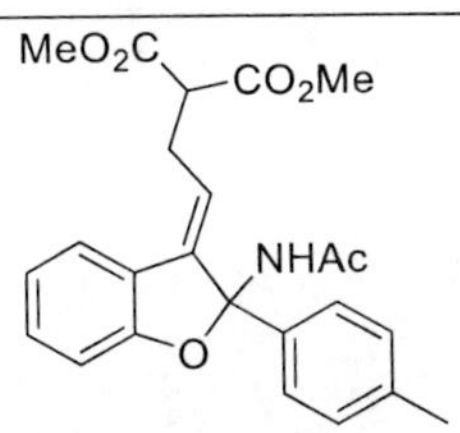	63%	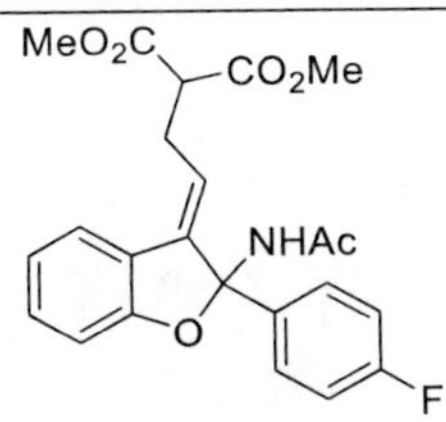	59%

Xu reported Cp*Rh(III)-catalyzed regioselective annulation of alkynylcyclopropanes 228 with N-aryloxyamides 227 via redox-neutral C-H/C-C activation, which generates a range of highly functionalized 2,3-dihydrobenzofurans 229 bearing an (E)-exocyclic carbon-carbon double bond and a tetrasubstituted carbon center (Table 36) [88]. Substitution at both the N-aryloxyamides 227 and alkynylcyclopropanes 228 was also applicable under the optimized conditions, as exemplified by the formation of products 229 in moderate yields. A series of para-, meta-substituted

N-aryloxyamides 227 with either electron-donating or -withdrawing groups (R2) participated well in the annulation, giving the corresponding products 229 good yields. The reaction constitutes the first example by using alkynylcyclopropanes 228 as coupling partners.

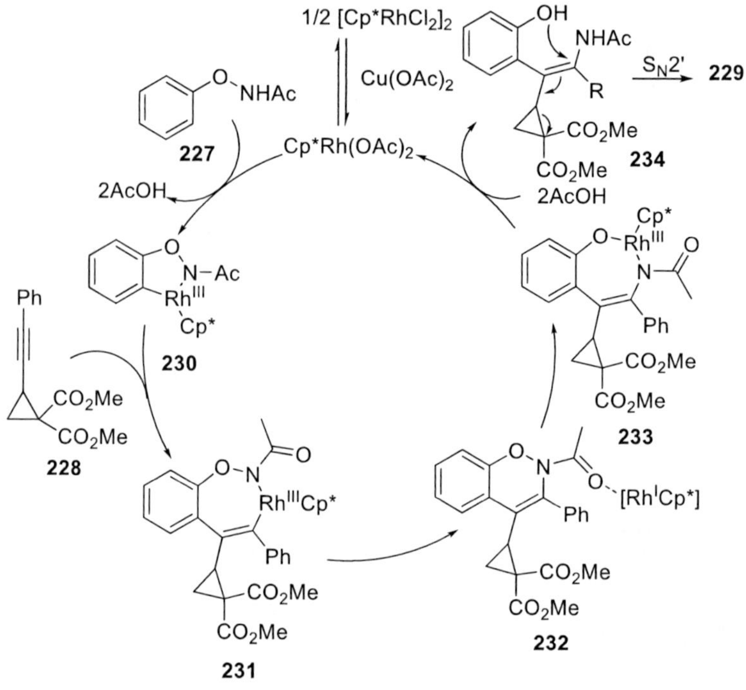

Scheme 24. Rhodium-catalyzed regioselective annulation of alkynylcyclopropanes with N-aryloxyamides.

The active catalytic species Cp*Rh(OAc)2 is formed which facilitates a facile ortho-C-H activation of 227 to afford a 5-membered rhodacycle intermediate 230 (Scheme 24). A regioselective migratory insertion of 230 into the alkyne of 228 forms a 7-membered rhodacycle 231 with the cyclopropane moiety away from rhodium center, probably due to the avoidance of the steric clash. Subsequent reductive elimination of 231 occurs to give a Rh(I) species 232, which undergoes oxidative addition to the N-O bond to generate the Rh(III) species 233. Finally, protonolysis for the demetalation of 233 gives the phenol precursor 234, which converts into the final product 229 via an intramolecular SN2′ ring opening of the vinylcyclopropane. However, another pathway featuring the conversion of the 7-membered rhodacycle 231 into a Rh(V) intermediate which undergoes reductive elimination to form the product 229 cannot be ruled out.

Table 37. Rh/Co relay catalyzed [3+2] C-H functionalization/annulation

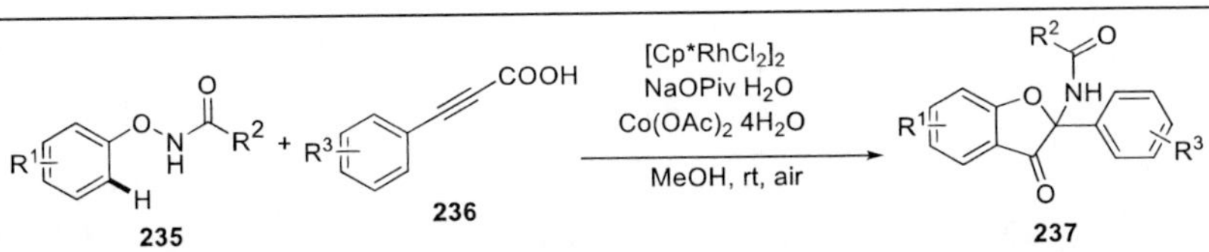

Product	Yield	Product	Yield
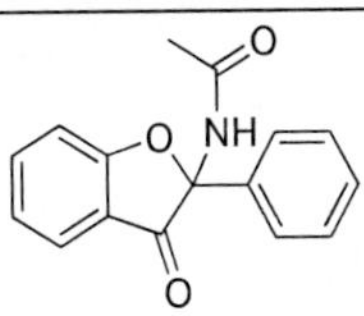	92%	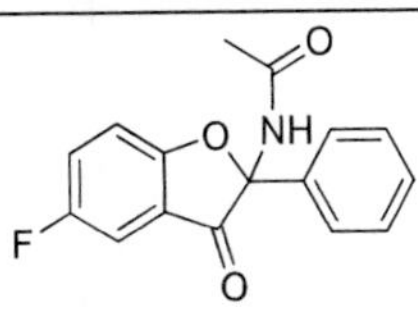	86%
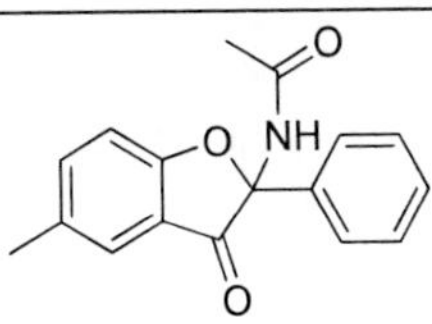	60%	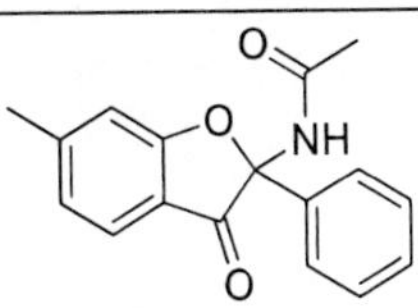	62%
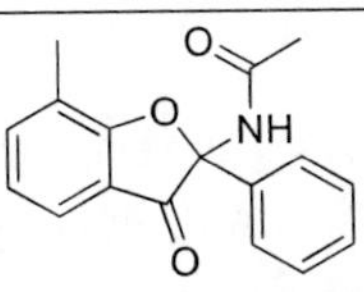	71%	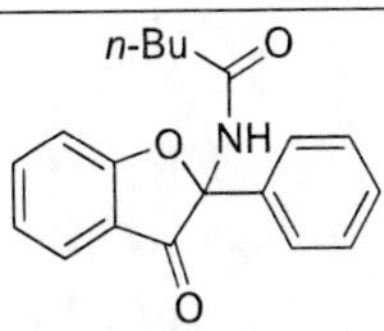	81%
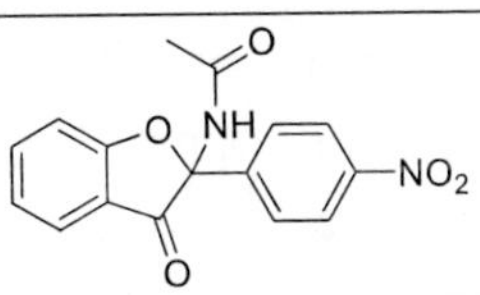	90%	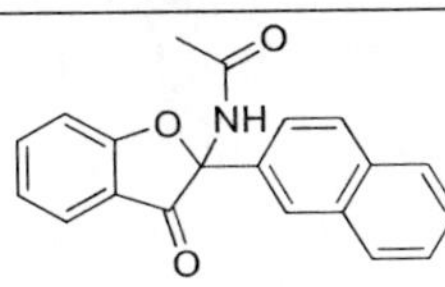	70%

Li reported the first Rh/Co relay catalyzed [3+2] C-H functionalization/annulation for the construction of benzofuran-3(2H)-one scaffolds 237 with a quaternary center from N-aryloxyacetamides 235 and propiolic acids 236 (Table 37) [89]. The authors investigated the substrate scope using various N-phenoxyacetamides 235 and propiolic acids 236, the resulting suggest that the N-phenoxyacetamides 235 bearing electron-withdrawing and electron-donating substituents at para-, meta- or ortho-position, the corresponding products 237 were isolated in moderate to good yields.

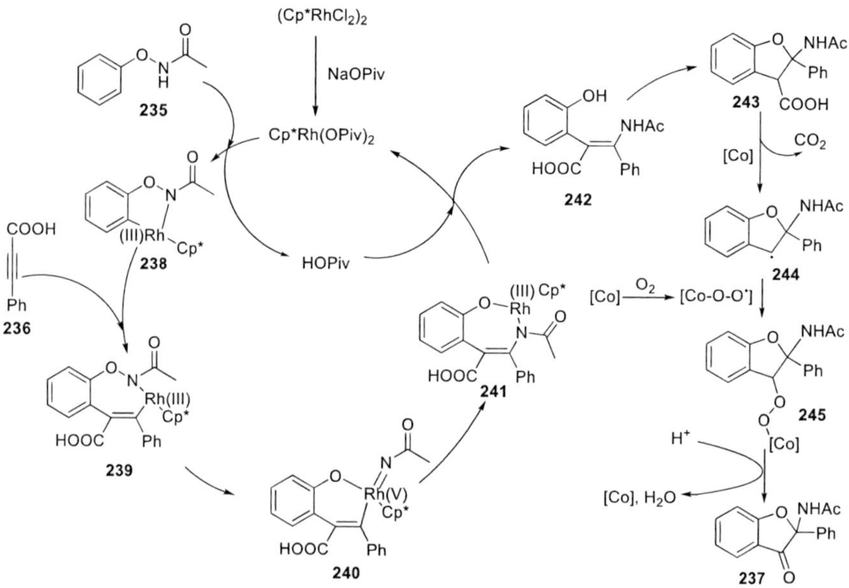

Scheme 25. Rh/Co relay catalyzed [3+2] C-H functionalization/annulation.

Then a possible mechanism for this reaction as shown, the coordination of 235 to the catalytically active species Cp*Rh(OPiv)2 from (Cp*RhCl2)2 and NaOPiv gives the rhodacycle complex 238 (Scheme 25). Next, an alkyne insertion of 236 to 238 provides intermediate 239. An intramolecular oxidative addition to the N-O bond gives the six-membered Rh(V) nitrenoid intermediate 240. Subsequent intramolecular nitrenoid migration produces a seven-membered Rh(III) intermediate 241. Then the protodemetalation of 241 provides the intermediate 242 along with the active catalyst Cp*Rh(OPiv)2. The cyclization of intermediate 242 provides 243, which undergoes a [Co]-mediated decarboxylation to generate the radical 244. The radical 244 traps the [Co–O–O•] species generated from [Co] under air to form hydroperoxide species 245, and 245 eventually provides the desired product 237.

Table 38. Rhodium-catalyzed cascade [3+2] redox-neutral annulation synthesis of benzofurans

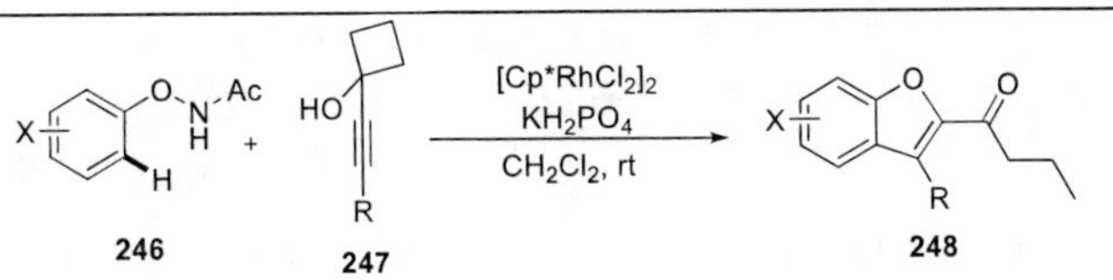

Product	Yield	Product	Yield
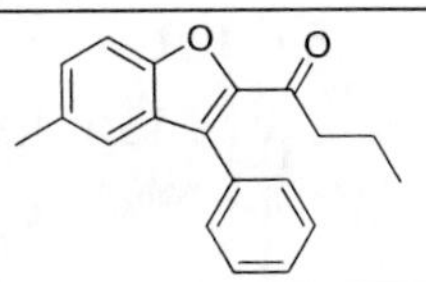	89%	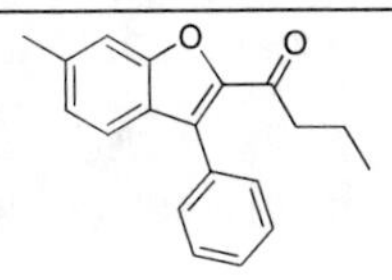	90%
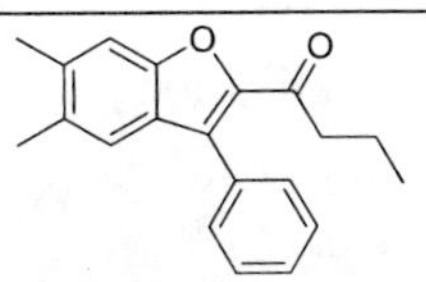	86%	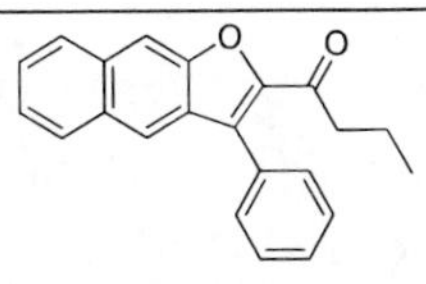	47%
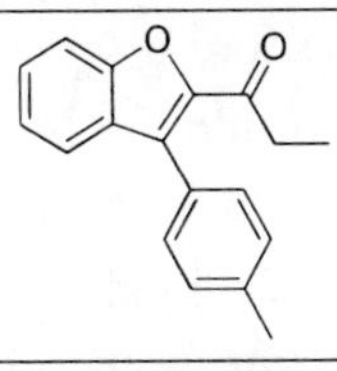	95%	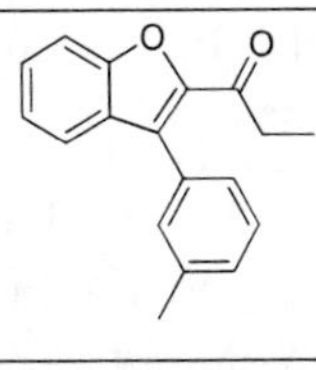	98%
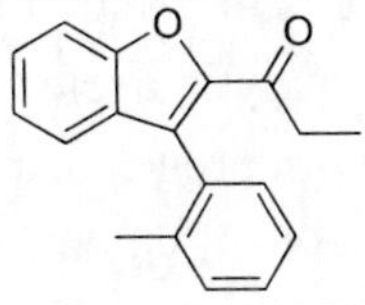	97%	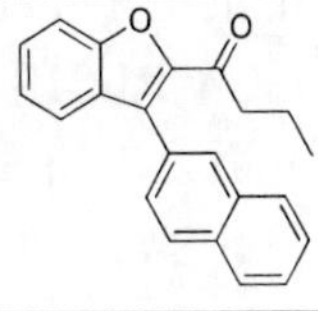	92%
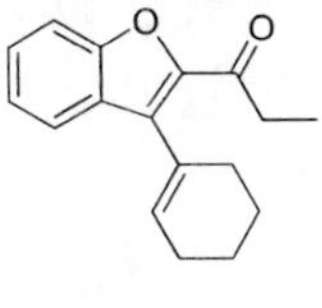	59%	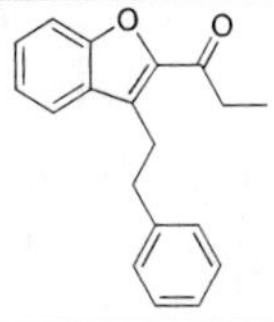	90%

Zhang and co-workers demonstrated Rh-(III)-catalyzed cascade [3+2] redox-neutral annulation and ring opening of N-aryloxyacetamides 246 with 1-alkynylcyclobutanols 247 for synthesis of benzofuran derivatives 248 (Table 38) [90]. Because alkynyl alcohols are widely used as important building blocks in synthetic chemistry [91-93]. In this reaction, -OH as the second traceless directing groups, O-NHAc served as both internal oxidant and directing group. This reaction has a broader substrate and mild reaction conditions. For example, 3,4-disubstituted and 2-naphthalenyl substrates were well-tolerated, and the reaction could be successfully applied to late-stage modification of a bioactive dopamine derivative. A wide range of 1-alkynylcyclobutanols 247 bearing different functionalities on the para-, meta-, or ortho-positions of the aromatic rings were well tolerated, giving the corresponding products in good to excellent yields.

Reduction of the product 248 was readily available using NaBH4 as a reductant, affording alcohol 249 in 94% yield. Epoxidation of 248 using m-CPBA as an oxidant, giving the epoxidation product 250 in 70% yield (Scheme 26). This transformation further proved versatility of the product. The possible reaction mechanism suggested the C-H activation via a concerted metalation-deprotonation (CMD) process. The regioselective migratory insertion of 247 into the C-Rh bond stemmed from the formation of hydrogen bonding between the dual directing groups to afford a seven-membered rhodacycle 252. The oxidative addition of Rh(III) into the N-O bond generated the Cp*Rh(V) nitrenoid 253, which afforded 254 through the reductive elimination and rhodium insertion into the C-C bond. The following metal protonation led to the formation of 248 and the regeneration of the active Cp*Rh(III) species for the next catalytic cycle.

Yi reported Ir(III)-catalyzed direct C-H alkynylation with arenes using commercially available TIPS-acetylene 256 as an efficient alkynylating reagent, using O-NHAc as an autocleavable oxidizing-directing group (Table 39) [94]. No desired product was deserved, when other terminal alkynes such as phenylacetylene and methyl propiolate instead of TIPS-acetylene 256, indicating that a TIPS substituent was crucial for this reaction.

The desilylation of 257 was easily achieved upon treatment with TBAF to provide 3-ethynylbiphenyl-4-ol 258, which could be used as a useful substrate for synthesis of highly attractive benzofuran derivatives 259 and 260 (Scheme 27).

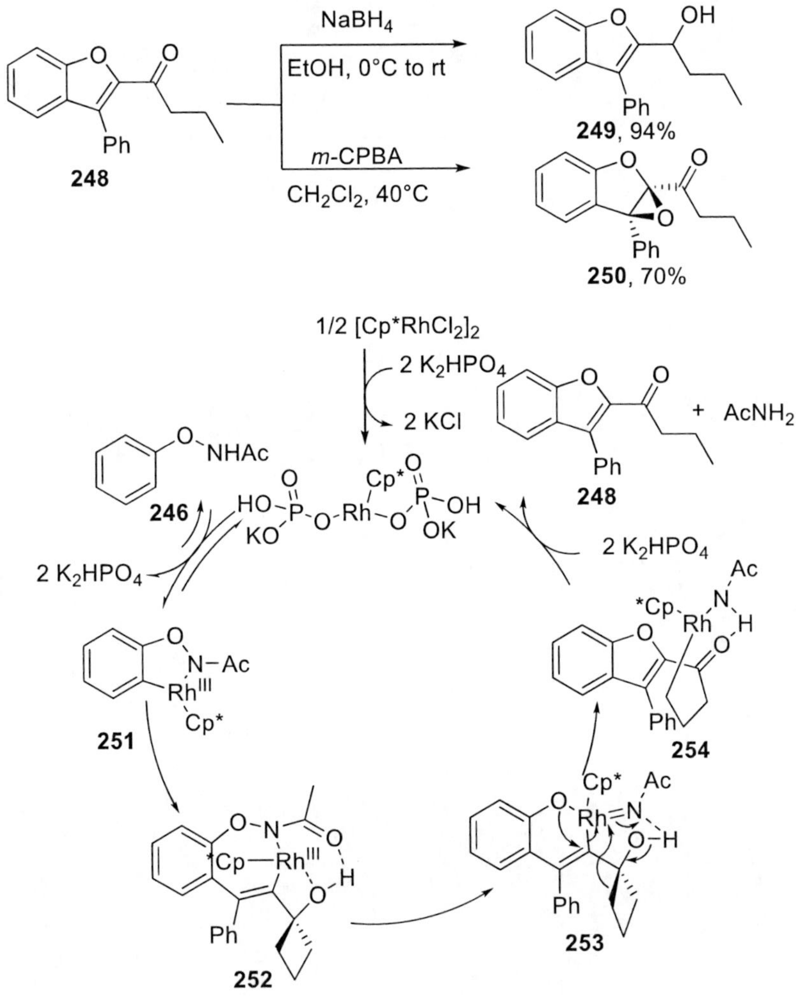

Scheme 26. Rhodium-catalyzed cascade [3+2] redox-neutral annulation synthesis of benzofurans and reduction/epoxidation.

Table 39. Iridium-catalyzed direct C-H alkynylation with arenes

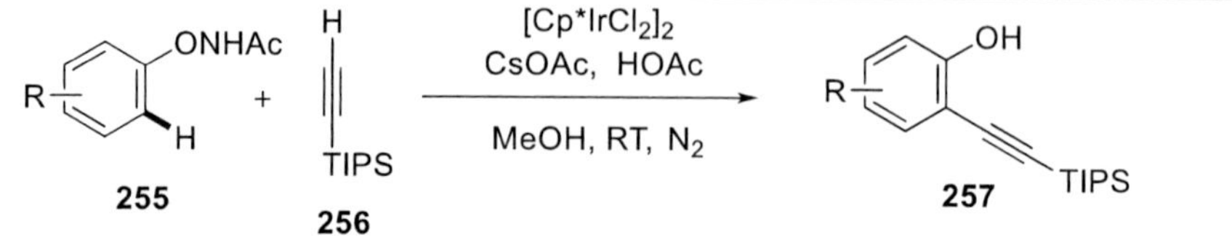

Product	Yield	Product	Yield
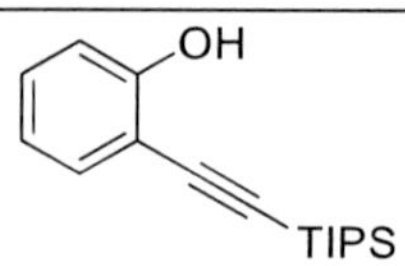	81%	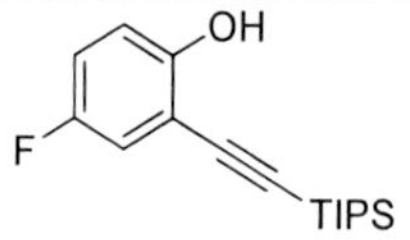	45%
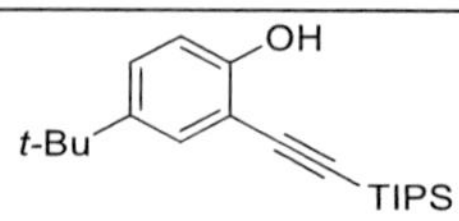	87%	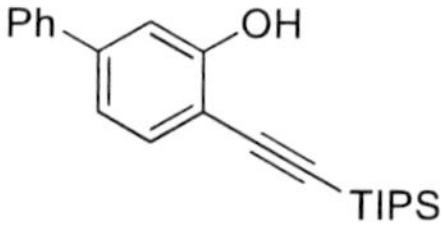	65%
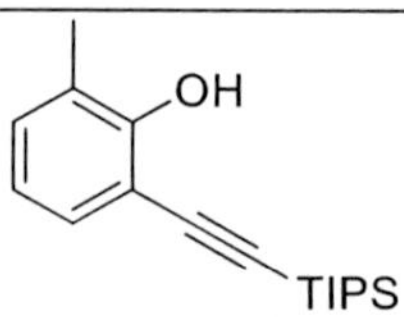	69%	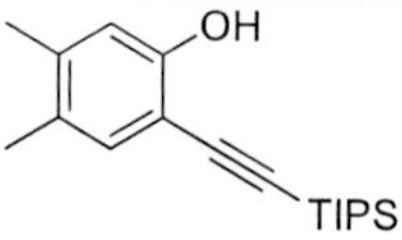	67%
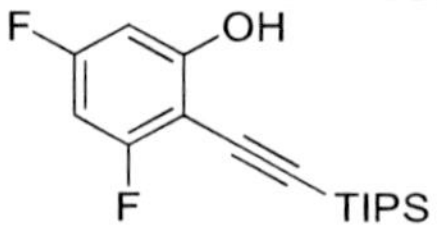	72%	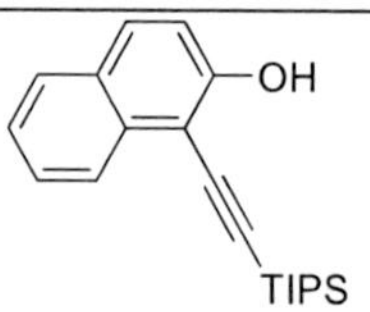	45%

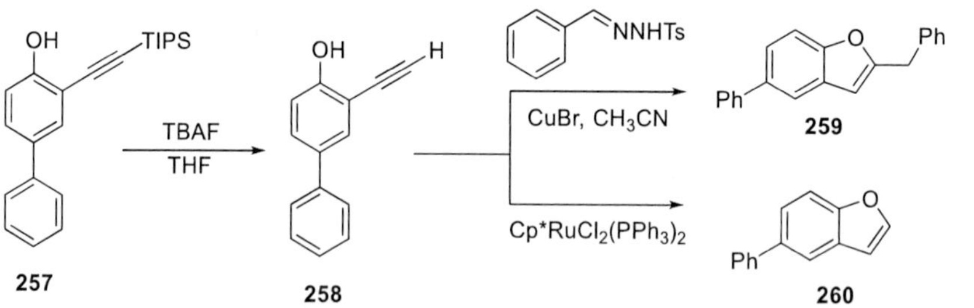

Scheme 27. Synthesis of benzofurans via annulation of 3-ethynylbiphenyl-4-ol.

Table 40. Iridium-catalyzed C-H annulation synthesis of benzofurans

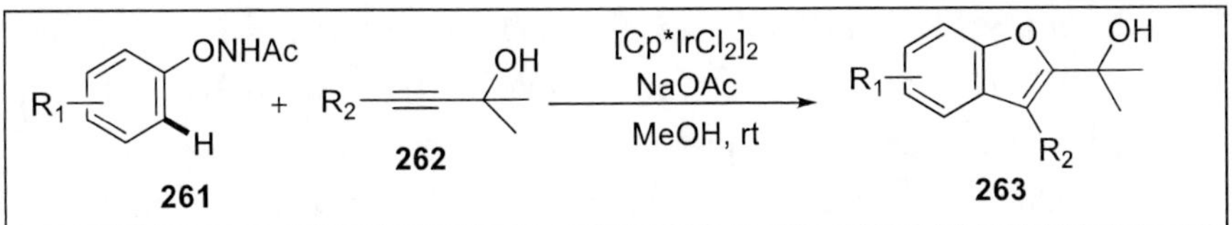

Product	Yield	Product	Yield
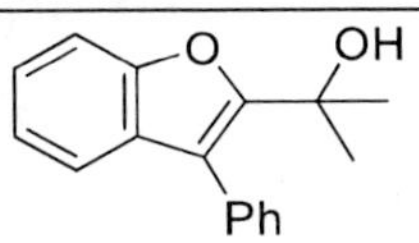	92%	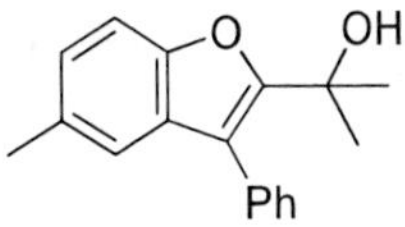	94%
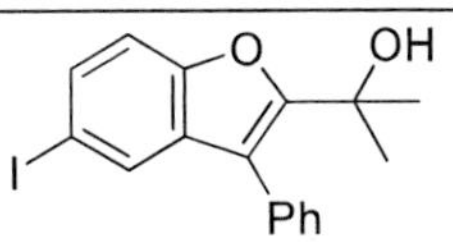	65%	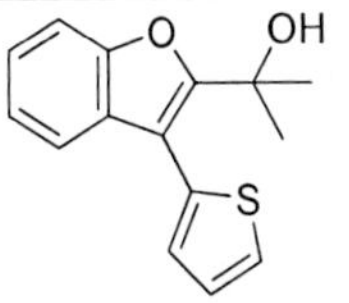	84%
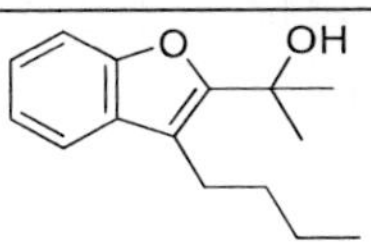	71%	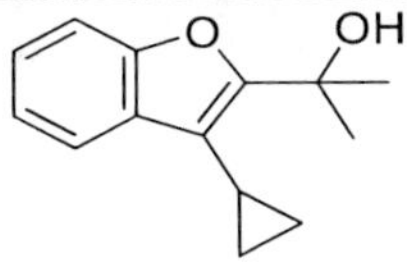	78%

Yi reported the first iridium(III)-catalyzed C-H annulation of N-phenoxyacetamides 261 with propargyl alcohols 262 for directed synthesis of benzofurans 263 (Table 40) [95]. Various N-phenoxyacetamides 261 bearing electron-withdrawing and electron-donating functional group were well tolerated in this reaction to give the desired benzofuran derivatives 263 in moderate to good yields with specific regioselectivity.

The control experiments and theoretical calculations suggested that the hydroxyl group in the tertiary propargyl alcohol substrate was found to be crucial for this transformation when the tertiary alcohol substrate was replaced by a steric similar tert-butyl-substituted alkyne or Boc-protected tertiary alcohol (Scheme 28). The hydroxyl group in the tertiary propargyl alcohol substrate was found to be crucial in controlling chemoselectivity by providing distinctive coordination modes with iridium catalysts. A plausible mechanism involving redox-neutral C-H activation and annulation is proposed. Cp*Ir(OAc)2 acted as active catalyst, followed by an irreversible C-H cleavage to yield the five-membered iridacycle intermediate 264. The following regioselective insertion of 262 generated intermediates 265 (major form) and 266 (minor form). The interaction mode in 265 potentially increases the reactivity by tuning the electronic property of the Ir center. Then, a plausible Ir(III)/Ir(V)/Ir(III) cycle was followed to yield the desired product 263 via an intramolecular oxidative addition/reductive elimination process (path a). Alternatively, another non-oxidative process could not be ruled out in which the intramolecular substitution occurred to form the C-O bond and break the O-N bond simultaneously (path b).

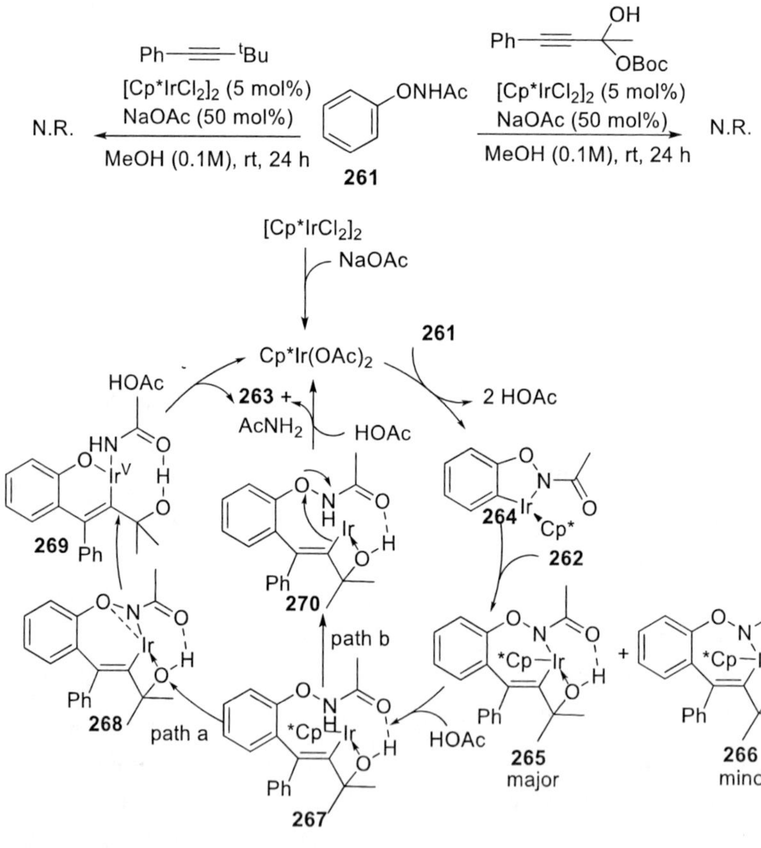

Scheme 28. Iridium-catalyzed C-H annulation synthesis of benzofurans.

Table 41. Rhodium-catalyzed redox-neutral cascade [3+2] annulation synthesis of benzofuran-2(*3H*)-ones

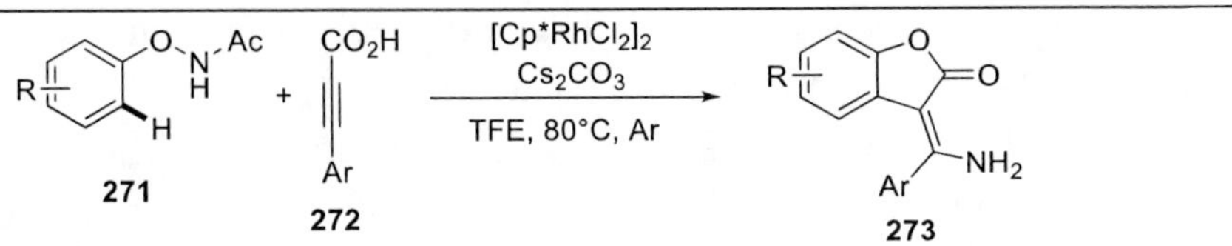

Product	Yield	Product	Yield
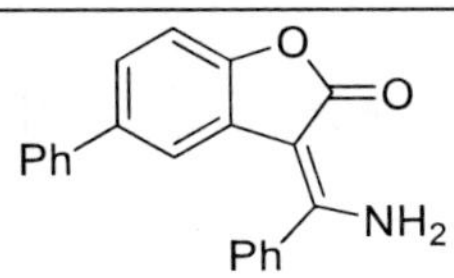	48%	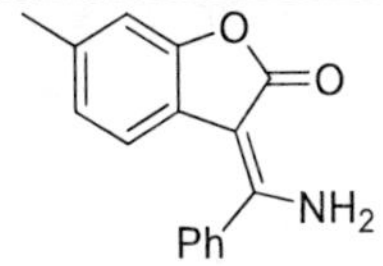	51%
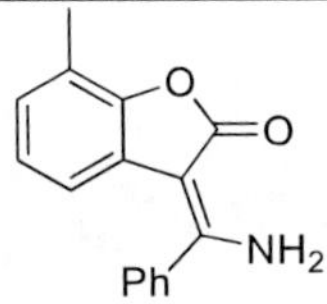	65%	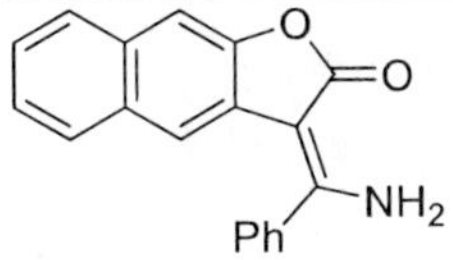	55%
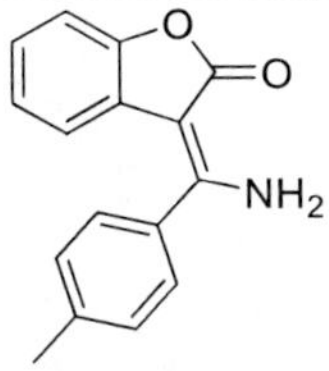	52%	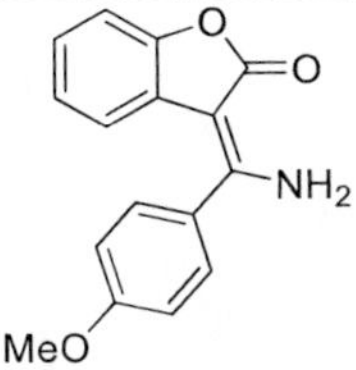	47%
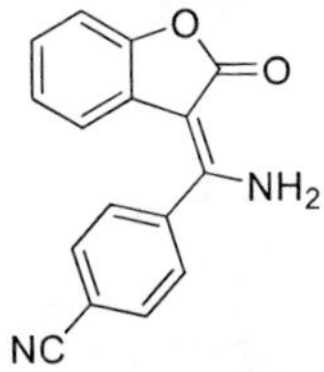	65%	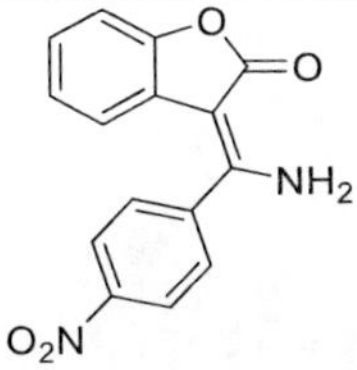	67%

Zhu reported the rhodium(III)-catalyzed redox-neutral cascade [3+2] annulation of N-aryloxyacetamides 271 with propiolic acids 272 for synthesis of benzofuran-2(3H)-ones 273 (Table 41) [96]. The reaction of N-aryloxyacetamides 271 and propiolic acids 272 did proceed leading to (Z)-benzofuran-2(3H)-one 273 in the presence of dimeric [Cp*RhC12]2 as a catalyst precursor, and Cs_2CO_3 as a base in TFE at 80°C for 12 h. It is interesting that no (E)-benzofuran-2(3H)-one and benzofuran-3(2H)-one were detected under these catalytic conditions. The selective formation of the Z isomer is likely due to the strong hydrogen-bonding between the amino proton and the oxygen atom of the carbonyl group. A wide range of substituted N-aryloxyacetamides 271 generally performed smoothly to provide the corresponding (Z)-3-aminomethylene benzofuran-2-one derivatives 273, when meta-substituted N-aryloxyacetamides were subjected to the reactions, yielding the expected products at sterically less hindered sites. A variety of differently substituted 3-(hetero)arylpropiolic acids 272 have been explored under the optimal reaction conditions. Electron-donating groups resulted in relatively lower reaction efficiency. In contrast, substrates bearing electron-withdrawing groups afforded the expected products in relatively higher yields.

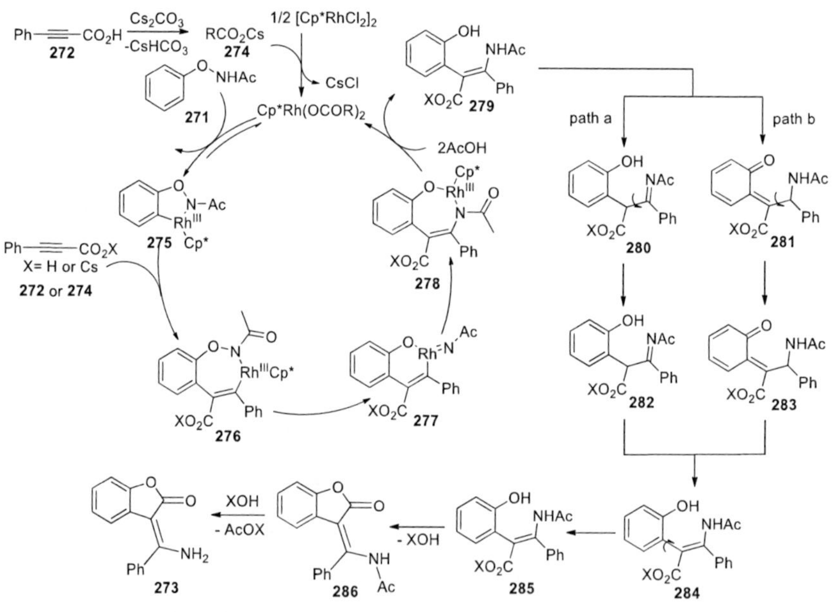

Scheme 29. Rhodium-catalyzed redox-neutral cascade [3+2] annulation synthesis of benzofuran-2(3H)-ones.

A plausible mechanism for the Cp*Rh(III)-catalyzed redox-neutral cascade [3+2] annulation reversible N-H deprotonation and C-H activation of 272 generated rhodacycle 275 (Scheme 29). The regioselective migratory insertion of 272 into the C-Rh bond gave 276. The formation of

278 was quite facile through oxidative addition of Rh into the O-N bond via Rh(V) nitrenoid 277 followed by reductive elimination. The protonation was involved to facilitate the generation of 279. The isomerization of the C-C double bond was achieved through enamine-imine (path a) or phenol-semiquinone (path b) tautomerization with promotion of the N-H…O type intramolecular interaction. The consecutive C-C single bond rotation, lactonization and hydrolysis afforded benzofuran-2-one derivative 273.

SUMMARY AND OUTLOOK

Over past years, significant progress has been made in Pd- or Rh-catalyzed benzofurans and their derivatives forming reactions through C-H bond functionalizations. Traditionally, the synthesis of benzofuran derivatives relies on the use of pre-functionalized starting precursors and the established methodologies were complicated and being limited by the scope of the substrates. Recent development in C-H activation involving the field of benzofuran synthesis has been achieved by using simple available starting materials, thereby avoiding wastes and thus providing a greener alternative. These methods are conceptually new and the mechanisms are unique. Most of the reactions proceed well to electron rich and electron poor substrates and most of the reactions follow a concerted metalation−deprotonation pathway.

However, a significant amount of work still needs to be done for the synthesis of benzofuran derivatives via C-H activation strategy. For example, stoichiometric oxidants such as silver and copper salts were required in Pd-catalyzed reactions and phenol derivatives were generated in some cases of Rh-catalyzed coupling reactions, which need additional steps to synthesize the final product. As a consequence, more environmental benign conditions through the use of air, electricity or visible light instead of metal salts need to be explored. Enantioselective synthetic reactions for such molecules via C-H bond activation are still in their infancy. We are excited to witness more developments in the growing field in the coming years.

Acknowledgments

This work was supported by the National Natural Science Foundation of China (21801152 and 21572110), the Natural Science Foundation of Shandong Province (ZR2019BB005 and ZR2019MB010).

References

[1] Okamoto, Y., Ojika, M. and Sakagami, Y. (1999). Iantheran A, a dimeric polybrominated benzofuran as a Na, K-ATPase inhibitor from a marine sponge, *Ianthella* sp. *Tetrahedron Letters*, 40, 507-510.

[2] Alvarezbuilla, J., Vaquero, J. J. and Barluenga, J. (2011). Modern heterocyclic chemistry. *Modern Heterocyclic Chemistry.*

[3] La Clair, J. J., Rheingold, A. L. and Burkart, M. D. (2011). Ganodone, a bioactive benzofuran from the fruiting bodies of *Ganoderma tsugae. Journal of Natural Products*, 74, 2045-2051.

[4] Simonetti, S. O., Larghi, E. L., Bracca, A. B. J. and Kaufman, T. S. (2013). Angular tricyclic benzofurans and related natural products of fungal origin. Isolation, biological activity and synthesis. *Natural Product Reports*, 30, 941-969.

[5] Miyaura, N. and Suzuki, A. (1995). Palladium-catalyzed cross-coupling reactions of organoboron compounds. *Chemical Reviews*, 95, 2457-2483.

[6] Jana, R., Pathak, T. P. and Sigman, M. S. (2011). Advances in transition metal (Pd,Ni,Fe)-catalyzed cross-coupling reactions using alkyl-organometallics as reaction partners. *Chemical Reviews*, 111, 1417-1492.

[7] Hartwig, J. F. (2010). Organotransition metal chemistry: from bonding to catalysis. *Organotransition Metal Chemistry: From Bonding to Catalysis.*

[8] Diederich, F. and Stang, P. J. (2004). Metal-catalyzed cross-coupling reactions. *Metal-Catalyzed Cross-Coupling Reactions.*

[9] Kakiuchi, F. and Murai, S. (2002). Catalytic C-H/olefin coupling. *Accounts of Chemical Research*, 35, 826-834.

[10] Daugulis, O., Do, H. Q. and Shabashov, D. (2009). Palladium- and Copper-catalyzed arylation of carbon-hydrogen bonds. *Accounts of Chemical Research*, 42, 1074-1086.

[11] Chen, X., Engle, K. M., Wang, D. H. and Yu, J. Q. (2009). Palladium(II)-catalyzed C-H activation/C-C cross-coupling reactions: versatility and practicality. *Angewandte Chemie-International Edition*, 48, 5094-5115.

[12] Ackermann, L. (2011). Carboxylate-assisted transition-metal-catalyzed C-H bond functionalizations: mechanism and scope. *Chemical Reviews*, 111, 1315-1345.

[13] Dey, A., Agasti S. and Maiti, D. (2016). Palladium catalysed *meta*-C-H functionalization reactions. *Organic & Biomolecular Chemistry*, 14, 5440-5453.

[14] Nakamura, I. and Yamamoto, Y. (2004). Transition-metal-catalyzed reactions in heterocyclic synthesis. *Chemical Reviews*, 104, 2127-2198.

[15] Blum, T. R., Zhu Y., Nordeen, S. A. and Yoon, T. P. (2014). Photocatalytic synthesis of dihydrobenzofurans by oxidative [3+2] cycloaddition of phenols. *Angewandte Chemie-International Edition*, 53, 11056-11059.

[16] Liu, L., Ji, X. Y., Dong, J. Y., Zhou, Y. B. and Yin, S. F. (2016). Metal-free oxidative annulation of 2-naphthols with terminal alkynes affording 2-arylnaphtho[2,1-*b*]furans. *Organic Letters*, 18, 3138-3141.

[17] Guo, X. W., Yu, R., Li, H. J. and Li, Z. P. (2009). Iron-catalyzed tandem oxidative coupling and annulation: an efficient approach to

construct polysubstituted benzofurans. *Journal of the American Chemical Society*, 131, 17387-17393.

[18] Kundu, D., Samim, M., Majee, A. and Hajra, A. (2011). Indium triflate-catalyzed coupling between nitroalkenes and phenol/naphthols: a simple and direct synthesis of arenofurans by a cyclization reaction. *Chemistry-An Asian Journal*, 6, 406-409.

[19] Ackermann, L. and Kaspar, L. T. (2007). $TiCl_4$-catalyzed indirect anti-markovnikov hydration of alkynes: application to the synthesis of benzo[*b*]furans. *The Journal of Organic Chemistry*, 72, 6149-6153.

[20] Anderson, K. W., Ikawa, T., Tundel, R. E. and Buchwald, S. L. (2006). The selective reaction of aryl halides with KOH: synthesis of phenols, aromatic ethers, and benzofurans. *Journal of the American Chemical Society*, 128, 10694-10695.

[21] Markina, N. A., Chen, Y. and Larock, R. C. (2013). Efficient microwave-assisted one-pot three-component synthesis of 2,3-disubstituted benzofurans under Sonogashira conditions. *Tetrahedron*, 69, 2701-2713.

[22] Willis, M. C., Taylor, D. and Gillmore, A. T. (2004). Palladium-catalyzed intramolecular *o*-arylation of enolates: application to benzo[*b*]furan synthesis. *Organic Letters*, 6, 4755-4757.

[23] Pei, T., Chen, C. Y., DiMichele, L. and Davies, I. W. (2010). Controlled synthesis of 2- and 3-substituted benzo[*b*]furans. *Organic Letters*, 12, 4972-4975.

[24] Zhao, J. and Larock, R. C. (2006). Synthesis of substituted carbazoles, indoles, and dibenzofurans by vinylic to aryl palladium migration. *The Journal of Organic Chemistry*, 71, 5340-5348.

[25] Thansandote, P. and Lautens, M. (2009). Construction of nitrogen-containing heterocycles by C-H bond functionalization. *Chemistry-A European Journal*, 15, 5874-5883.

[26] Wang, X. S., Lu, Y., Dai, H. X. and Yu, J. Q. (2010). Pd(II)-catalyzed hydroxyl-directed C-H activation/C−O cyclization: expedient construction of dihydrobenzofurans. *Journal of the American Chemical Society*, 132, 12203-12205.

[27] Xiao, B., Gong, T. J., Liu, Z. J., Liu, J. H., Luo, D. F., Xu, J. and Liu, L. (2011). Synthesis of dibenzofurans via palladium-catalyzed phenol-directed C-H activation/C-O cyclization. *Journal of the American Chemical Society*, 133, 9250-9253.

[28] Kuram, M. R., Bhanuchandra M. and Sahoo, A. K. (2013). Direct access to benzo[*b*]furans through Palladium-catalyzed oxidative annulation of phenols and unactivated internal alkynes. *Angewandte Chemie-International Edition*, 52, 4607-4612.

[29] Furstner, A. and Davies, P. W. (2005). Heterocycles by $PtCl_2$-catalyzed intramolecular carboalkoxylation or carboamination of alkynes. *Journal of the American Chemical Society*, 127, 15024-15025.

[30] Nakamura, I., Mizushima, Y. and Yamamoto, Y. (2005). Synthesis of 2,3-disubstituted benzofurans by platinum-olefin-catalyzed carboalkoxylation of *o*-alkynylphenyl acetals. *Journal of the American Chemical Society*, 127, 15022-15023.

[31] Yue, D. W., Yao, T. L. and Larock, R. C. (2005). Synthesis of 2,3-disubstituted benzo[*b*]furans by the palladium-catalyzed coupling of *o*-iodoanisoles and terminal alkynes, followed by electrophilic cyclization. *The Journal of Organic Chemistry*, 70, 10292-10296.

[32] Genin, E., Amengual, R., Michelet, V., Savignac, M., Jutand, A., Neuville, L. and Genet, J. P. (2004). A novel water-soluble *m*-TPPTC ligand: steric and electronic features-recent developments in Pd- and Rh-catalyzed C-C bond formations *Advanced Synthesis & Catalysis*, 346, 1733-1741.

[33] Li, C. L., Zhang, Y. C., Li, P. H. and Wang, L. (2011). Palladium-catalyzed oxidative cyclization of 3-phenoxyacrylates: an approach to construct substituted benzofurans from phenols. *The Journal of Organic Chemistry*, 76, 4692-4696.

[34] Kumar, M. P. and Liu, R. S. (2006). $Zn(OTf)_2$-catalyzed cyclization of proparyl alcohols with anilines, phenols, and amides for synthesis of indoles, benzofurans, and oxazoles through different annulation mechanisms. *The Journal of Organic Chemistry*, 71, 4951-4955.

[35] Wang, S. H., Li, P. H., Yu, L. and Wang, L. (2011). Sequential and one-pot reactions of phenols with bromoalkynes for the synthesis of (*Z*)-2-bromovinyl phenyl ethers and benzo[*b*]furans. *Organic Letters*, 13, 5968 -5971.

[36] Moritani, I. and Fujiwara, Y. (1967). Aromatic substitution of styrene-palladium chloride complex. *Tetrahedron Letters*, 8, 1119-1122.

[37] Fujiwara, Y., Moritani, I., Danno, S., Asano, R. and Teranishi, S. (1969). Aromatic substitution of olefins. VI. Arylation of olefins with palladium(II) acetate. *Journal of the American Chemical Society*, 91, 7166-7169.

[38] Fujiwara, Y., Asano, R., Moritani, I. and Teranishi, S. (1976). Aromatic substitution of olefins. XXV. Reactivity of benzene, naphthalene, ferrocene, and furan toward styrene, and the substituent effect on the reaction of monosubstituted benzenes with styrene. *The Journal of Organic Chemistry*, 41, 1681-1683.

[39] Zhang, H. M., Ferreira, E. M. and Stoltz, B. M. (2004). Direct oxidative Heck cyclizations: intramolecular Fujiwara–Moritani arylations for the synthesis of functionalized benzofurans and dihydrobenzofurans. *Angewandte Chemie-International Edition*, 43, 6144-6148.

[40] Ferreira, E. M. and Stoltz, B. M. (2003). Catalytic C-H bond functionalization with palladium(II): aerobic oxidative annulations of indoles. *Journal of the American Chemical Society*, 125, 9578-9579.

[41] Youn, S. W. and Eom, J. I. (2005). Facile construction of the benzofuran and chromene ring systems via Pd^{II}-catalyzed oxidative cyclization. *Organic Letters*, 7, 3355-3358.

[42] Tsuji, J. (2004). Transition metal reagents and catalysts. *Transition Metal Reagents and Catalysts.*

[43] Diederich, F. and Stang, P. J. (1998). Metal-catalyzed cross-coupling reactions. *Metal-Catalyzed Cross-Coupling Reactions.*

[44] Zhao, B. W. and Lu, X. Y. (2006). Cationic palladium(II)-catalyzed addition of arylboronic acids to nitriles. One-step synthesis of

benzofurans from phenoxyacetonitriles. *Organic Letters*, 8, 5987-5990.

[45] Zhao, B. W. and Lu, X. Y. (2002). Pd^{II}-catalyzed cyclization of alkynes containing aldehyde, ketone, or nitrile groups initiated by the acetoxypalladation of alkynes. *Angewandte Chemie-International Edition*, 41, 4343-4345.

[46] Zhao, B. W. and Lu, X. Y. (2006). Palladium(II)-catalyzed addition of arylboronic acid to nitriles. *Tetrahedron Letters*, 47, 6765-6768.

[47] Malamas, M. S., Sredy, J., Moxham, C., Katz, A., Xu, W., McDevitt, R., Adebayo, F. O., Sawicki, D. R., Seestaller, L., Sullivan, D. and Taylor, J. R. (2000). Novel benzofuran and benzothiophene biphenyls as inhibitors of protein tyrosine phosphatase 1B with antihyperglycemic properties. *Journal of Medicinal Chemistry*, 43, 1293-1310.

[48] Black, D. S., Craig, D. C., Kumar, N. and Rezaie, R. (2002). Acid-catalysed reactions of activated benzofuranylmethanols: formation of calixbenzofurans. *Tetrahedron*, 58, 5125-5134.

[49] Boehm, T. L. and Showalter, H. D. H. (1996). Development of a novel silyl ether linker for solid-phase organic synthesis. *The Journal of Organic Chemistry*, 61, 6498-6499.

[50] Smith, R. A., Chen, J., Mader, M. M., Muegge, I., Moehler, U., Katti, S., Marrero, D., Stirtan, W. G., Weaver, D. R., Xiao, H. and Carley, W. (2002). Solid-Phase synthesis and investigation of benzofurans as selective estrogen receptor modulators. *Bioorganic & Medicinal Chemistry Letters*, 12, 2875-2878.

[51] Park, K. K. and Jeong, J. (2005). Facile synthesis of regio-isomeric naphthofurans and benzodifurans. *Tetrahedron*, 61, 545-553.

[52] Carrer, A., Brinet, D., Florent, J. C., Rousselle, P. and Bertounesque, E. (2012). Palladium-catalyzed direct arylation of polysubstituted benzofurans. *The Journal of Organic Chemistry*, 77, 1316-1327.

[53] Dwight, T. A., Rue, N. R., Charyk, D., Josselyn, R. and DeBoef, B. (2007). C-C bond formation via double C-H functionalization: aerobic oxidative coupling as a method for synthesizing heterocoupled biaryls. *Organic Letters*, 9, 3137-3139.

[54] Yang, M. Y., Jiang, X. Y., Shi, W. J., Zhu, Q. L. and Shi, Z. J. (2015). Direct Lactonization of 2-Arylacetic Acids through Pd(II)-Catalyzed C-H Activation/C-O Formation. *Organic Letters*, 15, 690-693.

[55] Sharma, U., Naveen, T., Maji, A., Manna, S. and Maiti, D. (2013). Palladium-catalyzed synthesis of benzofurans and coumarins from phenols and olefins. *Angewandte Chemie-International Edition*, 52, 12669-12673.

[56] Dehmlow, H., Aebi, J. D., Jolidon, S., Ji, Y. H., von der Mark, E. M., Himber, J. and Morand, O. H. (2003). Synthesis and structure-activity studies of novel orally active non-terpenoic 2,3-oxidosqualene cyclase inhibitors. *Journal of Medicinal Chemistry*, 46, 3354-3370.

[57] Greve, H., Meis, S., Kassack, M. U., Kehraus, S., Krick, A., Wright A. D. and Konig, G. M. (2007). New iantherans from the marine sponge *ianthella quadrangulata*: novel agonists of the $P2Y_{11}$ receptor. *Journal of Medicinal Chemistry*, 50, 5600-5607.

[58] Ribeiro-Rodrigues, R., dos Santos, W. G., Oliveira, A. B., Snieckus, V., Zani, C. L. and Romanha, A. J. (1995). Growth inhibitory effect of naphthofuran and naphthofuranquinone derivatives on *Trypanosoma cruzi* epimastigotes. *Bioorganic & Medicinal Chemistry Letters*, 5, 1509-1512.

[59] Agasti, S., Sharma, U., Naveen, T. and Maiti, D. (2015). Orthogonal selectivity with cinnamic acids in 3-substituted benzofuran synthesis through C-H olefination of phenols. *Chemical Communications*, 51, 5375-5378.

[60] Neely, J. M. and Rovis, T. (2014). Rh(III)-catalyzed decarboxylative coupling of acrylic acids with unsaturated oxime esters: carboxylic acids serve as traceless activators. *Journal of the American Chemical Society*, 136, 2735-2738.

[61] Cornella, J., Righi, M. and Larrosa, I. (2011). Carboxylic acids as traceless directing groups for formal *meta*-selective direct arylation. *Angewandte Chemie-International Edition*, 50, 9429-9432.

[62] Agasti, S., Maity, S., Szabo K. J. and Maiti, D. (2015). Palladium-catalyzed synthesis of 2,3-disubstituted benzofurans: an approach towards the synthesis of deuterium labeled compounds. *Advanced Synthesis & Catalysis*, 357, 2331-2338.

[63] Satoh, T. and Miura, M. (2010). Oxidative coupling of aromatic substrates with alkynes and alkenes under rhodium catalysis. *Chemistry-A European Journal*, 16, 11212-11222.

[64] Colby, D. A., Bergman, R. G. and Ellman, J. A. (2010). Rhodium-catalyzed C-C bond formation via heteroatom-directed C-H bond activation. *Chemical Reviews*, 110, 624-655.

[65] Song, G. Y., Wang F. and Li, X. W. (2012). C-C, C-O and C-N bond formation via rhodium(III)-catalyzed oxidative C-H activation. *Chemical Society Reviews*, 41, 3651-3678.

[66] Zhu, R. Y., Wei, J. B. and Shi, Z. J. (2013). Benzofuran synthesis via copper-mediated oxidative annulation of phenols and unactivated internal alkynes. *Chemical Science*, 4, 3706-3711.

[67] Zeng, W., Wu, W. J., Jiang, H. F., Huang, L. B., Sun, Y. D., Chen Z. W. and Li, X. W. (2013). Facile synthesis of benzofuransvia copper-catalyzed aerobic oxidative cyclization of phenols and alkynes. *Chemical Communications*, 49, 6611-6613.

[68] Moon, Y., Kim, Y., Hong, H. and Hong, S. (2013). Synthesis of heterocyclic-fused benzofuransvia C-H functionalization of flavones and coumarins. *Chemical Communications*, 49, 8323-8325.

[69] Huang, H. W., Ji, X. C., Wu, W. Q. and Jiang, H. F. (2015). Transition metal-catalyzed C-H functionalization of *N*-oxyenamine internal oxidants. *Chemical Society Reviews*, 44, 1155-1171.

[70] Zhu, R. Y., Farmer, M. E., Chen, Y. Q. and Yu, J. Q. (2016). A simple and versatile amide directing group for C-H functionalizations. *Angewandte Chemie-International Edition*, 55, 10578-10599.

[71] Zhu, C., Wang, C. Q. and Feng, C. (2018). Recent advance in transition-metal-catalyzed oxidant-free 4+1 annulation through C-H bond activation. *Tetrahedron Letters*, 5, 430-437.

[72] Wang, Z. R., Xie, P. P. and Xia, Y. Z. (2018). Recent progress in Ru(II)-catalyzed C-H activations with oxidizing directing groups. *Chinese Chemical Letters*, 29, 47-53.

[73] Liu, G. X., Shen, Y. Y., Zhou, Z. and Lu, X. Y. (2013). Rhodium(III)-catalyzed redox-neutral coupling of *N*-phenoxyacetamides and alkynes with tunable selectivity. *Angewandte Chemie-International Edition*, 52, 6033-6037.

[74] Zhou, Z., Liu, G. X., Shen Y. Y. and Lu, X. Y. (2014). Synthesis of benzofurans via ruthenium-catalyzed redox-neutral C-H functionalization and reaction with alkynes under mild conditions. *Organic Chemistry Frontiers*, 1, 1161-1165.

[75] Chen, Y., Wang, D. Q., Duan, P. P., Ben, R., Dai, L., Shao, X. R., Hong, M., Zhao, J. and Huang, Y. (2014). A multitasking functional group leads to structural diversity using designer C-H activation reaction cascades. *Nature Communications*, 5, 4610.

[76] Duan, P. P., Yang, Y. F., Ben, R., Yan, Y. Y., Dai, L., Hong, M., Wu, Y. D., Wang, D. Q., Zhang X. H. and Zhao J. (2014). Palladium-catalyzed benzo[*d*]isoxazole synthesis by C-H activation/[4+1] annulation. *Chemical Science*, 5, 1574-1578.

[77] Hu, F. D., Xia, Y., Ye, F., Liu, Z. X., Ma, C., Zhang, Y. and Wang, J. B. (2014). Rhodium(III)-catalyzed *ortho* alkenylation of *N*-phenoxyacetamides with *N*-tosylhydrazones or diazoesters through C-H activation. *Angewandte Chemie-International Edition*, 53, 1364-1367.

[78] Zhou, J., Shi, J. J., Liu, X. L., Jia, J. L., Song, H. C., Xu H. E. and Yi, W. (2015). Rh(III)-catalyzed and alcohol-involved carbenoid C-H insertion into *N*-phenoxyacetamides using α-diazomalonates. *Chemical Communications*, 51, 5868-5871.

[79] Zhou, Z., Liu, G. X., Chen, Y. and Lu, X. Y. (2015). Cascade synthesis of 3-alkylidene dihydrobenzofuran derivatives via rhodium(III)-catalyzed redox-neutral C-H functionalization/ cyclization. *Organic Letters*, 17, 5874-5877.

[80] Li, Y., Tang, Y. H., He, X., Shi, D. D., Wu, J. and Xu, S. L. (2017). Rhodium(III)-catalyzed annulative carbooxygenation of 1,1-

disubstituted alkenes triggered by C-H activation. *Chemistry-A European Journal*, 23, 7453-7457.

[81] Wu, Y. X., Chen, Z. Q., Yang, Y. X., Zhu, W. L. and Zhou B. (2018). Rh(III)-catalyzed redox-neutral unsymmetrical C-H alkylation and amidation reactions of *N*-phenoxyacetamides. *Journal of the American Chemical Society*, 140, 42-45.

[82] Yi, W., Chen, W. J., Liu, F. X. M., Zhong, Y. T., Wu, D., Zhou, Z. and Gao, H. (2018). Rh(III)-catalyzed and solvent-controlled chemoselective synthesis of chalcone and benzofuran frameworks via synergistic dual directing groups enabled regioselective C-H functionalization: a combined experimental and computational study. *ACS Catalysis*, 8, 9508-9519.

[83] Pan, J. L., Xie, P. P., Chen, C., Hao, Y., Liu, C., Bai, H. Y., Ding, J., Wang, L. R., Xia, Y. Z. and Zhang, S. Y. (2018). Rhodium(III)-catalyzed redox-neutral cascade [3+2] annulation of *N*-Phenoxyacetamides with propiolates via C-H functionalization/ isomerization/lactonization. *Organic Letters*, 20, 7131-7136.

[84] Wang, Y. B., Chen, Y., Yang, Y. X. and Zhou, B. (2018). A Rh(III)-catalyzed redox-neutral C-H alkylation reaction with allylic alcohols by using a traceless oxidizing directing group. *Organic Chemistry Frontiers*, 5, 1844-1847.

[85] Li, Y., Shi, D. D., Tang, Y. H. He, X. and Xu, S. L. (2018). Rhodium(III)-catalyzed redox-neutral C-H activation/annulation of *N*-aryloxyacetamides with alkynyloxiranes: synthesis of highly functionalized 2,3-dihydrobenzofurans. *The Journal of Organic Chemistry*, 83, 9464-9470.

[86] Zhang, Y., He, Y., Li, L. S., Ji, M. M., Li, X. Z. and Zhu, G. G. (2018). Synthesis of polyaryl-substituted olefins via a Rh(III)-catalyzed one-pot reaction using *N*-phenoxyacetamides, ketones, and hydrazines. *The Journal of Organic Chemistry*, 83, 2898-2903.

[87] Wang, H. N., Wang, B. Q. and Li, B. (2017). Synthesis of 3-arylbenzofuran-2-ylphosphines via rhodium-catalyzed redox-neutral C-H activation and their applications in palladium-catalyzed cross-

coupling of aryl chlorides. *The Journal of Organic Chemistry*, 82, 9560-9569.

[88] Li, Y., Shi, D. D., He, X., Wang, Y. Z., Tang, Y. H., Zhang, J. J. and Xu, X. L. (2019). Redox-Neutral annulation of alkynylcyclopropanes with *N*-aryloxyamides via rhodium(III)-catalyzed sequential C-H/C-C activation. *The Journal of Organic Chemistry*, 84, 1588-1595.

[89] Yuan, W. K., Zhu, M. H., Geng, R. S., Ren, G. Y., Zhang, L. B., Wen, L. R. and Li, M. (2019). Construction of benzofuran-3(*2H*)-one scaffolds with a quaternary center via Rh/Co relay catalyzed C-H functionalization/annulation of *N*-aryloxyacetamides and propiolic acids. *Organic Letters*, 21, 1654-1658.

[90] Pan, J. L., Liu, C., Chen, C., Liu, T. Q., Wang, M., Sun, Z. L. and Zhang, S. Y. (2019). Dual directing-groups-assisted redox-neutral annulation and ring opening of *N*-Aryloxyacetamides with 1-alkynylcyclobutanols via rhodium(III)-catalyzed C-H/C-C activations. *Organic Letters*, 21, 2823-2827.

[91] Zhou, W., Mei, Y. L., Li, B., Guan, Z. Y. and Deng, Q. H. (2018). Synthesis of *β*-alkyl 2-hydroxychalcones by rhodium-catalyzed coupling of *N*-phenoxyacetamides and nonterminal propargyl alcohols. *Organic Letters*, 20, 5808-5812.

[92] Markham, J. P., Staben, S. T. and Toste, F. D. (2005). Gold(I)-catalyzed ring expansion of cyclopropanols and cyclobutanols. *Journal of the American Chemical Society*, 127, 9708-9709.

[93] Jones, C., Nguyen, Q. and Driver, T. G. (2014). Dirhodium(II) carboxylate catalyzed formation of 1,2,3-trisubstituted indoles from styryl azides. *Angewandte Chemie-International Edition*, 53, 785-788.

[94] Zhou, J., Shi, J. J., Qi, Z. S., Li, X. W., Xu, H. E. and Yi, W. (2015). Mild and efficient Ir(III)-catalyzed direct C-H alkynylation of *N*-phenoxyacetamides with terminal alkyne. *ACS Catalysis*, 5, 6999-7003.

[95] Chen, W. J., Liu, F. X. M., Gong, W. C., Zhou, Z., Gao, H., Shi, J. J., Wu, B. and Yi, W. (2018). Hydroxyl group-prompted and iridium(III)-catalyzed regioselective C-H annulation of *N*-phenoxyacetamides with propargyl alcohols. *Advanced Synthesis & Catalysis*, 360, 2470-2475.

[96] Pan, J. L., Liu, T. Q., Chen, C., Li, Q. Z., Jiang, W., Ding, T. M., Yan, Z. Q. and Zhu, G. D. (2019). Rhodium(III)-catalysed cascade [3+2] annulation of *N*-aryloxyacetamides with 3-(hetero) arylpropiolic acids: synthesis of benzofuran-2(*3H*)-ones. *Organic & Biomolecular Chemistry*, 17, 8589-8600.

In: Benzofuran: Production and Applications ISBN: 978-1-53617-873-9
Editor: Murilo Souza Barros

Chapter 4

BENZOFURANS SCAFFOLDS: APPLICATIONS IN ANTICANCER DRUGS DISCOVERY AND DEVELOPMENT

Jorddy Neves Cruz*[1,*]*, Kauê Santana da Costa*[2]*,
Marcia Moraes Cascaes*[3]*,
Fernanda Wariss Figueiredo Bezerra*[4]*,
Ingryd Nayara de Farias Ramos*[5]*,
Mozaniel Santana de Oliveira*[1]*,
***Eloisa Helena de Aguiar Andrade*[1]
***and Cleydson Breno Rodrigues dos Santos*[6]

[1]Botany Coordinating, Paraense Emilio Goeldi Museum, Belém, PA, Brazil
[2]Institute of Biodiversity, Federal University of Western Pará, Santarém, Pará, Brazil
[3]Program of Post-Graduation in Chemistry, Federal University of Pará, Belém, Pará, Brazil
[4]Faculty of Food Engineering, Federal University of Pará, Belém, Pará, Brazil

[*] Corresponding Author's E-mail: jorddynevescruz@gmail.com.

[5]Institute of Biological Sciences, Federal University of Pará, Belém, Pará, Brazil
[6]Laboratory of Modeling and Computational Chemistry, Federal University of Amapá, Macapá, AP, Brazil

ABSTRACT

In this review, we exhibit a general view of aspects related to cancer initiation and progress, as well as, the applications of benzofurans as structural frameworks for the development of new drugs against cancer. Using recent studies in the scientific literature, we discuss the cytotoxic activities of compounds containing benzofurans as structural core against different cancer cell lines and their applications to the design and discovery of new benzofuran derivatives with biological activities against interesting molecular targets for cancer therapy.

Keywords: benzofurans, cancer, anticancer activity, privileged scaffold

1. BENZOFURAN AS MULTIFUNCTIONAL SCAFFOLD WITH PHARMACOLOGICAL AND BIOLOGICAL APPLICATIONS

Benzofurans are heterocyclic compounds isolated from different natural origins, but also could be obtained from synthetic methods. Compound containing benzofuran ring as a structural core exhibit a wide range of pharmacological and biological applications (Khodarahmi, Asadi, Hassanzadeh, et al. 2015; Bhargava and Rathore 2017). Recently, their molecular frameworks have drawn special attention as an interesting scaffold to develop new drugs against different molecular targets (Zhu and Hou 2020; Marriott et al. 2012; Lee et al. 2007). Currently, benzofurans derivatives have been used in the treatment of different human diseases, and among these compounds we can cite some FDA-approved drugs, such as trioxsalen, 8-methoxypsoralen, and miodarone applied in the vitiligo, cancer, and arrhythmia treatment, respectively (Miao et al. 2019).

Benzofuran and their derivatives exhibit a broad spectrum of biological activities, such as antifungal (Masubuchi et al. 2003; Ryu et al. 2010), anti-leishmanicidal (MacIel-Rezende et al. 2013), antioxidant (Miceli et al. 2018), anticonvulsant (Kamal et al. 2014), anti-HIV (Rida et al. 2006), analgesic (Rádl et al. 2000), anti-bacterial (Yempala et al. 2013; Jiang et al. 2011; J. Liu et al. 2012), anti-hyperlipidemic (Sashidhara et al. 2013), antiplasmodial (Sundaram, Ganesan, and Murugesan 2012), and anticancer activities (Napiórkowska et al. 2019). Regarding the anticancer activity, different studies have identified cytotoxic activity of different benzofuran derivatives from natural and synthetic origin against cancer cell lines (Chen et al. 2016; El-Nakkady et al. 2012; Abd El-Karim et al. 2015), and in some cases, these activities show better efficiencies than the approved drugs (Wang et al. 2013). In addition, benzofuran scaffolds have emerged as an interesting pharmacophore and structural core to design and discovery new anticancer drugs that are active against different approved molecular targets relevant for cancer initiation and progression, such as Pim-1 kinase (Xiang et al. 2011; Aouidate et al. 2018), topoisomerase I and II (Lee et al. 2007), Mcl1 (Marimuthu and Singaravelu 2019), farnesyl transferase (Asoh et al. 2009), aromatase (Khodarahmi, Asadi, Farrokhpour, et al. 2015), and sigma receptors (Marriott et al. 2012).

2. Cancer: An Overview

2.1. Initiation and Progression

Cancer is a wide term used to identify a set of more than one hundred diseases, which are characterized as a disorder in the growth and proliferation of cells that escape from the normal control mechanisms. This process requires several steps and leads to the progressive accumulation of molecular changes, involving the activation of proto-oncogenes in oncogenes, resulting in uncontrolled cell growth and dysregulation of tumor suppressor genes, related to inhibition of tumor growth and DNA repair genes, which act on gene regulation (Bommer 2014). Cancer is classified as a genetic disease, where the transformation of normal cells into cancer cells involves a series of mutations, as well as, genetic and epigenetic changes, which leads to the persistence and survival of cells with an advantage of selective growth in a specific microenvironment, thus culminating in a transformed phenotype, malignancy and the ability to form metastases (Luzzatto and Pandolfi 2015; Todd, Groundwater, and Gill 2017). This cascade defines oncogenesis and confers well-defined characteristics to cells, such as sustaining proliferative signaling, resisting cell death (apoptosis), evading growth suppressors, replicative immortality, inducing the new blood vessels development's (angiogenesis), invasion and metastasis activation, evading immune destruction, tumor-promoting inflammation and genomic instability (Hanahan and Weinberg 2011).

The development of cancer cells is related directly to the intrinsic and extrinsic factors acting on cells, even as the organism's response to these changes, observing huge levels of complexity in relation to the factors responsible for the initiation and development of cancer disease (Todd, Groundwater, and Gill 2017; Loohuis et al. 2014). Carcinogenesis can result from the action of different factors, such as environmental (chemical, and physical factors) and genetic factors, or to the combination of both and consists of integrated stages, such as initiation, promotion, progression, and metastasis (Peters and Gonzalez, n.d.). These genetic changes can result in dysregulation of biochemical signaling pathways associated with cell proliferation, survival, and differentiation. The promotion stage is considered a relatively slow and reversible process, in which the modified cells are transformed into malignant cells and accumulate multiple new genetic and epigenetic alterations (Siddiqui et al. 2015).

Progression is the phase between a pre-malignant lesion and the development of invasive cancer; it is the final stage of neoplastic transformation where occurs the cell proliferation, as well as, genetic and phenotypic changes. This process involves a rapid increase in the size of the tumor; with cells undergoing more mutations with invasive and metastatic potential. Finally, metastasis involves the dissemination of cancer cells from the primary site to other parts of the body through the bloodstream or the lymphatic system, constituting the systemic state of neoplasia (Siddiqui et al. 2015; Sonnenschein and Soto 2016).

2.2. Epidemiology

In the last decades, the advancement in the research of the molecular basis of cancer occurred with remarkable upgrades in the theoretical/practical understanding of its origin, however, cancer still arises as a prevalent disease with high mortality rates (Siegel, Miller, and Jemal 2019). Incidence is growing rapidly worldwide and the reasons are complex, but reflect the aging and growth of the population, just as changes in the prevalence and distribution of the main risk factors of cancer (Bray et al. 2018). According to the International Agency for Research on Cancer (IARC) and the World Health Organization (WHO), there were 18.1 million new cases of cancer in 2018 worldwide and the corresponding estimates for total cancer deaths were 9.6 million (Torre et al. 2016). Considering 2030, new cancer cases are expected to increase to 21.7 million and deaths to 13 million (Torre et al. 2016; Fidler, Soerjomataram, and Bray 2016).

It is known that almost half of the cases and more than half of cancer deaths in the world occurred in Asia in 2018, due to this region having almost 60% of the global population. Europe responds for 23.4% of total cancer cases and 20.3% of cancer deaths, although it represents only 9% of the global population. Followed by the 21% incidence in the Americas and 14.4% of mortality worldwide. In contrast to the other regions, the proportions of cancer deaths in Asia (57.3%), and Africa (7.3%) are higher than the incidence proportions (48.4% and 5.8%, respectively), as a consequence of the epidemiology of cancer disease in these regions (Bray et al. 2018).

2.3. Current Treatment

Currently, cancer is considered as one of the biggest public health problem worldwide, and it is considered the most challenging disease in recent decades. A major amount of resources have been devoted to cancer for researches related to prevention and treatment (Adjiri, n.d.). Radiotherapy is one of the main cancer treatment modalities, as it has excellent control over the tumor, preserving normal tissues and less systemic effects. However, in some cases with resistance, there is a failure in the treatment, so that it is not possible to achieve a complete remission, with the manifestation of tumor recurrence, increased potential for cell migration and invasion and consequently distant metastases (Allen, Her, and Jaffray 2017). In contrast, the surgery is an effective treatment method for tumors before they reach the stage of invasion or metastasis, but it is not efficient when cancer has already migrated to other regions of the body (Wyld, Audisio, and Poston 2015). Currently, chemotherapy is the most used form of treatment and consists of the application of cytotoxic substances, which generally result in serious side effects for the patients. The number of drugs used against cancer is increasing, targeting a range of molecular targets that have been shown to play a functional role in tumor progression. However, personalized treatments, on a practical level, are complex to design, because of the variable character and heterogeneity of cancer cells. In addition, the cost of this treatment strategy can also indicate an objection (Zugazagoitia et al. 2016).

Despite the several cases of complete remission of cancer described in the literature, there are still challenges mainly due to the complexity of the disease, which involves different genetic and epigenetic changes and the influence of many environmental factors that affect these changes. Thus, it is extremely important to research for new compounds, and strategies for treatment and prevention of cancer (Yuan et al. 2017), (de Castro Sant' Anna et al. 2018).

3. Benzofurans and Their Derivatives with Anticancer Activity

3.1. Synthetic Origin

Benzofurans from synthetic origin have shown different biological activities against cancer cell lines (Nevagi, Dighe, and Dighe 2015; Ezzatzadeh and Hossaini 2019). Studies have reported anticancer activity of compounds with benzofuran rings as principal structural scaffolds, such as cercosporamide, hydroxy-3-methyl-6-phenylbenzofuran-2-carboxylic acid ethyl ester derivatives, benzofuran-2-carboxamides, egonol, moracin (Naik et al. 2015; Bazin et al. 2013).

Abdelhafez et al. (2004) reported that synthetic chalcone and thiopyrimidine benzofuran derivatives showed inhibitory *in vitro* activity against the *vascular endothelial* growth factor *receptor* (VEGFR-2) and a significative *in vivo* against prostate cancer cells (Abdelhafez et al. 2014). In addition, the compound 1-(6-hydroxy-4-methoxybenzofuran-5-yl)-3-(4-nitrophenyl)prop-2-en-1-one showed the best *in vitro* activity against VEGFR-2 when compared with the sorafenib drug. Furthermore, the correlation between biological activity and molecular docking results demonstrated that the analyzed compounds are proposed to act inhibiting the kinase pathway, considered essential to the development of many types of cancers.

Recently, a study investigated the antiproliferative activity of derivative compounds synthesized from 3-metil-1-benzofurane (cercosporamide analogs) against non-small cell lung cancer cell lines and demonstrated that three compounds exhibited cytotoxic activity in nanomolar concentration (Bazin et al. 2013). Similarly, El-Nakkady et al. (2012) demonstrated that benzofurans derivatives obtained from visnaginone, a visnagine derivative, showed cytotoxicity against hepatocytes cancer cell lines (HEPG2) when compared with two well-known anticancer drugs, the 5-fluorouracil, and doxorubicin (El-Nakkady et al. 2012).

Wang et al. (2013) performed the synthesis of a series of novel hybrid compounds of imidazole scaffold-based 2-benzylbenzofurans, and identified a satisfactory *in vitro* cytotoxic activity against different human tumor cell lines, such as MCF-7 (breast cell carcinoma) and SMMC-7721 (myeloid liver carcinoma) when compared with cisplatin a broad-spectrum drug against cancer cells (Wang et al. 2013). In addition, the authors attributed the modulation of cytotoxic activity to the existence of the benzimidazole ring and the replacement of the imidazolyl-3 position by a naphthylacyl or 4-methoxyphenacil groups.

3.2. Natural Origin

Benzofurans and their derivatives of natural origin have been isolated from different organisms, in special from plants. Regarding to plants these compounds are widely distributed in angiosperms plants families, such as Asteraceae, Lauraceae, Liliaceae, Rutaceae, Fabaceae and Meliaceae.

Zhou et al. (Zhou et al. 2013) using the MTT assay reported the cytotoxic activity of six new benzofurans neolignanes isolated from the soluble fraction of *Aristolochia fordiana* (Aristolochiaceae) leaves against three human cancer cell lines (MCF-7, HepG2, and MG-63). These authors identified that the presence of the methoxide group increased the cytotoxic activity of these compounds, and the double bond in the furan ring in C-7/C-8 position increased the anticancer activity when compared with the other compounds lacking these structural changes. In addition, the authors concluded that the aldehyde group also increased the cytotoxicity of the tested compounds.

Manna et al. (Manna et al. 2010) reported that a new benzofuran lignin, named Benfur(5-{(E)-2-carboxyvinyl}-2-(4-hydroxy-3-methoxyphenyl)-7-methoxy-2,3-dihydro-1-benzofuran-3-carboxylic acid) isolated from the *Dracaena* resin could be used as a chemopreventive and chemotherapic agents against the malignant T cells due to its inductive action to the apoptosis in the p53-positive cells.

Similarly, Liu et al. (Q. Liu et al. 2010b), isolated different benzofurans from *Ligularia veitchiana* root (2,5-diacetyl-6-methoxybenzofuran, 6-hydroxytremetone, and 2-isopropenyl-6-acetyl-8-methoxy1,3-benzodioxin-4-one) and tested these compounds against two well-known cell lines: human leukemia cells (HL-60) and human cervical carcinoma cells (HeLa). The authors identified a moderated cytotoxicity and inhibition of cell growth and indicated that the bioactivity of these compounds is probably related to 1,4-dioxilanoxi radical attached to the benzofuran ring that acts as an alkylation center.

The benzofuran 1-(6-hydroxy-2-isopropenyl-1-benzofuran-5-yl)-1-ethanone isolated from the root of *Petasites hybridus* showed cytotoxic and apoptotic activities against human breast cancer MCF-7 cells (Soleimani et al. 2015). The compound assigned as (1R,2R,3S,3aR,8bS)-4‴-{[(2‴R,4‴R)-4‴-[(S)-1,2-dihydrox-yethyl]-3-hydroxy-1,4-dioxan-2-yl]oxy}-1,8b-hydroxy-8-me-thoxy-3a-(4-methoxyphenyl)-3-phenyl-2,3,3a,8b-tetrahydro-1H-cyclopenta[b]benzofuran-2-carboxamide isolated from an aqueous extract of a large-scale recollection of *Aglaia perviridis* roots exhibited cytotoxic activity against a human colon cancer (HT-29) and human prostate cancer cell line (PC-3) (Agarwal et al. 2019).

The cyclopenta[b]-benzofuran derivatives, silvestrol, and episilvestrol, isolated from stems of *Aglaia stellatopilosa* (Meliaceae) showed significant cytotoxic activity against human cancer cell lines, such as HT-29,17 MCF-7, and NCI-H460 (Othman et al. 2015). Interestingly, the silvestrol obtained from fruits and twigs of *Aglaia foveolate* (Meliaceae) was investigated against melanoma cell lines (MDA-MB-435) (Chen et al. 2016). In this study, the authors verified that the compound strongly inhibited malignant cell growth and induced cell death, and this activity was attributed to the induction of apoptosis mediated by early autophagy. Corroborating with these results previous studies have demonstrated that silvestrol acts as an inhibitor of protein synthesis in MDA-MB-231 breast and PC-3 prostate cancer cell lines and the benzofuran ring is essential for this properties (Cencic et al. 2009).

Natural benzofurans, such as benzofuran-chalcanes millpuline E and millpuline F isolated from the extract of the leaves of *Millettia pulchra*

(Benth) could be potentially important natural cancer chemoprevention agents (Dunn et al. 2016). The compound bergapten, obtained the extract of the roots of *Dorstenia kameruniana*, displayed cytotoxicity against the sensitive CCRF-CEM and multidrug-resistant CEM/ADR5000 leukemia cells, toward the development of novel drugs to treat cancer (Adem et al. 2018).

The compounds methyl rocaglate, 4'-demethoxy-3',4'-merhylene-dioxy-methyl rocaglate, 4'-demethoxy-3',4'-methylenedioxyrocaglaol, 1-oxo-4'-demethoxy-3',4'-methyenedioxyrocaglaol and 1-*o*-formyl-4'-demethoxy-3',4'-methylenedioxy-methyl rocaglate isolated from the stems and fruits of *Aglaia elliptica* (Meliaceae) showed potent cytotoxic activity when evaluated against different human cancer cell lines (Cui et al. 1997). Similarly, the benzonfuran calophione A isolated from roots of *Tephrosia calophylla* (Leguminosae) exhibited significant cytotoxicity against cancer cell lines (Ganapaty et al. 2009).

Three arylbenzofurans (ebenfurans I, II and III) isolated from *Onobrychis ebenoides* (Fabaceae) showed cytotoxic activity against human cancer cell lines (Katsanou et al. 2007). Similarly, the compound isoneorautenol isolated from the *Erythrina excelsa* (Fabaceae) showed significant cytotoxicity toward sensitive and drug-resistant cancer cell lines (Kuete et al. 2014). The benzofuran derivative 1-(6-hydroxy-2-isopropenyl-1-benzofuran-5-yl)-1-ethanone was isolated from the roots of *Petasites hybridus* (Asteraceae) and showed moderate inhibitory activity *in vitro* against human breast cancer MCF-7 cells (Khaleghi et al. 2011). The isolated compounds 2,5-diacetyl-6-methoxybenzofuran and 6-hydroxytremetone from the roots of *Ligularia veitchiana* (Asteraceae), also exhibited cytotoxic activity against HL-60, SMMC-7721, and HeLa cancer cell lines (Q. Liu et al. 2010a).

The benzofuran lignans, such as herpetosiol A and herpetosiol B isolated from the ethyl acetate extract of the stems of *Herpetospermum pedunculosum* (Cucurbitaceae) exhibited anti-proliferative effects against four tumor cell lines: MDA-MB-231 (breast carcinoma), SGC7901 (human gastric carcinoma), A549 (lung carcinoma), and HepG2 (hepatocellular carcinoma) with IC_{50} ranging from 1.7 ± 0.1 to 32.6 ± 1.1 μM (Ma et al.

2019). The compound obovatinal isolated from leaves of *Persea obovatifolia* (Lauraceae) showed significant cytotoxic activities against P-388, KB16, A549 and HT-29 cancer cell lines (Tsai et al. 1996).

REFERENCES

Abd El-Karim, Somaia S., Manal M. Anwar, Neama A. Mohamed, Tamer Nasr, and Samia A. Elseginy. 2015. "Design, Synthesis, Biological Evaluation and Molecular Docking Studies of Novel Benzofuran-Pyrazole Derivatives as Anticancer Agents." *Bioorganic Chemistry* 63 (December): 1-12. https://doi.org/10.1016/j.bioorg.2015.08.006.

Abdelhafez, Omaima M., Kamelia M. Amin, Hamed I. Ali, Mohamed M. Abdalla, and Eman Y. Ahmed. 2014. "Design, Synthesis and Anticancer Activity of Benzofuran Derivatives Targeting VEGFR-2 Tyrosine Kinase." *RSC Advances* 4 (23): 11569-79. https://doi.org/10.1039/c4ra00943f.

Adem, Fozia A., Victor Kuete, Armelle T. Mbaveng, Matthias Heydenreich, Albert Ndakala, Beatrice Irungu, E. Thomas, and Abiy Yenesew. 2018. "Fitoterapia Cytotoxic Benzylbenzofuran Derivatives from Dorstenia Kameruniana" 128 (February): 26-30. https://doi.org/10.1016/j.fitote. 2018.04.019.

Adjiri, Adouda. n.d. "Identifying and Targeting the Cause of Cancer Is Needed to Cure Cancer." *Oncology and Therapy* 4. https://doi.org/10.1007/s40487-015-0015-6.

Agarwal, Garima, James R. Wilson, Steven J. Kurina, Gerardo D. Anaya-eugenio, Tran N. Ninh, Joanna E. Burdette, Djaja D. Soejarto, et al. 2019. "Structurally Modified Cyclopenta[b]Benzofuran Analogues Isolated from Aglaia Perviridis." Review-article. *Journal of Natural Products* 82: 2870-77. https://doi.org/10.1021/acs.jnatprod.9b00631.

Allen, Christine, Sohyoung Her, and David A. Jaffray. 2017. "Radiotherapy for Cancer: Present and Future." *Advanced Drug Delivery Reviews*. Elsevier B. V. https://doi.org/10.1016/j.addr.2017.01.004.

Aouidate, Adnane, Adib Ghaleb, Mounir Ghamali, Abdellah Ousaa, M'barek Choukrad, Abdelouahid Sbai, Mohammed Bouachrine, and Tahar Lakhlifi. 2018. "3D QSAR Studies, Molecular Docking and ADMET Evaluation, Using Thiazolidine Derivatives as Template to Obtain New Inhibitors of PIM1 Kinase." *Computational Biology and Chemistry* 74 (June): 201-11. https://doi.org/10.1016/j.compbiolchem.2018.03.008.

Asoh, Kohsuke, Masami Kohchi, Ikumi Hyoudoh, Tatsuo Ohtsuka, Miyako Masubuchi, Kenichi Kawasaki, Hirosato Ebiike, et al. 2009. "Synthesis and Structure-Activity Relationships of Novel Benzofuran Farnesyltransferase Inhibitors." *Bioorganic and Medicinal Chemistry Letters* 19 (6): 1753-57. https://doi.org/10.1016/j.bmcl.2009.01.074.

Bazin, Marc Antoine, Lizeth Bodero, Christophe Tomasoni, Bénédicte Rousseau, Christos Roussakis, and Pascal Marchand. 2013. "Synthesis and Antiproliferative Activity of Benzofuran-Based Analogs of Cercosporamide against Non-Small Cell Lung Cancer Cell Lines." *European Journal of Medicinal Chemistry* 69: 823-32. https://doi.org/10.1016/j.ejmech.2013.09.013.

Bhargava, Sangeeta, and Deepti Rathore. 2017. "Synthetic Routes and Biological Activities of Benzofuran and Its Derivatives: A Review." *Letters in Organic Chemistry* 14 (6): 381-402. https://doi.org/10.2174/1570178614666170321125853.

Bommer, U., Kara L. Perrow. 2014. "Cancer Biology: Molecular and Genetic Basis." In: *Cancer Biology: Molecular and Genetic Basis*, edited by Olver, I., Sabesan, S., 2-20.

Bray, Freddie, Jacques Ferlay, Isabelle Soerjomataram, Rebecca L. Siegel, Lindsey A. Torre, and Ahmedin Jemal. 2018. "Global Cancer Statistics 2018: GLOBOCAN Estimates of Incidence and Mortality Worldwide for 36 Cancers in 185 Countries." *CA: A Cancer Journal for Clinicians* 68 (6): 394-424. https://doi.org/10.3322/caac.21492.

Castro Sant' Anna, Carla de, Alberto Gomes Ferreira Junior, Paulo Soares, Fabricio Tuji, Eric Paschoal, Luiz Cláudio Chaves, and Rommel Rodriguez Burbano. 2018. "Molecular Biology as a Tool for the

Treatment of Cancer." *Clinical and Experimental Medicine*. Springer-Verlag Italia s.r.l. https://doi.org/10.1007/s10238-018-0518-1.

Cencic, Regina, Marilyn Carrier, Gabriela Galicia-Vázquez, Marie-Eve Bordeleau, Rami Sukarieh, Annie Bourdeau, Brigitte Brem, et al. 2009. "Antitumor Activity and Mechanism of Action of the Cyclopenta[b]Benzofuran, Silvestrol." *PloS One* 4 (4): e5223. https://doi.org/10.1371/journal.pone.0005223.

Chen, Wei Lun, Li Pan, A. Douglas Kinghorn, Steven M. Swanson, and Joanna E. Burdette. 2016. "Silvestrol Induces Early Autophagy and Apoptosis in Human Melanoma Cells." *BMC Cancer* 16 (17): 1-10. https://doi.org/10.1186/s12885-015-1988-0.

Cui, Baoliang, Heebyung Chai, Thawatchai Santisuk, Vichai Reutrakul, Norman R Farnsworth, and Royal Forest. 1997. *Novel Cytotoxic LH-Cyclopenta[6]Benzofuran Lignans From* 53 (52): 17625-32.

Dunn, Benth, Wenli Wang, Ning Li, Jian Wang, Gang Chen, Renbin Huang, Weihong Zhao, Jiayuan Li, and Yingying Si. 2016. "Phytochemistry Bioactive Benzofuran-Chalcanes as Potential NQO1 Inducers From." *Phytochemistry* 131: 107-14. https://doi.org/10.1016/j.phytochem.2016.09.005.

El-Nakkady, Sally S., Hanaa F. Roaiah, Weam S. El-Serwy, Abdel Mohsen Soliman, Sherein I. Abd El-Moez, and Adel A. H. Abdel-Rahman. 2012. "Antitumor and Antimicrobial Activities of Some Hetero Aromatic Benzofurans Derived from Naturally Occurring Visnagin." *Acta Poloniae Pharmaceutica - Drug Research* 69 (4): 645-55.

Ezzatzadeh, Elham, and Zinatossadat Hossaini. 2019. "Green Synthesis and Antioxidant Activity of Novel Series of Benzofurans from Euparin Extracted of Petasites Hybridus." *Natural Product Research* 33 (11): 1617-23. https://doi.org/10.1080/14786419.2018.1428598.

Fidler, Miranda M., Isabelle Soerjomataram, and Freddie Bray. 2016. "A Global View on Cancer Incidence and National Levels of the Human Development Index." *International Journal of Cancer* 139 (11): 2436-46. https://doi.org/10.1002/ijc.30382.

Ganapaty, Seru, Guttula Veera, Kantha Srilakshmi, Steve Thomas, Hafizur Rahman, Hartmut Laatsch, and Reto Brun. 2009. "Phytochemistry Cytotoxic Benzil and Coumestan Derivatives from Tephrosia Calophylla." *Phytochemistry* 70 (1): 95-99. https://doi.org/10.1016/j.phytochem.2008.10.009.

Hanahan, Douglas, and Robert A. Weinberg. 2011. "Hallmarks of Cancer: The next Generation." *Cell* 144 (5): 646-74.

Jiang, Xizhen, Wenlu Liu, Wei Zhang, Faqin Jiang, Zhe Gao, Hao Zhuang, and Lei Fu. 2011. "Synthesis and Antimicrobial Evaluation of New Benzofuran Derivatives." *European Journal of Medicinal Chemistry* 46 (8): 3526-30. https://doi.org/10.1016/j.ejmech.2011.04.053.

Kamal, Mehnaz, Ashok Shakya, Mohamed Ahsan, and Talha Jawaid. 2014. "Synthesis, Anticonvulsant and Neurotoxicity Evaluation of Some Newer N-(2-Benzoylbenzofuran-3-Yl)-3-(Substituted)-Propanamide Analogs." *Central Nervous System Agents in Medicinal Chemistry* 13 (3): 159-65. https://doi.org/10.2174/187152491366613112216082 8.

Katsanou, Efrosini S., Maria Halabalaki, Nektarios Aligiannis, Sofia Mitakou, Alexios-leandros Skaltsounis, Xanthippi Alexi, Harris Pratsinis, and Michael N. Alexis. 2007. *Cytotoxic Effects of 2-Arylbenzofuran Phytoestrogens on Human Cancer Cells : Modulation by Adrenal and Gonadal Steroids* ☋ 104: 228-36. https://doi.org/10.1016/j.jsbmb.2007.03.028.

Khaleghi, Fatemeh, Laily Bin, Faramarz Rostami, Wan Ahmad, Mohammad A. Khalilzadeh, Brian Skelton, and Mohamed Makha. 2011. "Phytochemistry Letters A New Bioactive Compound from the Roots of Petasites Hybridus." *Phytochemistry Letters* 4 (3): 254-58. https://doi.org/10.1016/j.phytol.2011.04.009.

Khodarahmi, Ghadamali, Parvin Asadi, Hossein Farrokhpour, Farshid Hassanzadeh, and Mohammad Dinari. 2015. "Design of Novel Potential Aromatase Inhibitors via Hybrid Pharmacophore Approach: Docking Improvement Using the QM/MM Method." *RSC Advances* 5 (71): 58055-64. https://doi.org/10.1039/c5ra10097f.

Khodarahmi, Ghadamali, Parvin Asadi, Farshid Hassanzadeh, and Elham Khodarahmi. 2015. "Benzofuran as a Promising Scaffold for the Synthesis of Antimicrobial and Antibreast Cancer Agents: A Review." *Journal of Research in Medical Sciences* 20 (11): 1094-1104. https://doi.org/10.4103/1735-1995.172835.

Kuete, Victor, Louis P. Sandjo, Guy M. N. Kwamou, Benjamin Wiench, Augustin E. Nkengfack, and Thomas Efferth. 2014. "Activity of Three Cytotoxic Isoflavonoids from Erythrina Excelsa and Erythrina Senegalensis (Neobavaisoflavone, Sigmoidin H. and Isoneorautenol) toward Multi-Factorial Drug Resistant Cancer Cells." *Phytomedicine* 21 (5): 682-88. https://doi.org/10.1016/j.phymed.2013.10.017.

Lee, Yeun Kyung, Mi Hee Woo, Chul Ho Kim, Youngsoo Kim, Sung Ho Lee, Byeong Seon Jeong, Hyun Wook Chang, and Jong Keun Son. 2007. "Two New Benzofurans from Gastrodia Elata and Their DNA Topoisomerases I and II Inhibitory Activities." *Planta Medica* 73 (12): 1287-91. https://doi.org/10.1055/s-2007-981619.

Liu, Jingbao, Faqin Jiang, Xizhen Jiang, Wei Zhang, Jingjing Liu, Wenlu Liu, and Lei Fu. 2012. "Synthesis and Antimicrobial Evaluation of 3-Methanone-6-Substituted-Benzofuran Derivatives." *European Journal of Medicinal Chemistry* 54 (August): 879-86. https://doi.org/10.1016/j.ejmech.2012.05.013.

Liu, Quan, Ling Shen, Tong-tong Wang, Chao-jun Chen, Wei-yan Qi, and Kun Gao. 2010a. "Novel Modified Furanoeremophilane-Type Sesquiterpenes and Benzofuran Derivatives from Ligularia Veitchiana." *Food Chemistry* 122 (1): 55-59. https://doi.org/10.1016/j.foodchem.2010.01.069.

Liu, Quan, Ling Shen, Tong Tong Wang, Chao Jun Chen, Wei Yan Qi, and Kun Gao. 2010b. "Novel Modified Furanoeremophilane-Type Sesquiterpenes and Benzofuran Derivatives from Ligularia Veitchiana." *Food Chemistry* 122 (1): 55-59. https://doi.org/10.1016/j.foodchem.2010.01.069.

Loohuis, Loes Olde, Giulio Caravagna, Alex Graudenzi, Daniele Ramazzotti, Giancarlo Mauri, Marco Antoniotti, and Bud Mishra. 2014. "Inferring Tree Causal Models of Cancer Progression with

Probability Raising." Edited by Lars Kaderali. *PLoS ONE* 9 (10): e108358. https://doi.org/10.1371/journal.pone.0108358.

Luzzatto, Lucio, and Pier Paolo Pandolfi. 2015. "Causality and Chance in the Development of Cancer." *New England Journal of Medicine* 373 (1): 84-88. https://doi.org/10.1056/NEJMsb1502456.

Ma, Yingxiong, Hong Wang, Rui Wang, Fancheng Meng, Zhaoyue Dong, Guowei Wang, Xiaozhong Lan, Hong Quan, Zhihua Liao, and Min Chen. 2019. "Cytotoxic Lignans from the Stems of Herpetospermum Pedunculosum." *Phytochemistry* 164 (September 2018): 102-10. https: //doi.org/10.1016/j.phytochem.2019.05.004.

MacIel-Rezende, Claudia Mara, Letícia De Almeida, Éderson D. Martin Costa, Francieli Ribeiro Pires, Karina Ferreira Alves, Cláudio Viegas Junior, Danielle Ferreira Dias, Antônio Carlos Doriguetto, Marcos José Marques, and Marcelo Henrique Dos Santos. 2013. "Synthesis and Biological Evaluation against Leishmania Amazonensis of a Series of Alkyl-Substituted Benzophenones." *Bioorganic and Medicinal Chemistry* 21 (11): 3114-19. https://doi.org/10.1016/j.bmc.2013.03.045.

Manna, Sunil K., Julie S. Bose, Vijay Gangan, Nune Raviprakash, Thota Navaneetha, Pongali B. Raghavendra, Banaganapalli Babajan, Chitta S. Kumar, and Swatantra K. Jain. 2010. "Novel Derivative of Benzofuran Induces Cell Death Mostly by G2/M Cell Cycle Arrest through P53-Dependent Pathway but Partially by Inhibition of NF-KB." *Journal of Biological Chemistry* 285 (29): 22318-27. https://doi.org/10.1074/jbc.M110.131797.

Marimuthu, Parthiban, and Kalaimathy Singaravelu. 2019. "Unraveling the Molecular Mechanism of Benzothiophene and Benzofuran Scaffold-Merged Compounds Binding to Anti-Apoptotic Myeloid Cell Leukemia 1." *Journal of Biomolecular Structure and Dynamics* 37 (8): 1992-2003. https://doi.org/10.1080/07391102.2018.1474805.

Marriott, Karla Sue C., Andrew Z. Morrison, Misty Moore, Olarongbe Olubajo, and Leonard E. Stewart. 2012. "Synthesis of N-Phenyl-N-(3-(Piperidin-1-Yl)Propyl)Benzofuran-2-Carboxamides as New Selective

Ligands for Sigma Receptors." *Bioorganic and Medicinal Chemistry* 20 (23): 6856-61. https://doi.org/10.1016/j.bmc.2012.09.044.

Masubuchi, Miyako, Hirosato Ebiike, Ken Ichi Kawasaki, Satoshi Sogabe, Kenji Morikami, Yasuhiko Shiratori, Shinji Tsujii, et al. 2003. "Synthesis and Biological Activities of Benzofuran Antifungal Agents Targeting Fungal N-Myristoyltransferase." *Bioorganic and Medicinal Chemistry* 11 (20): 4463-78. https://doi.org/10.1016/S0968-0896(03)00429-2.

Miao, Yu-hang, Yu-heng Hu, Jie Yang, Teng Liu, Jie Sun, and Xiao-jing Wang. 2019. "Natural Source, Bioactivity and Synthesis of Benzofuran Derivatives." *RSC Advances* 9 (47): 27510-40. https://doi.org/10.1039/c9ra04917g.

Miceli, Martina, Elia Roma, Paolo Rosa, Marta Feroci, M. Antonietta Loreto, Daniela Tofani, and Tecla Gasperi. 2018. "Synthesis of Benzofuran-2-One Derivatives and Evaluation of Their Antioxidant Capacity by Comparing DPPH Assay and Cyclic Voltammetry." *Molecules* 23 (4): 710. https://doi.org/10.3390/molecules23040710.

Naik, Ravi, Dipesh S. Harmalkar, Xuezhen Xu, Kyusic Jang, and Kyeong Lee. 2015. "Bioactive Benzofuran Derivatives: Moracins A. Z. in Medicinal Chemistry." *European Journal of Medicinal Chemistry* 90: 379-93. https://doi.org/10.1016/j.ejmech.2014.11.047.

Napiórkowska, Mariola, Marcin Cieślak, Julia Kaźmierczak-Barańska, Karolina Królewska-Golińska, and Barbara Nawrot. 2019. "Synthesis of New Derivatives of Benzofuran as Potential Anticancer Agents." *Molecules* 24 (8): 1529. https://doi.org/10.3390/molecules24081529.

Nevagi, Reshma J., Santosh N. Dighe, and Satish N. Dighe. 2015. "Biological and Medicinal Significance of Benzofuran." *European Journal of Medicinal Chemistry* 97 (1): 561-81. https://doi.org/10.1016/j.ejmech.2014.10.085.

Othman, Nuraqilah, Li Pan, Michele Mejin, Julian C. L. Voong, Hee-byung Chai, Caroline M. Pannell, A. Douglas Kinghorn, and Tiong C. Yeo. 2015. *Cyclopenta[b]Benzofuran and Secodammarane Derivatives from the Stems of Aglaia Stellatopilosa.* https://doi.org/10.1021/acs. jnatprod.5b00810.

Peters, Jeffrey M., and Frank J. Gonzalez. n.d. *The Evolution of Carcinogenesis*. https://doi.org/10.1093/toxsci/kfy184.

Rádl, Stanislav, Petr Hezký, Petr Konvička, and Ivan Krejčí. 2000. "Synthesis and Analgesic Activity of Some Substituted 1-Benzofurans and 1-Benzothiophenes." *Collection of Czechoslovak Chemical Communications* 65 (7): 1093-1108. https://doi.org/10.1135/cccc2000 1093.

Rida, Samia M., Soad A. M. El-Hawash, Hesham T. Y. Fahmy, Aly A. Hazza, and Mostafa M. M. El-Meligy. 2006. "Synthesis and *In Vitro* Evaluation of Some Novel Benzofuran Derivatives as Potential Anti-HIV-1, Anticancer, and Antimicrobial Agents." *Archives of Pharmacal Research* 29 (1): 16-25. https://doi.org/10.1007/BF02977463.

Ryu, Chung Kyu, Ae Li Song, Jung Yoon Lee, Jung An Hong, Joo Hee Yoon, and Aram Kim. 2010. "Synthesis and Antifungal Activity of Benzofuran-5-Ols." *Bioorganic and Medicinal Chemistry Letters* 20 (22): 6777-80. https://doi.org/10.1016/j.bmcl.2010.08.129.

Sashidhara, Koneni V., Ram K. Modukuri, Ravi Sonkar, K. Bhaskara Rao, and Gitika Bhatia. 2013. "Hybrid Benzofuran-Bisindole Derivatives: New Prototypes with Promising Anti-Hyperlipidemic Activities." *European Journal of Medicinal Chemistry* 68 (October): 38-46. https://doi.org/10.1016/j.ejmech.2013.07.009.

Siddiqui, Imtiaz A., Vanna Sanna, Nihal Ahmad, Mario Sechi, and Hasan Mukhtar. 2015. "Resveratrol Nanoformulation for Cancer Prevention and Therapy." *Annals of the New York Academy of Sciences* 1348 (1): 20-31. https://doi.org/10.1111/nyas.12811.

Siegel, Rebecca L., Kimberly D. Miller, and Ahmedin Jemal. 2019. "Cancer Statistics, 2019." *CA: A Cancer Journal for Clinicians* 69 (1): 7-34. https://doi.org/10.3322/caac.21551.

Soleimani, Afsane, Jahanbakhsh Asadi, Faramarz Rostami-charati, and Roghaye Gharaei. 2015. *High Cytotoxicity and Apoptotic Effects of Natural Bioactive Benzofuran Derivative on the MCF-7 Breast Cancer Cell Line*, 505-13.

Sonnenschein, Carlos, and Ana M. Soto. 2016. "Carcinogenesis Explained within the Context of a Theory of Organisms." *Progress in Biophysics and Molecular Biology* 122 (1): 70-76. https://doi.org/10.1016/j.pbiomolbio.2016.07.004.

Sundaram, Ravikumar, Ramanathan Ganesan, and Gnanadesigan Murugesan. 2012. "*In Vitro* Antiplasmodial Activity of Spiro Benzofuran Compound from Mangrove Plant of Southern India." *Asian Pacific Journal of Tropical Medicine* 5 (5): 358-61. https://doi.org/10.1016/S1995-7645(12)60059-5.

Todd, Adam, Paul W. Groundwater, and Jason H. Gill. 2017. "Cellular and Molecular Basis of Cancer." In: *Anticancer Therapeutics*, 39-80. Chichester, UK: John Wiley & Sons, Ltd. https://doi.org/10.1002/9781118696194.ch1.3.

Torre, Lindsey A., Rebecca L. Siegel, Elizabeth M. Ward, and Ahmedin Jemal. 2016. *Global Cancer Incidence and Mortality Rates and Trends-An Update*. https://doi.org/10.1158/1055-9965. EPI-15-0578.

Tsai, Ian Lih, Chih Feng Hsieh, Chang Yih Duh, and Ih Sheng Chen. 1996. "Cytotoxic Neolignans from Persea Obovatifolia." *Phytochemistry* 43 (6): 1261-63. https://doi.org/10.1016/S0031-9422(96)00509-2.

Wang, Xue Quan, Lan Xiang Liu, Yan Li, Cheng Jun Sun, Wen Chen, Liang Li, Hong Bin Zhang, and Xiao Dong Yang. 2013. "Design, Synthesis and Biological Evaluation of Novel Hybrid Compounds of Imidazole Scaffold-Based 2-Benzylbenzofuran as Potent Anticancer Agents." *European Journal of Medicinal Chemistry* 62: 111-21. https://doi.org/10.1016/j.ejmech.2012.12.040.

Wyld, Lynda, Riccardo A. Audisio, and Graeme J. Poston. 2015. "The Evolution of Cancer Surgery and Future Perspectives." *Nature Reviews Clinical Oncology*. Nature Publishing Group. https://doi.org/10.1038/nrclinonc.2014.191.

Xiang, Yibin, Bradford Hirth, Gary Asmussen, Hans Peter Biemann, Kimberly A. Bishop, Andrew Good, Maria Fitzgerald, et al. 2011. "The Discovery of Novel Benzofuran-2-Carboxylic Acids as Potent Pim-1 Inhibitors." *Bioorganic and Medicinal Chemistry Letters* 21 (10): 3050-56. https://doi.org/10.1016/j.bmcl.2011.03.030.

Yempala, Thirumal, Jonnalagadda Padma Sridevi, Perumal Yogeeswari, Darmarajan Sriram, and Srinivas Kantevari. 2013. "Design, Synthesis and Antitubercular Evaluation of Novel 2-Substituted-3H-Benzofuro Benzofurans via Palladium-Copper Catalysed Sonagashira Coupling Reaction." *Bioorganic and Medicinal Chemistry Letters* 23 (19): 5393-96. https://doi.org/10.1016/j.bmcl.2013.07.048.

Yuan, Renyikun, Wen Sun, Liu Xin, and Jin-Jian Lu. 2017. *Natural Products to Prevent Drug Resistance in Cancer Chemotherapy: A Review Tanshinones and Diethyl Blechnics with Anti-Inflammatory and Anti-Cancer Activities from Salvia Miltiorrhiza Bunge (Danshen) View Project Total Tanshinones-Induced Apoptosis And.* https://doi.org/10.1111/nyas.13387.

Zhou, Zhong Bo, Jian Guang Luo, Ke Pan, Si Ming Shan, Wei Zhang, and Ling Yi Kong. 2013. "Bioactive Benzofuran Neolignans from Aristolochia Fordiana." *Planta Medica* 79 (18): 1730-35. https://doi.org/10.1055/s-0033-1351100.

Zhu, Naiqiang, and Jingyi Hou. 2020. "Exploring the Mechanism of Action Xianlingubao Prescription in the Treatment of Osteoporosis by Network Pharmacology." *Computational Biology and Chemistry* 85 (April): 107240. https://doi.org/10.1016/j.compbiolchem.2020.107240.

Zugazagoitia, Jon, Cristiano Guedes, Santiago Ponce, Irene Ferrer, Sonia Molina-Pinelo, and Luis Paz-Ares. 2016. "Current Challenges in Cancer Treatment." *Clinical Therapeutics* 38 (7). https://doi.org/10.1016/j.clinthera.2016.03.026.

INDEX

A

acetic acid, 61, 148
acetone, 43
acetylation, 74, 75
acetylcholine, 17
acid, 7, 61, 67, 86, 93, 100, 103, 105, 106, 112, 116, 118, 133, 135, 142, 148, 167, 191
acrylic acid, 192
active compound, 74
activity of benzofurans, 21
acylation, 61, 133
aggregation, 17, 78, 92, 94
alcohols, 34, 63, 120, 122, 153, 160, 165, 166, 178, 181, 189, 195, 196, 197
alkenes, 155, 158, 193, 195
alkylation, 11, 12, 158, 159, 160, 165, 166, 195, 207
allergic reaction, 72
amino, 10, 158, 184
amyloid beta, 17
analgesic, 6, 201
angiogenesis, 99, 202
antibiotic, 70, 75, 76
anticancer, vi, 22, 26, 199, 200, 201, 205, 206, 209, 215, 216, 217
anti-cancer, 60
anticancer activity, 22, 200, 201, 205, 206, 209
anticancer drug, 201, 205
antioxidant, 6, 18, 69, 72, 73, 83, 84, 85, 86, 100, 201
antitumor, 6, 79, 99
antitumor agent, 99
apoptosis, 92, 202, 206, 207
aromatic compounds, 3
aromatic rings, 165, 178
asymmetric synthesis, 61, 62, 96, 100

B

bacteria, 2, 21, 102, 106, 107, 113
bacterial strains, 18
benign, 10, 118, 185
benzene, 4, 13, 32, 43, 134, 190
benzofuran, v, vii, ix, 2, 3, 4, 5, 6, 7, 8, 9, 10, 11, 13, 17, 18, 21, 22, 23, 24, 25, 26, 27, 33, 63, 78, 96, 100, 103, 112, 117, 118, 125, 127, 128, 129, 131, 133, 137,

138, 140, 141, 142, 145, 146, 148, 150, 153, 154, 160, 162, 164, 170, 175, 178, 179, 181, 183, 184, 185, 186, 190, 191, 192, 193, 195, 196, 197, 200, 201, 205, 206, 207, 208, 209, 210, 211, 212, 213, 214, 215, 216, 217, 218
bioassay, 72
bioavailability, 3
biological activities, vii, viii, ix, 2, 6, 17, 31, 33, 34, 60, 69, 70, 75, 79, 118, 141, 200, 201, 205
biological control, 110
biologically active compounds, viii, 78, 117, 153
biomarkers, 18
biomolecules, 6
biosynthesis, 34, 63, 88
biotic factor, 4
blood vessels, 202
bloodstream, 203
bonding, 13, 162, 178, 184, 187
bonds, 131, 154, 158, 171
Brazil, 1, 5, 31, 76, 101, 102, 105, 109, 199, 200
breast cancer, 207, 208
breast carcinoma, 208
building blocks, 75, 178

C

cancer, vii, ix, 17, 60, 71, 73, 74, 200, 201, 202, 203, 204, 205, 206, 207, 208, 209, 210, 211, 212, 213, 214, 216, 217, 218
cancer cells, 202, 204, 206
cancer death, 203
cancer therapy, ix, 200
carbon, 10, 32, 97, 100, 128, 131, 142, 152, 153, 155, 167, 173, 187
carbon atoms, 142
carboxylic acid, 137, 141, 142, 145, 192, 205, 206
carboxylic acids, 137, 142, 145, 192
carcinoma, 74, 92, 206, 207, 208
cascades, 149, 151, 194
catalysis, 132, 187, 193
catalyst, 7, 9, 10, 11, 13, 62, 67, 79, 119, 120, 124, 127, 129, 131, 135, 137, 138, 140, 149, 154, 161, 166, 176, 181, 183
catalytic system, 153, 171
C-C, 13, 62, 79, 131, 146, 173, 178, 185, 187, 189, 191, 193, 196
cell culture, 88
cell cycle, 86
cell death, 78, 202, 207
cell line, ix, 73, 74, 200, 201, 205, 206, 207, 208
C-H functionalization, v, vii, ix, 117, 118, 119, 138, 145, 146, 148, 149, 152, 153, 154, 155, 156, 160, 161, 162, 163, 164, 175, 176, 187, 191, 193, 194, 195, 196
chemical, 4, 5, 6, 7, 20, 32, 34, 38, 49, 71, 82, 101, 109, 111, 113, 202
chemical reactions, 7
chemical structures, 5, 32, 34, 49
chemical synthesis, 32, 34
chemicals, 3, 78
chiral catalyst, 61, 62
chlorobenzene, 7
cholinesterase, 88
cholinesterase inhibitors, 88
cleavage, 97, 134, 154, 181
climate, 4
closure, 12, 60, 61, 62
composition, 4, 20, 104, 109, 110, 111, 113, 114, 115, 116
compounds, vii, viii, ix, 2, 3, 4, 6, 7, 9, 10, 15, 17, 18, 19, 21, 31, 32, 33, 34, 35, 36, 37, 60, 69, 70, 71, 72, 73, 74, 75, 76, 78, 79, 80, 82, 85, 87, 94, 95, 98, 101, 158, 170, 186, 193, 200, 204, 205, 206, 207, 208
condensation, 118, 133
configuration, 61, 81, 96, 97, 165

constituents, 38, 72, 82, 83, 84, 85, 86, 87, 90, 92, 93, 94, 95, 110, 111
construction, 120, 129, 131, 133, 145, 165, 175, 188, 190
coordination, 161, 176, 181
copper, 10, 15, 115, 185, 193
cytotoxicity, 74, 77, 78, 89, 92, 97, 205, 206, 207, 208

D

derivatives, vii, viii, ix, 4, 6, 7, 9, 10, 11, 17, 34, 69, 76, 78, 87, 88, 95, 96, 101, 104, 108, 110, 112, 114, 117, 118, 119, 124, 133, 135, 142, 146, 150, 154, 155, 160, 170, 178, 179, 181, 184, 185, 192, 194, 200, 201, 205, 206, 207
dihydrobenzofuran neolignan, v, vii, viii, 31, 32, 36, 37, 38, 39, 44, 45, 46, 47, 48, 49, 50, 51, 52, 53, 54, 55, 56, 57, 58, 59, 60, 68, 69, 70, 80, 82, 83, 84, 85, 90, 94, 98, 99, 102, 103, 112, 113, 114
diseases, 3, 17, 21, 73, 76, 100, 200, 201
distribution, 20, 38, 80, 203
diversity, viii, 2, 4, 17, 34, 51, 63, 74, 79, 194
DNA repair, 201
dopamine, 178
drugs, vii, ix, 2, 3, 6, 17, 19, 77, 78, 200, 201, 204, 208

E

electron, 7, 34, 65, 119, 120, 128, 129, 134, 142, 148, 154, 158, 164, 171, 174, 176, 181, 184, 185
elucidation, 101
enantioselective synthesis, 63, 97
encapsulation, vii, viii, 2, 19
environmental factors, 204
enzymes, 67, 69, 70
epigenetic alterations, 202
equilibrium, 20, 66, 67, 119
ester, 13, 14, 17, 60, 75, 77, 149, 205
ethers, 98, 125, 129, 188, 190
ethyl acetate, 17, 43, 69, 208
extracts, 2, 3, 17, 18, 21, 49, 76, 87, 93, 101, 102, 103, 108, 111, 115

F

families, 37, 38, 206
formation, 13, 21, 60, 61, 62, 65, 66, 69, 79, 92, 124, 136, 139, 142, 145, 146, 168, 173, 178, 183, 184, 191, 193, 196
fruits, 17, 18, 43, 72, 81, 85, 86, 88, 89, 91, 95, 110, 207, 208
functionalization, 118, 119, 125, 129, 134, 138, 145, 146, 149, 152, 153, 154, 155, 156, 160, 161, 162, 163, 164, 175, 176, 187, 188, 190, 191, 193, 194, 195, 196
furan, 4, 9, 17, 60, 61, 89, 166, 188, 190, 206

G

genetic disease, 202
genetic factors, 202
genomic instability, 202
glucose, 44, 45, 46, 48, 50
glucoside, 85, 86, 92
glycoside, 82, 89, 94
growth, 88, 201, 203, 207

H

heart failure, 17
hepatocellular carcinoma, 208
hepatocytes, 89, 205
human, 73, 74, 115, 200, 206, 207, 208
human leukemia cells, 207

hydrogen, 161, 178, 184, 187
hydrogen bonds, 187
hydrogenation, 75
hydrolysis, 17, 185
hydroxyl, 18, 35, 49, 62, 66, 72, 73, 75, 76, 78, 120, 129, 160, 161, 181, 188
hydroxyl groups, 35, 73
hypothesis, 37, 79

I

in vitro, 18, 73, 75, 77, 87, 104, 110, 111, 114, 205, 206, 208
in vivo, 71, 87, 111, 205
inflammation, 18, 71, 72, 87, 100, 202
inflammatory mediators, 70
inflammatory responses, 85
inhibition, 18, 71, 78, 79, 89, 201, 207
inhibitor, 84, 186, 207
initiation, vii, ix, 200, 201, 202
insertion, 60, 96, 100, 119, 129, 145, 152, 154, 161, 166, 174, 176, 178, 182, 184, 194
ionization, 101, 103, 104, 108, 112, 114, 115
isolation, 34, 37, 49, 60, 73, 79, 85, 90
isomerization, 63, 96, 128, 164, 185, 195

K

ketones, 13, 132, 133, 165, 168, 170, 195

L

ligand, 98, 123, 124, 138, 189
light, 4, 13, 15, 19, 185
lignans, 32, 33, 63, 65, 79, 81, 82, 83, 84, 85, 86, 88, 89, 93, 94, 96, 97, 98, 208
lignin, 33, 98, 206
lymphatic system, 203

M

macrophages, 17, 71, 74
mass spectrometry, 101, 103, 104, 108, 112, 114, 115
materials, 119, 170, 185
metabolites, vii, viii, 2, 4, 31, 34, 84
methanol, 5, 43, 49, 67, 72, 147, 148, 160
methodology, 6, 9, 63, 69, 128
methylation, 74, 75, 76
migration, 119, 120, 176, 188, 204
molecular oxygen, 72
molecular structure, 4
molecules, viii, 2, 3, 6, 9, 21, 35, 117, 119, 125, 185
mortality, 75, 203
motor activity, 75

N

natural compound, 75
natural product, viii, 2, 3, 4, 6, 15, 25, 32, 33, 80, 82, 83, 84, 85, 86, 87, 90, 93, 94, 95, 96, 98, 100, 101, 105, 109, 110, 111, 114, 117, 141, 153, 186, 209, 211, 218
neolignans, vii, viii, 18, 31, 32, 35, 36, 39, 44, 45, 46, 47, 48, 49, 50, 51, 52, 53, 54, 55, 56, 57, 58, 59, 60, 70, 79, 80, 81, 82, 83, 84, 85, 87, 88, 89, 90, 91, 92, 93, 94, 95, 96, 98, 99, 100, 102, 103, 112, 113, 114
neurodegenerative disorders, 73
neutral, 138, 146, 149, 154, 155, 158, 160, 164, 165, 167, 168, 173, 177, 179, 181, 183, 184, 194, 195, 196
nitric oxide, 17, 18, 70, 71, 73, 82, 84, 87, 93
nitric oxide synthase, 18, 70
nuclear magnetic resonance, 102
nucleus, vii, 2, 3, 5, 6, 8, 17, 21

O

oil, 5, 71, 104, 109, 110, 111, 114, 115, 116
olefins, 128, 139, 140, 168, 169, 170, 190, 192, 195
organic compounds, 101
oxidation, 34, 62, 63, 65, 72, 74, 98, 122, 137, 140, 151, 162
oxidative stress, 82
oxygen, 19, 49, 66, 94, 119, 124, 131, 142, 184

P

palladium, v, vii, ix, 10, 13, 15, 23, 29, 30, 100, 117, 118, 119, 120, 121, 122, 123, 124, 125, 126, 127, 128, 129, 130, 131, 132, 133, 134, 135, 136, 137, 138, 139, 140, 141, 143, 145, 172, 186, 187, 188, 189, 190, 191, 192, 193, 194, 195, 218
pathogenesis, 79, 100
pathogens, 76, 104, 108, 115
pathway, vii, viii, 31, 34, 37, 82, 129, 133, 134, 136, 152, 175, 185, 205
phenol, 8, 9, 11, 13, 35, 61, 65, 100, 122, 123, 124, 125, 127, 130, 138, 140, 145, 166, 174, 185, 188, 189
phenolic compounds, 85
phenylbenzofuran, 8, 9, 33, 205
physicochemical properties, 3, 19
plants, viii, 4, 31, 34, 36, 80, 206
platelet aggregation, 90
preparation, iv, viii, 11, 32, 97
privileged scaffold, 200
pro-inflammatory, 70, 87
proliferation, 70, 84, 201, 202
prostate cancer, 205, 207
protection, 100, 122
protein synthesis, 207

Q

quinone, 125, 140, 142

R

radicals, 2, 6, 8, 21, 64
reaction mechanism, 65, 137, 166, 178
reaction medium, 20, 160
reaction rate, 64
reaction temperature, 151
reactions, 11, 15, 61, 65, 67, 71, 74, 79, 96, 100, 103, 105, 112, 113, 115, 118, 128, 132, 146, 153, 162, 164, 170, 171, 184, 185, 186, 187, 190, 191, 195
reactive oxygen, 18, 72
reactivity, 125, 132, 134, 165, 182
reagents, 13, 63, 190
receptor, 98, 192, 205
regioselectivity, 64, 137, 149, 152, 160, 171, 181
rhodium, v, vii, ix, 100, 117, 118, 146, 147, 148, 149, 150, 151, 152, 153, 154, 155, 156, 157, 159, 160, 161, 162, 163, 164, 165, 166, 167, 168, 169, 170, 173, 174, 177, 178, 179, 183, 184, 193, 194, 195, 196, 197
room temperature, 69, 168
ruthenium, 11, 149, 194

S

scope, 11, 123, 128, 134, 138, 158, 165, 166, 167, 176, 185, 187
secondary metabolism, 4
selective estrogen receptor modulator, 191
selectivity, 2, 21, 76, 119, 148, 165, 192, 194
silver, 63, 65, 98, 151, 185
skeleton, 60, 63, 150, 153

solution, 10, 19, 20, 65
solvents, 7, 19, 148, 162
species, 2, 4, 17, 18, 35, 37, 39, 40, 42, 43, 49, 60, 72, 81, 90, 110, 122, 123, 124, 131, 132, 133, 137, 140, 142, 145, 148, 154, 158, 162, 166, 174, 176, 178
structural changes, 206
structural modifications, 60
structure, viii, 2, 3, 17, 21, 31, 62, 67, 77, 80, 94, 102, 192
substitution, 119, 148, 155, 182, 190
Sun, v, 6, 9, 22, 24, 26, 29, 82, 94, 95, 117, 193, 196, 215, 217, 218
synthesis, v, vii, viii, 2, 6, 7, 9, 10, 11, 12, 13, 14, 15, 17, 19, 23, 24, 25, 26, 27, 29, 30, 32, 33, 34, 44, 45, 46, 47, 48, 49, 50, 51, 52, 53, 54, 55, 56, 57, 58, 59, 60, 61, 62, 63, 64, 67, 68, 69, 79, 80, 96, 97, 98, 99, 100, 101, 102, 107, 109, 117, 118, 120, 122, 124, 125, 126, 127, 128, 129, 130, 131, 133, 136, 137, 139, 140, 141, 142, 143, 144, 145, 146, 148, 149, 150, 151, 152, 154, 155, 156, 157, 160, 161, 162, 163, 166, 167, 168, 169, 171, 177, 178, 179, 180, 181, 182, 183, 184, 185, 186, 187, 188, 189, 190, 191, 192, 193, 194, 195, 196, 197, 206, 207, 209, 210, 211, 212, 213, 214, 215, 216, 217
synthetic methods, 60, 170, 200

T

target, viii, 117, 119, 122, 128, 134
techniques, 6, 19, 34, 51, 79
temperature, 10, 20, 64, 67, 154
transformation, 20, 119, 120, 122, 124, 125, 128, 133, 136, 142, 153, 158, 170, 178, 181, 202
treatment, 17, 78, 116, 122, 141, 178, 200, 204
trifluoroacetate, 68
triphenylphosphine, 15
tuberculosis, 17, 76, 93
tumor, 74, 201, 202, 204, 206, 208
tumor growth, 201
tumor progression, 204

V

vascular endothelial growth factor, 205

W

water, 5, 9, 10, 13, 15, 19, 35, 43, 67, 88, 189
workers, 36, 60, 61, 62, 64, 65, 66, 67, 69, 70, 74, 77, 78, 119, 146, 148, 152, 154, 158, 164, 177
worldwide, viii, 2, 6, 23, 203, 204
wound healing, 69, 84

Y

yield, 7, 11, 13, 60, 62, 63, 67, 69, 122, 129, 132, 134, 135, 138, 149, 150, 151, 153, 154, 158, 162, 166, 178, 181